SUSTAINABLE DEVELOPMENT IN CREATIVE INDUSTRIES: EMBRACING DIGITAL CULTURE FOR HUMANITIES

This book provides the thoughtful writings of a selection of authors illustrating a central concept: Sustainable Development in Creative Industries, which utilizes a monetary equilibrium addressing issues, particularly those associated with the use of an integrated area in cyberspace and physical space, and their effect on the creative industries. 15 universities from Asia and Europe have participated in the 9th Bandung Creative Movement, where this topic was explored. Sustainability issues are now at the forefront of progress.

The book covers four main areas. The first section, entitled "Art, Culture, and Society," delves into the various sectors that contribute to building a more sustainable environment, including the arts and culture. Whereas "Design and Architecture" is referring to cutting-edge practices in the fields of manufacturing, transportation, interior design, and building construction. The third section "Technology and New Media" delves into the transformation of technology into a new medium for the development of the creative industries. The final section, "management and Business," discusses an innovative perspective on the state of the market and management in the sector.

Anyone interested in the intersection of creative industries, sustainability, and digital cultures would benefit intellectually from reading this book.

PROCEEDINGS OF THE 9TH BANDUNG CREATIVE MOVEMENT INTERNATIONAL CONFERENCE ON CREATIVE INDUSTRIES (BCM 2022), BANDUNG, INDONESIA, 1 SEPTEMBER 2022

Sustainable Development in Creative Industries: Embracing Digital Culture for Humanities

Edited by

Dyah Ayu Wiwid Sintowoko, Idhar Resmadi, Hanif Azhar, Ganjar Gumilar and Taufiq Wahab
Universitas Telkom, Indonesia

First published 2023
by Routledge
4 Park Square, Milton Park, Abingdon, Oxon OX14 4RN
and by Routledge
605 Third Avenue, New York, NY 10158
e-mail: enquiries@taylorandfrancis.com
www.routledge.com – www.taylorandfrancis.com

Routledge is an imprint of the Taylor & Francis Group, an informa business

© 2023 selection and editorial matter, Dyah Ayu Wiwid Sintowoko, Idhar Resmadi, Hanif Azhar, Ganjar Gumilar & Taufiq Wahab; individual chapters, the contributors

The right of Dyah Ayu Wiwid Sintowoko, Idhar Resmadi, Hanif Azhar, Ganjar Gumilar, & Taufiq Wahab to be identified as the author[/s] of the editorial material, and of the authors for their individual chapters, has been asserted in accordance with sections 77 and 78 of the Copyright, Designs and Patents Act 1988.

The Open Access version of this book, available at www.taylorfrancis.com, has been made available under a Creative Commons Attribution-Non Commercial-No Derivatives 4.0 license.

Although all care is taken to ensure integrity and the quality of this publication and the information herein, no responsibility is assumed by the publishers nor the author for any damage to the property or persons as a result of operation or use of this publication and/or the information contained herein.

Library of Congress Cataloging-in-Publication Data
A catalog record has been requested for this book

ISBN: 978-1-032-44503-8 (hbk)
ISBN: 978-1-032-44504-5 (pbk)
ISBN: 978-1-003-37248-6 (ebk)

DOI: 10.1201/9781003372486

Typeset in Times New Roman
by MPS Limited, Chennai, India

Table of contents

Preface	xi
Scientific Committee	xiii
Batik Bomba: Symbolic interaction through artwork *S. Alam, A. Budiman, D. Hidayat & Sunarto*	1
Female students' perceptions on the comfort of the classroom and dormitory bedroom, in Insan Sejahtera Integrated Islamic Boarding Middle School Sumedang *T. Cardiah, S. Salsabila Az-zahra, R. Firmansyah & N. Laksitarini*	7
Design of a website-based competency improvement program at Sanggar Batik Katura Cirebon *G.M. Putra & F. Ciptandi*	14
Thrift shop-based business rebranding practice, Case study: Andys_store24 *G.P.S. Basten & R. Febriani*	19
The physical and philosophical values of the Mappalette Bola tradition and its potential in industrial development design earthquake resistant house foundation construction *S.D.S. Rusdy & F. Ciptandi*	24
Marketing strategy of *Tipa-Tipa* (souvenirs from Lake Toba) with SWOT method *J.B.W. Jawak & I. Wirasari*	30
Designing Markaz Domba strategy business using design thinking approach *K.M. Syahid & I. Wirasari*	37
The implementation of health protection-based design on Telkom University dormitory building *D. Murdowo, V. Haristianti & R.H.W. Abdulhadi*	42
Redesign coffee packaging for sustainable local community development *W. Swasty, A. Mustikawan & F.E. Naufalina*	48
The mosque rest area circulation system with a design thinking approach *D.P. Pangestu & D.W. Soewardikoen*	54
Personal values vs UTAUT factors on mobile gamers in Indonesia *R. Aulia, A. Bismo & R. Amansyah*	59
The efforts to develop Forest Farmer Group (FFG) business through designing visual identity packaging of mangrove coffee products in Indramayu *D.M. Septiani & I. Wirasari*	65
Exploring ideas of sustainable workplace through visual storytelling: A case study of #KomikCeritaIndah *D. Apsari & P. Maharani*	70

How SME-scaled branding agencies creates a high-quality yet affordable brand identity design for SMEs in Indonesia　76
R.A. Siswanto & J.B. Dolah

Philosophical potential of *Ngalokat Cai* ritual for *Cimahi* city branding design based on sustainable development goals　81
D.S. Pratiwi & M.I.P. Koesoemadinata

Zine on sharia investment basics for college students　87
S. Desintha, A.T.Z. Abdalloh & S. Hidayat

Fish consumption culture in Manado as a consideration to a web design　91
R.B. Adji, D.W. Soewardikoen & Ilhamsyah

Study of digital branding as media strategic toward metaverse　96
G.A. Prahara & J. Adler

Identification of building around the Bogor market as forming the identity of the Chinatown area in Bogor City　102
H. Anwar, R. Hambali & T. Mulya Raja

Tjap Go Meh animation process as a form of acculturation of Indonesian culture through edutainment media　107
I. Wirasari, D. Aditya, A.E. Adi, N.D. Nugaraha, S. Salayanti & S. Anis

Preliminary study into the Genshin impact's aesthetics: The sustainability of visual culture through the character design　113
L. Agung & L.R. Wiwaha

The phenomenon of digital wedding invitations: Its potential and cultural shift in Surakarta　118
S.A. Sari, I. Azhari & M.I.P. Koesoemadinata

Computer generated photography: Still image to moving image　123
A.P. Zen, C.R. Yuningsih & I.M. Miraj

The legal status of metaverse law and its implementation in modern era　129
A.P. Zen & I.M. Miraj

Street furniture: Design and material studies on public facilities　135
T.Z. Muttaqien, R.H.W. Abdulhadi & T.M. Raja

Study of second-hand clothing upcycle with mixed media techniques in the context of sustainable fashion case study: Bazooqu Brand　140
D.Y.P. Longi & M. Rosandini

Reliefs at the Desa lan Puseh Temple in Sudaji Village, North Bali, image of Bali tourism during the Dutch colonial period　144
I.D.A.D. Putra & S. Abdullah

Cultural influences toward society's view to young adult BTS fans (army) in Indonesia and its influence on fans consumption　149
A. Syafikarani, F. Balqis, A.R. Alissa & I.P. Narwasti

The value of Robo-Robo tradition as design inspiration in public spaces in the new normal era　154
U. Nafi'ah & M.I.P. Koesoemadinata

Gender-based design translation: An actor-network theory of MKS Shoes *I. Resmadi, R.P. Bastari & R.A. Siswanto*	159
Descriptive analysis of graphic layout in interior design catalog *C. Chalik, Andrianto & A.S.M. Atamtajan*	164
Constructing meaning on Tahilalats comics based on Instagram comments section *P. Aditia & A.N. Fadilla*	170
Digital handmade: A craftsmanship shifts in block printing surface textile design *M.S. Ramadhan & A. Widiandari*	175
Eyeglasses as a functional fashion accessory for generation Z in urban Indonesia: Recommendation for sustainable development *E. Mayor & I. Santosa*	180
Interior visual identity: Customer interest in interior elements in a retail store *T.M. Raja, K.P. Amelia, D.D. Prameswari & Y. Sabrani*	186
Analysis of library design in the digital era *F. Mahdiyah & R. Machfiroh*	192
Digital unification: The sustainability of Indonesian NFT communities through digital content *R.P. Bastari & I. Resmadi*	196
Website news media Ayobandung.com user experience *A.D. Pinasti & D.W. Soewardikoen*	202
Subverted narrative: On Bandung civic identity within Mufti Priyanka's monography *G. Gumilar*	208
Sustainable modest fashion design based on consumer needs *C. Tifany Fahira & M. Rosandini*	214
Strategy to maintain the existence of the museum in era 4.0 through the virtual Museum of Sultan Syarif Kasim in Bengkalis Regency *M.R. Kurniawan & T. Hendiawan*	220
Analysis of startup business promotion mistakes and solutions through digital media *T.F.L. Adino & M. Wardaya*	226
Suroboyo UNESA bus stop design with the iconic design approach *A.M. Nurkamilah, D.D.A. Utami, I.F. Fauzia & I. Sudarisman*	231
The influence of visual visibility on the behaviour of visitors in the cinema *Imtihan Hanom, M.D.D. Ar-Rasyid & A.M. Auliarahman*	237
Promoting animation technology in preserving and reintroducing *Wayang Cina Jawa* to younger generations *A. Lionardi, D.K. Aditya & I.G.A. Rangga Lawe*	242

Small classroom design adaptation as a response to teaching and learning activities in the new normal conditions *A. Farida, R. Fawwaz & I. Subagio*	247
Adaptive reuse strategy at Maison Teraskita Hotel Bandung *A.D. Purnomo, N. Laksitarini & L.C. Lase*	252
Process on public service advertisement of COVID-19 *S. Nurbani & Y.A. Barlian*	257
Implementation of the AISAS method in Waroeng Soejo's frozen food promotion strategy *E. Mentari, I. Wirasari & Ilhamsyah*	263
Behavioral patterns and interaction of backpacker visitors in limited areas at snooze Hostel Yogyakarta *N. Laksitarini, A.D. Purnomo & L.C. Lase*	268
Form of dynamic identity in restaurant interior design *E.A. Wismoyo, T.M. Raja & V. Haristianti*	273
Investigating consumers' attitude toward food souvenir packaging color *W. Swasty & M. Mustafa*	279
Folklore-based art performances in digital art space as a form of revitalizing local wisdom in the metaverse era *R. Rachmawanti & C.R. Yuningsih*	285
Nighttime attraction based on of Gedung Sate area *R.H.W. Abdulhadi, H. Anwar & I.Z. Budiono*	290
The development of the virtual environment and its impact on interior designers and architects. Case study: Zaha Hadid Architects *V. Haristianti & D. Murdowo*	296
Recommendations for restructuring urban development in Kemang area *N.A. Hapsoro & K.P. Amelia*	302
Extended reality: How digital technology transformed in film festivals? *D.A.W. Sintowoko, H. Azhar & H. Humaira*	307
Destination branding strategy *Bitombang* old village as alternate tourism destination in Selayar, Sulawesi Selatan *M. Andhyka Satria Putra & F. Ciptandi*	311
Exploring sustainable fashion market in Indonesia. Case study: *Sukkha Citta, Setali Indonesia,* and *TukarBaju* *R. Febriani*	316
Portable handle push-up bar design: How could plastic waste encourage healthy living? *H. Azhar, D.A.W. Sintowoko, Akhmadi, P. Viniani & M. Bashori*	321
Surround sound system matrix in Reaper for learning module *A.A. Anwar*	327

Succupedia: Augmented reality-driven horticultural book and its potential on becoming a metaverse object — 332
A.P. Budi & A. Harditya

The effect of digital marketing through social media is increasing online consumer trust in a pandemic period — 338
Christian & F.V. Kurniawan

The application of contemporary artistic theme in Aloft Jakarta Wahid Hasyim Hotel — 343
T. Sarihati, M.A. Alfarizy, R. Firmansyah, A.N.S. Gunawan & D.G. Dijiwa

The application of the philosophy and technical style of Batuan Bali painting to the modern narratives and process — 349
D.K. Aditya & I.D.A. Dwija Putra

La Kakao: Promoting small enterprises through visual storytelling based on the interpretation of local wisdom — 354
S.M.B. Haswati, K.S. Ahada, M. Buana & R.R.F. Ramli

The role of compact furniture design in the efficiency of floor area used of micro residence. Case study: Loft-style apartments in Surabaya — 360
Mahendra Nur Hadiansyah

Investigating commercial interior space presented in tourist Instagram posts — 365
A. Kusumowidagdo & M. Rahadiyanti

IRES marketing public relations strategy to raise children's environmental awareness through *KOMIK SOTA* — 370
P.N. Larasaty, M. Lemona, N.P. Pangaribuan, D. Widowati & M.B. Simanjuntak

@kiossahabat.id as social media marketing by Asperger individuals — 376
S.U. Suskarwati, E.A. Puruhito, A.H. Simanjuntak, A.J.C. Sagala & B.R. Satria

Forging ASEAN's cultural identity through digital museum diplomacy — 381
B. Riyanto, A.H. Assegaf, Y.W. Kurniawan, C. Mawuntu & G. Aulia

Dance as the medium to develop interpersonal communication for persons with Down syndrome in a digital era — 387
M.Y. Cobis, G.A. Mulyosantoso, K. Syahna, M. Wiguna & K.I. Septiani

The inclusive concept for designing four categories of disability education facilities — 393
A. Akhmadi, S.T. Virgana, A.P. Yuniati & H. Azhar

Utilization of planar material for shell structure with digital tectonics approach — 399
S.E. Indrawan

Systematic literature review: Games for social skills development in people with Autism Spectrum Disorder (ASD) — 405
O.D. Hutagaol, Y. Rahayu, P. Nova, M. Rafly & A. Sultani

Development of Tasikmalaya Kerancang solder design embroidery motifs for women's resort wear clothing — 410
J. Evan, M.Y. Tanzil & Y.K.S. Tahalele

Utilization of waste fabric for ready to wear deluxe unisex clothing design with fabric manipulation technique on Kamisado brand *C.A.P. Kurniadi, D.M.W. Githapradana & Soelistyowati*	415
Strategy for revitalizing city parks in the city of Bandung, West Java, Indonesia. Case study: City Hall Park *I. Sudarisman, M. Mustafa & M. Hafizal*	420
Reflection of sustainability concept on the use of resources in the 19th-century plantation *R. Wulandari & L. Nuralia*	426
Digital skills in education: Perspective from teaching capabilities in technology *R.R. Wulan, D.A.W. Sintowoko, I. Resmadi & Y. Siswantini*	432
Author index	437

Preface

On September 1, 2022, at Telkom University in Bandung, Indonesia, the 9th annual Bandung Creative Movement (BCM) conference took place with the theme "Sustainable Development in Creative Industries Embracing Digital Culture for Humanities." This topic, which employs a monetary balance to address challenges, will be discussed at the 9th BCM. These issues, in particular, concern the usage of an integrated sustainability area in cyberspace and physical space and the impact this has on the creative industries.

Sustainability will become an integral part of the creative industry's future growth in several key areas. At the conference, participants discussed

- Arts, Culture, and Society
- Design and Architecture
- Technology and New Media
- Management and Business

Issues pertaining to the cultural industries and digital technology were also considered as a solution to the sustainability in Creative industries. How does the creative community and the realm of digital technology react to and offer solutions to the global challenges posed by the sustainable development goals? The keynote speakers that came from various backgrounds and different countries are:

1. Prof. Lia Vilahur Chiaraviglio (EU ERAM University of Girona)
2. Prof. Elisabetta Lazzaro (UCA Business School for the Creative Industries)
3. Dr Mohd Asyiek Bin Mat Desa (University of Science Malaysia)
4. Dr. Arini Arumsari (Telkom University)
5. Dewa Made Weda Githapradana, S.Tr.Ds., M.Sn. (Ciputra University)
6. Mikhael Yulius Cobis, M.Si., M.M. (LSPR Communication and Business Institute)

Eighty papers, both group and individual efforts, were presented throughout the simultaneous sessions. In this open-access proceeding, all papers are organized into one of four categories. We hope that the 8th BCM will provide readers with a valuable overview of the digital creative world.

The Editors

Sustainable Development in Creative Industries: Embracing Digital Culture for Humanities –
Sintowoko et al (eds)
© 2023 the Editor(s), ISBN 978-1-032-44503-8
Open Access: www.taylorfrancis.com, CC BY-NC-ND 4.0 license

Scientific Committee

Chairman

Hanif Azhar, S.T., M.Sc.

Reviewers:

1. Dr. Ranang Agung Sugihartono S.Pd., M.Sn. *Indonesia Institute of the Arts, Surakarta*
2. Dr. Desy Nurcahyanti, S.Sn., M.Hum. *Sebelas Maret University*
3. Dr. Astrid Kusumowidagdo, S.T., M.M. *Ciputra University*
4. Dr. Mumtaz Mokhtar *Universiti Teknologi MARA UiTM, Malaysia*
5. Nadia Sigi Prameswari, S.Sn., M.Sn. *Universitas Negeri Semarang*
6. Jejen Jaelani, S.S., M.Hum. *Padjadjaran University*
7. Fadhly Abdillah, M.Ds *Pasundan University*
8. Prof. Dr. Aan Komariah, M.Pd. *Indonesia University of Education*
9. Rahayu Budhi Handayani, S.Sn., M.Ds. *Ciputra University*
10. Dr. Iwan Gunawan, M.Sn. *Jakarta Institute of the Arts*
11. Dr. Moh. Isa Pramana K, S.Sn., M.Sn *Telkom University*
12. Dandi Yunindar S.Sn., M.Ds *Telkom University*
13. Dr. Didit Widiatmoko Soewardikoen, M.Sn. *Telkom University*
14. Dr. Ira Wirasari, S.Sos., M.Ds. *Telkom University*
15. Ratri Wulandari, ST., M.T. *Telkom University*
16. Setiamurti Rahardjo, ST., M.T. *Telkom University*
17. Lingga Agung, S.I.Kom., M. Sn *Telkom University*
18. Wirania Swasty, S.Ds., M.AB. *Telkom University*
19. Idhar Resmadi, S.Ikom, M.T. *Telkom University*
20. Dr. Ranti Rachmawanti. M.Hum *Telkom University*
21. Andreas Rio Adriyanto, SE., M.Eng. *Telkom University*
22. Santi Salayanti, S.Sn., M.Sn. *Telkom University*
23. Aulia Ibrahim Yeru, S.Ds., M.Sn. *Telkom University*
24. Dr. Ully Irma Maulina Hanafiah, S.T., M.T. *Telkom University*
25. Dyah Ayu Wiwid Sintowoko, S.Sn., M.A. *Telkom University*
26. Dr. Djoko Murdowo, MBA *Telkom University*

Editors:

Dyah Ayu Wiwid Sintowoko, S.Sn., M.A.
Idhar Resmadi, S.Ikom, M.T.
Hanif Azhar, S.T., M.Sc.
Ganjar Gumilar, S.Sn., M.Sn.
Taufiq Wahab, S.Sn., M.Sn.

Telkom University
School of Creative Industries

Batik Bomba: Symbolic interaction through artwork

S. Alam & A. Budiman
Universitas Telkom, Bandung, Indonesia

D. Hidayat
Universiti Sains Malaysia, Malaysia

Sunarto
Universitas Negeri Semarang, Semarang, Indonesia
ORCID ID: 0000-0002-3136-1437

ABSTRACT: Batik Bomba has several dimensions such as ideas, behaviours, and artefacts. Batik Bomba is not only used for clothing needs but in it has a symbolic value that has meaning in the traditions of the people of Sulawesi Tengah. Batik Bomba not only has economic needs but Batik Bomba was created because of the influence of the surrounding environment, both natural, social and cultural so that there is an interaction between tradition and the environment in an image. Batik Bomba as the artwork is related to the socio-cultural conditions that became the background for the creation of the artwork. Symbolic interaction in Batik Bomba emphasizes the relationship between symbols and interactions, and the core of this approach is the individual view. This symbolic interaction assumes that each individual in himself has the essence of culture, interacts in the social midst of his community, and produces the meaning of "thoughts" that are collectively agreed upon. The symbolic meanings in Batik Bomba are depicts the natural environment and community traditions of Sulawesi Tengah.

Keywords: artwork, batik bomba, symbolic interaction

1 INTRODUCTION

Batik is still maintained and preserved, even by one of the United Nations agencies in charge of culture, namely UNESCO has designated batik as Humanity Heritage for Oral and Intangible Culture (Masterpieces of the Oral and Intangible Heritage of Humanity) since October 2, 2009. This fact gives hope about the sustainability of batik as an artwork that has high values with various variations, as well as increasing public and government attention to managing the preservation of traditional arts. This indicates that batik has been regarded as a cultural icon of the nation which has its uniqueness as well as symbols and deep philosophy. Furthermore, batik is not only considered a culture originating from Indonesia but is also recognized as a representation of the intangible culture of universal humanity.

Batik is one of the human creations that contain philosophical values, has character and artistic value, and has been part of Indonesian culture for a long time. As a cultural icon, batik is local wisdom that contains a very high historical value (Widagdo *et al.* 2021). Batik is a certain kind of fabric that is specially made with distinctive motifs, which are immediately recognized by the general public (Lokaprasidha 2017). Meanwhile, according to Widagdo *et al.* (2021) "Batik is a craft that has a high artistic value and has been part of Indonesian culture".

Almost every region in Indonesia has its characteristics and uniqueness in batik. In Palu, Sulawesi Tengah, there is also a typical batik, namely Batik Bomba. Although not as famous as Solo, Yogyakarta and Pekalongan, Batik Bomba craftsmen continue to explore to develop batik motifs and materials. Not only to be used as clothing but also to be used as decoration or also as a household appliance.

Efforts to understand the culture of Batik Bomba must of course include efforts to understand the existence of symbolic ideas in the batik object itself. As stated by Gamble that batik is a cultural object that includes traditions, and the environment in which humans and communities live and interact around them (Gamble 2004). This is because batik does not exist in a vacuum, but becomes the embodiment of human activities and ideas around it.

The use of Batik Bomba by the people of Palu is adjusted to the circumstances or an event, the adjustment to the use of Batik Bomba lies in the motifs in this batik. Batik Bomba with Taiganja motif, for example, this batik is only used at the time of the proposal; this indicates that a motif has a message to be conveyed to the public so that there is a symbolic communication and interaction in the community. Although the use of motifs as symbols in this interaction is intentionally and consciously carried out and has been much attached to the Kaili tribal community, many people do not realize that they are applying the theory of symbolic interaction in their daily lives. Therefore, this paper examines the symbolic interaction in Batik Bomba.

2 RESEARCH METHODS

The method used is a documentation study by examining various theories and research results about Batik Bomba. Sources of data come from publications both through international journals and national journals with the keywords bomba batik and the traditions of the people of Central Sulawesi. The strategy used in searching for data is to search for research reports, research articles, conceptual articles and books through the google search site on the internet.

Batik Bomba as an object of study was analyzed using the symbolic interaction theory which was first introduced by George Herbert Mead in 1934 at the University of Chicago, United States (Ritzer 2021). Symbolic interaction is one approach to discussing social interaction. According to Mead, social interaction in society occurs in two main forms, namely; 1) sign conversation (non-symbolic interaction); and; 2) the use of important symbols (symbolic interaction). The statement asserts that the emphasis on symbolic interaction is on the context of symbols because people try to understand the meaning or intent of an action performed by one another.

3 RESULT AND DISCUSSION

Symbolic interaction is a communication activity or exchange of symbols that are given meaning (Carter & Fuller 2016). Symbolic interaction teaches that humans interact with each other all the time, they share meanings for certain terms and actions and understand events in certain ways (Littlejohn & Foss 2011). Symbolic interaction exists because the basic ideas informing meaning come from the human mind (Mind) about the self (Self), and its relationship in social interaction and the ultimate aim is to mediate, and interpret meaning in the society (Society) where the individual stays.

Batik Bomba is related to the socio-cultural conditions that are the background of the creative process of the art. Batik motifs also appear as a result of socio-cultural changes or evolutionary cultural acculturation. However, the socio-cultural nuances involve agents of change, which in this context are the artists who make batik motifs. The symbolic interaction in Batik Bomba emphasizes the relationship between symbols and interactions, and the core of this approach is the individual view. This symbolic interaction assumes that each individual in himself has the essence of culture, interacts in the social midst of his community, and produces the meaning of "thoughts" that are collectively agreed upon.

Humans can create and manipulate symbols. This ability is needed for communication and interaction between individuals and between communities so that symbolic interaction is born. Based on George Herbert Mead's three main concepts about symbolic interaction, the symbolic interaction aspects contained in Batik Bomba are:

3.1 *Batik Bomba is the result of the mind process (mind)*

Bomba means "openness" and "togetherness", which is the name of the typical batik of Palu, Sulawesi Tengah. This is following the attitude of the Palu community, which is open to other communities or does not close themselves off to jointly advance their region. The tradition of Batik Bomba and weaving in Palu has been going on for a long time, but when this tradition began, valid literacy has not found. In the past, the process of making Batik Bomba started with making silk threads which were then woven into a cloth then made batik with ink material from tree sap using *canting* whose prints were made of wood.

Batik Bomba motif emerged from a thought process to distinguish it from other motifs to highlight its own identity as a form of cultural creation and innovation while taking inspiration from the characteristics of Sulawesi Tengah. In the process of the birth of the Batik Bomba motif, the motif makers respond to the socio-cultural situation in Sulawesi Tengah in a symbolic situation, they respond to the environment including physical objects (objects) and social objects (the behaviour of the people of Sulawesi Tengah) in Batik Bomba. As stated by Mead that thinking is a process in which individuals interact with themselves by using meaningful symbols. Through the process of interaction with oneself, the individual chooses which of the stimuli directed at him will be responded to.

3.2 *Batik Bomba is self reflection (self)*

3.2.1 *Gesture*

Each motif in Batik Bomba describes the uniqueness of Sulawesi Tengah, both in the natural environment and in the socio-cultural environment (Nuraedah & Bakri 2017). Batik Bomba is considered a complement to traditional ceremonies that have magical powers. The main motifs of Batik Bomba are the "*taiganja*" motif and the "*sambulugana*" motif as wedding ceremony equipment which have certain mythology and give special meaning.

Batik Bomba is a social response that occurs because of Batik Bomba craftsmen. The motif in Batik Bomba is a gesture or gesture. As stated by Mead that gestures are movements of the first organism that act as specific stimuli that evoke (socially) appropriate responses in the second organism (Ritzer 2021).

3.2.2 *Significant symbols*

A culture consists of ideas, symbols and values as a result of human work and behaviour, so it is not an exaggeration to say that humans are symbolic creatures. Humans think, feel and act with symbolic expressions. These symbolic expressions are characteristic of humans that clearly distinguish them from animals, so humans are called animal symbolicum or symbolic animals (Cassirer 1966).

Symbols complete all aspects of human life which include aspects of culture, including behaviour and knowledge. Batik Bomba as a work of human art has elements that reflect symbols of the people of Sulawesi Tengah. The symbol is reflected in the name of the batik motif, and the role and use of batik cloth. Batik Bomba is not just a painting written on cloth using canting. This is because the motif written on a piece of Batik Bomba cloth has the meaning to be conveyed. The symbolic meaning contained in Batik Bomba motif is discussed as follows.

1) Batik Bomba Describes the Natural Environment of Sulawesi Tengah

Sulawesi Tengah has abundant natural potential, both flora and fauna. Various flora that is used as a source of stylization for Batik Bomba motif is the flora that is often found in Sulawesi Tengah such as the Moringa leaf motif and the bamboo motif. One of the many faunas in Sulawesi Tengah is the Maleo Bird, so the maleo bird is distilled into one of the typical motifs of Batik Bomba.

a. Moringa Leaves Motif   b. Motif Bamboo   c. Maleo Bird Motif

Figure 1. Batik Bomba motifs sourced from flora and fauna. a. Moringa Leaves Motif b. Motif Bamboo c. Maleo Bird Motif (Source: UPT PKB-PNFI Sulawesi Tengah)

2) Batik Bomba Describes the Tradition of the People of Sulawesi Tengah

Various kinds of traditions exist in Sulawesi Tengah, one of which is the "*Sambulugana*" tradition. "*Sambulugana*" is a procession for proposing to the bride-to-be by bringing various kinds of offerings such as areca nut and betel nut, as well as giving a dowry in the form of a Taiganja pendant (Andiwa 2016). This tradition then inspired the birth of the "Taiganja" motif which became the main motif in Batik Bomba. Taiganja symbolizes the status of its owner, which is obtained by a certain customary and

Figure 2. Taiganja pendant and Taiganja motif. (Source: Museum Negeri Sulawesi Tengah)

generative procedure only. Thus, the ownership of Taiganja is not only based on the ability to obtain financing, but also on the customary and generative authority it has.

This decorative ornament from Taiganja is a symbol of fertility, prosperity and warding off evil. Women who come from high social strata usually wear Taiganja. so it is only suitable for use in traditional activities or wedding parties. Taiganja motifs symbolize fertility, prosperity, and immunity from adversity.

3.3 Batik Bomba is a social process in society (society)

Batik Bomba is not only used for ceremonial activities but can also be used for formal and non-formal events. The role of local governments in developing batik bomba is strongly supported by other parties such as batik entrepreneurs and the community. Batik Bomba entrepreneurs are very involved in developing this Batik Bomba by increasing work relations and collaborating with parties who can help develop and promote batik. In addition to the parties above, the community also contributes to the development of batik by wearing Batik Bomba.

The efforts that have been made by the local government in developing Batik Bomba (Oktaria 2018) include:

1) The local government has provided initial and development capital assistance to Batik Bomba craftsmen
2) The local government has required Batik Bomba to be worn by all government employees on certain days
3) Batik Bomba has been included as one of the skills taught in schools
4) Batik Bomba entrepreneurs actively promote both through print and electronic media.

The enthusiasm of the people of Sulawesi Tengah in preserving batik is a response to the government's concern in fighting for Indonesian batik, which has received confirmation from UNESCO. The intangible cultural values of batik are related to manufacturing rituals, artistic expressions, decorative symbolism, and regional cultural identity. Batik making, which begins with a special ritual, aims to provide an in-depth aesthetic and philosophical value to batik. The inauguration of batik by UNESCO is not solely based on batik itself but is more based on the aesthetic value contained in batik. The legitimacy of batik as a Global Cultural Heritage by UNESCO is expected to make a positive multidimensional contribution to the people of Indonesia in particular and the world community in general.

4 CONCLUSION

Batik Bomba motif emerged from a thought process to distinguish it from other motifs to highlight its own identity as a form of cultural creation and innovation while taking inspiration from the characteristics of Sulawesi Tengah. Batik Bomba as an artwork has elements that reflect symbols in the people of Sulawesi Tengah. The symbol is reflected in the batik motif, the role and the use of batik cloth. The symbolic meanings in Batik Bomba are; 1) to describe the natural environment, and 2) to describe the traditions of the community. The widespread use of batik by the community will give rise to new industries engaged in batik which will ultimately lead to sustainable social and economic development and the creation of new job opportunities.

REFERENCES

Andiwa. (2016). Makna Simbolik Sambulugana Pada Upacara Perkawinan Suku Kaili (Suatu Kajian Hermeunetika). *Bahasantodea*, 4(3), 28–34. http://jurnal.untad.ac.id/jurnal/index.php/Bahasantodea/article/view/13168

Carter, M. J., & Fuller, C. (2016). Symbols, Meaning, and Action: The Past, Present, and Future of Symbolic Interactionism. *Current Sociology*, 64(6), 931–961. https://doi.org/10.1177/0011392116638396

Cassirer, E. (1966). *An Essay on Man: An Introduction to a Philosophy of Human Culture*. London: Yale University Press.

Gamble, C. (2004). *The Basic Archeology*. New York: Routledge.

Littlejohn, S. W., & Foss, K. A. (2011). *Theories of Human Communication* (10th ed.). Waveland Press, Inc.

Lokaprasidha, P. (2017). The History of Batik and The Development of Kampung Batik Kauman as a Local to International Tourism Destination. *Journal of Tourism and Creativity*, *1*(1), 39–48. https://jurnal.unej.ac.id/index.php/tourismjournal/article/view/13796

Nuraedah, & Bakri, M. (2017). Klasifikasi Motif Kain Tradisional Batik Bomba Kaili. *Seminar Nasional Sistem Informasi UNMER Malang*, *1*(14 September), 715–723. https://jurnalfti.unmer.ac.id/index.php/senasif/article/view/97

Oktaria, M. D. (2018). Pengolahan Motif Pada Buya Bomba Dengan Teknik Digital Printing. *ATRAT: Jurnal Seni Rupa*, *6*(1), 13–24. https://doi.org/http://dx.doi.org/10.26742/atrat.v6i1.573

Ritzer, G. (2021). *Sociological Theory* (11th ed.). New York: McGraw-Hill Companies, Inc.

Widagdo, J., Ismail, A. I., & Alwi, A. binti. (2021). Study of the Function, Meaning, and Shape of Indonesian Batik From Time To Time. *Proceedings of the ICON ARCCADE 2021: The 2nd International Conference on Art, Craft, Culture and Design (ICON-ARCCADE 2021)*, 1–7. https://doi.org/10.2991/assehr.k.211228.001

Female students' perceptions on the comfort of the classroom and dormitory bedroom, in Insan Sejahtera Integrated Islamic Boarding Middle School Sumedang

T. Cardiah, S. Salsabila Az-zahra, R. Firmansyah & N. Laksitarini
Telkom University, Bandung, Indonesia

ORCID ID: 0000-0002-6211-8560

ABSTRACT: Insan Sejahtera Integrated Islamic Middle School is projected to create tough, achieving, and Islamic-minded students. To achieve this, the learning process must be supported by the comfort of a space in the perceptions of its users. This study aims to determine user perceptions of classrooms and dormitory rooms reviewed through 16 variables as a measure of room comfort. Research using descriptive and qualitative methods proves that in dormitory rooms, discomfort arises from the lack of artificial ventilation that makes the room feel hot. Meanwhile, classroom discomfort arises due to external noise and high temperatures when the weather is hot. The results of the study using a 'Likert scale' analysis showed that user perceptions of the rooms still do not meet the ideal level of comfort because there are still many comfort factors that have not been fulfilled optimally. The level of comfort of the room is an aspect of further research.

Keywords: boarding school, classrooms, dormitory rooms, perception, room comfort

1 INTRODUCTION

Insan Sejahtera Integrated Islamic Boarding Middle School Sumedang, which is located in Kampung Toga Housing Area Block G No.1, South Sumedang District, Sumedang Regency, West Java Province, is one of the private Islamic education institutions at the middle school level. It has been under the auspices of the Mitra Insan Sejahtera Foundation since 2016 and has been accredited by A. This school aims to create tough, achieving, and Islamic-minded students. The programs offered by this school include 2 pattern programs, namely full-day school and boarding school. To support the school in functioning properly as a place for teaching and learning, ideally, the classrooms should be able to meet the expectations of the comfort of teachers and students (Cardiah & Sudarisman 2019). In addition, dormitory facilities with shared rooms during school certainly have an influence on the learning process of students, each of whom has a different personality and traits. (Firmansyah *et al.* 2020). However, based on the obtained data, the user's perception of classrooms and dormitory rooms still does not meet the ideal level of comfort. Therefore, this study aims to determine user perceptions of classrooms and dormitory rooms reviewed through variables of benchmarks as a measure of room comfort with a focus on case studies at Insan Sejahtera Integrated Islamic Boarding Middle School Sumedang.

Figure 1. Classrooms.

Figure 2. Female dormitory rooms.

2 RESEARCH METHODS

The method used in this research is the descriptive qualitative and quantitative analysis where the results of this study are about the perceptions of female students and teachers, as users of classrooms and dormitory rooms, regarding the level of comfort there. The method of descriptive qualitative for explaining the results, along with a new understanding of the phenomenon is used in the study (Raco 2018).

2.1 *Methods of data collecting*

Primary data were obtained through data collection methods using field survey observations in dorm rooms and classrooms, and through questionnaires with teacher and student

respondents. This questionnaire is related to the theory of the 'content validation' test in which the responses and perspectives of respondents regarding the variables asked in the questionnaire are then analyzed and the results of the analysis will be used as research samples.

2.2 *Methods of data analysis.*

Data analysis of the questionnaire results used a 'Likert scale' analysis to assess the achievement of room comfort from the perspective of users based on predetermined benchmarks for room comfort needs, including natural lighting, artificial lighting, natural ventilation, artificial ventilation, temperature, external noise, internal noise, scent, cleanliness, safety, humidity, the visual beauty of space, functional space, security of space, furniture used, appearance of furniture form, and furniture's functionality.

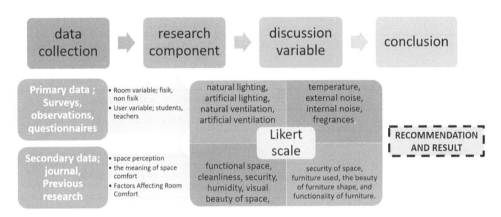

Figure 3. Research methods.

3 RESULT AND DISCUSSION

The analysis was carried out by measuring the comfort level of classrooms and dormitory rooms based on the scale, that is, uncomfortable, less comfortable, neutral, quite comfortable, and very comfortable. The respondent's opinion optional filling column is also included to find out the reason for filling the room comfort scale. The data were obtained from the results of questionnaires filled out by female students of Insan Sejahtera Integrated Islamic Boarding Middle School Sumedang. There are 16 variables as a measure of comfort in the dorm room, and these variables are taken from various journals including the GBCI standard, which discusses the comfort of the room, then adjusted according to the study's requirements.

3.1 *Classroom*

Table 1 shows the analysis results of the classroom comfort level from the perspective of female students as room users, with 16 benchmarks for room comfort requirements. The results of the questionnaire from 30 students at SMP IT Insan Sejahtera Boarding School will be explained in Table 1.

Table 1. Classroom comfort level analysis.

No.	Benchmarks	U	LC	N	QC	VC	Brief Description
1.	Natural lighting	0%	0%	0%	33.3%	66.7%	Many windows installed so a lamp isn't needed since it's already bright outside.
2.	Artificial lighting	0%	0%	0%	58.3%	41.7%	Bright lights used provides extra lighting when the weather is cloudy.
3.	Natural ventilation	0%	0%	8.3%	50%	41.7%	Windows installed can be opened for fresh air to enter so the room didn't feel stuffy.
4.	Artificial ventilation	16.7%	16.7%	41.7%	8.3%	16.7%	No fan or air conditioner.
5.	Temperature	0%	16.7%	25%	41.7%	16.7%	During the day, the class feels hot and stuffy.
6.	External noise	8.3%	16.7%	16.7%	41.7%	16.7%	Distracted by vehicle noises in a quiet and peaceful area.
7.	Internal noise	0%	8.3%	15%	33.3%	33.3%	Depends on each class. Mostly quiet because students are focused on studying.
8.	Scent	0%	0%	16.7%	41.7%	41.7%	Smells nice from the air freshener.
9.	Cleanliness	0%	8.3%	0%	50%	41.7%	Always in a clean condition.
10.	Humidity	0%	8.3%	16.7%	33.3%	41.7%	Good. Air inlets are present.
11.	Room visual beauty	0%	0%	16.7%	25%	58.3%	The visual beauty of the space is very simple. There are no decorations in the classroom.
12.	Room functionality	0%	0%	8.3%	25%	66.7%	The room is quite spacious and comfortable for studying.
13.	Room security	0%	0%	33.3%	25%	41.7%	After school, the classrooms are locked by the school security guard.
14.	Furniture used	0%	0%	16.7%	41.7%	41.7%	Very good and sturdy. No damage to the furniture in the classroom.
15.	Furniture visual beauty	0%	0%	25%	16.7%	58.3%	Chairs and tables look simple and beautiful.
16.	Furniture functionality	0%	0%	16.7%	33.3%	50%	Chairs and tables are in good condition and very suitable to use.

Notes: U = Uncomfortable, LC = Less Comfortable, N = Neutral, QC = Quite Comfortable, VC = Very Comfortable.

User's Perceptions of Insan Sejahtera Integrated Islamic Boarding Middle School Classrooms
In general, students' perceptions of classrooms are considered sufficient to provide room comfort because the facilities provided are still functioning properly (Mandulangi et al. 2016). Students' perceptions of comfort in the most uncomfortable classrooms are 16.7% artificial ventilation and 8.3% external noise. Meanwhile, the most comfortable perception is natural lighting and functional space, which is 66.7%. Classrooms are considered to be very clean so that they are comfortable to study with sufficient sunlight and without the artificial lights if it is not cloudy or rainy. Several windows in the classroom make the classroom free from stuffy (Bokoharjo 2017). The comfort of the learning process in the classroom can not only be measured based on students' perceptions of the interior of the classroom but also on classmates who learn together, and the way or process of learning in the classroom itself which is

considered very conducive with obedient students adhering to school regulations (Ma'nun 2020). Factors that can affect the comfort of the room are room circulation, natural forces or climate around the room environment, room noise, aromas or odors in the room, room shape, room security, room cleanliness, room beauty, and room lighting (Ashadi & Anisa 2017).

3.2 *Dormitory room*

Student dormitories should be thermally comfortable so that students stay healthy, both physically and psychologically (Kirani *et al.* 2016). Table 2 shows the analysis results of the dormitory room comfort level based on the perception of boarding female students as room

Table 2. Dormitory room comfort level analysis.

No.	Bench-marks	U	LC	N	QC	VC	Brief Description
1.	Natural lighting	0%	17.7%	25%	50%	8.3%	Windows in the rooms let sunlight in, but some are blocked by cupboards.
2.	Artificial lighting	0%	25%	8.3%	25%	41.7%	Very bright. Completely different from natural lighting.
3.	Natural ventilation	0%	8.3%	33.3%	8.3%	50%	The environment outside is cool but stuffy on inside.
4.	Artificial ventilation	25%	16.7%	41.7%	8.3%	8.3%	No fan or air conditioner.
5.	Temperature	0%	16.7%	41.7%	25%	16.7%	Uncertain.
6.	External noise	0%	0%	25%	25%	50%	Peaceful and quiet. The location is far from the main road; less noisy.
7.	Internal noise	0%	16.7%	8.3%	41.7%	33.3%	Came from the boarding resident.
8.	Scent	0%	0%	33.3%	25%	41.7%	Smells good from the air freshener.
9.	Cleanliness	0%	16.7%	16.7%	33.35	33.%	Always clean due to the dorm rules.
10.	Humidity	0%	8.3%	16.7%	41.7%	33.3%	Quite humid and cold.
11.	Room visual beauty	0%	9.3%	50%	25%	16.7%	Not so appealing. No decorations. Some furniture are damaged.
12.	Room functionality	0%	8.3%	8.3%	50%	33.3%	Cramped but still works well and is comfortable.
13.	Room security	0%	0%	58.3%	16.7%	24%	There is a key for every dormitory, also there is a security guard for the housing area, so it is secured.
14.	Furniture used	0%	05	25%	33.3%	41.7%	Some used sturdy, new furniture, but others have many damaged ones.
15.	Furniture visual beauty	0%	8.3%	41.7%	16.7%	33.3%	Considered quite simple and standard.
16.	Furniture functionality	0%	0%	16.7%	50%	33.3%	Furniture functions well even if some are damaged.

Notes: U = Uncomfortable, LC = Less Comfortable, N = Neutral, QC = Quite Comfortable, VC = Very Comfortable.

users, with 16 benchmarks for room comfort requirements. The results of the questionnaire with 30 students at SMP IT Insan Sejahtera Boarding School are explained in Table 2.

User's Perceptions of Insan Sejahtera Integrated Islamic Boarding Middle School Dorm Room.

Bedrooms that have a capacity of at least four people per room are generally considered comfortable enough to be used for activities, especially relaxing. The students' perception of the comfort of the dorm room was the highest percentage at 50% Very Comfortable is the use of natural air and external noise. and 50% Quite Comfortable is Natural Lighting. The most uncomfortable perception is artificial ventilation, which is 25%. However, the room feels hot because of the lack of openings and there is no artificial ventilation such as air conditioning or fans. Comfort is divided into two groups: physical comfort and psychological comfort. Comfort in a dorm room based on room interior factors, namely a conducive complex environment and friendly neighbors, as well as the proximity of roommates who share in all situations (Nurhidayat *et al.* 2021). Physical comfort is formed from several aspects including space comfort, visual comfort, thermal comfort (temperature, and audial/sound comfort (Aienna *et al.* 2016), while psychological comfort is an individual or personal feeling where this psychological comfort means a feeling condition, and thoughts of someone who expresses the individual's level of satisfaction with the surrounding environment so that information about the quality of psychological comfort for each individual needs to involve the process of feeling the sensation of comfort itself personally field (Gunawan & Ananda 2017).

4 CONCLUSION

Based on the perception of female students, in both dorm rooms and classrooms, **the most uncomfortable feeling (U)** is the use of artificial ventilation, while the **Very Comfortable (VC)** is natural ventilation and natural lighting. The dorm rooms are considered quite comfortable due to the beautiful environment outside and the dormitory facilities are quite complete and functional. The discomfort in the dorm room is temporary and comes from the weather conditions outside the building which makes the room feel hot. As for the comfort classroom, the perceived comfort is quite good, mainly because the class is always clean and bright so the learning atmosphere is quite focused and conducive. Classroom discomfort comes from the heat when the weather conditions are hot and the distraction from motorized vehicles passing by the school environment which is quiet. In conclusion, in general, the comfort of dorm rooms and classrooms is quite comfortable but cannot meet the ideal level of comfort because there are still many comfort factors that have not been maximally met as the results of the data analysis that has been obtained above. The recommendation for further research is that comparative studies are needed in different objects/locations, to prove the perception of the Most Uncomfortable and Very Comfortable.

REFERENCES

Aienna, Adyatma, S., & Arisanty, D. (2016). Kenyamanan Termal Ruang Kelas di Sekolah Tingkat SMA Banjarmasin Timur. *Jurnal Pendidikan Geografi*, *3*(3), 1–12. http://eprints.unlam.ac.id/1914/1/volume 3 nomor 3_a.pdf

Ashadi, A., & Anisa, A. (2017). Konsep Disain Rumah Sederhana Tipe Kecil Dengan Mempertimbangkan Kenyamanan Ruang. *NALARs*, *16*(1), 1. https://doi.org/10.24853/nalars.16.1.1–14

Cardiah, T., & Sudarisman, I. (2019). *Full Day School Education Concept as Forming Characteristics of Interior Space.*

Firmansyah, R., Ismail, S., Utaberta, N., Yuli, G. N., & Shaari, N. (2020). *Student's Perception of Common Rooms in Daarut Tauhid Tahfidz Islamic Boarding School, Bandung*. 192(EduARCHsia 2019), 86–89. https://doi.org/10.2991/aer.k.200214.012

Gunawan, & Ananda, F. (2017). Sekolah Menengah Umum. *Jurnal Inovtek Polbeng*, *7*(2), 98–103.

Kirani, F. F., Astrini, W., & Iyati, W. (2016). Evaluasi Desain Asrama Siswa dalam Aspek Kenyamanan Termal pada Unit Pelaksana Teknis (UPT) SMA Negeri Olahraga (SMANOR) Jawa Timur. *Jurnal Mahasiswa Jurusan Arsitektur*, *4*(4).

Ma'nun, L. (2020). *Perancangan Sekolah Berasrama Yayasan Cahaya Aceh Dengan Pendekatan Open Building*.

Mandulangi, L., Rondonuwu, D. M., & Rate, J. Van. (2016). Waterscape Architecture. *Freshwater Quality: Defining the Indefinable?*, 423–427.

Nurhidayat, A. R., Wulandari, R., Yuniati, A. P., Telkom, U., Ruang, P., & Remaja, K. (2021). *Perancangan Interior Smp Al Ma'Soem Islamic Boarding School*. *8*(4), 1682–1687.

Raco, J. (2018). *Metode Penelitian Kualitatif: Jenis, Karakteristik dan Keunggulannya*. https://doi.org/10.31219/osf.io/mfzuj

Design of a website-based competency improvement program at Sanggar Batik Katura Cirebon

G.M. Putra & F. Ciptandi
Telkom University, Bandung, Indonesia

ABSTRACT: Katura Batik Cirebon is a business engaged in the sale of Cirebon typical batik with various types of materials such as silk, semi-silk, and sanwos cotton. Now, the system runs where consumers still have to come directly to the store to purchase the product. Constraints faced by the store today are difficult to market or sell its products outside the city or region. The purpose of this study is to design an e-commerce system to support the sales process which can make reservations online and provide information up to date. This research uses a qualitative case study method of analysis that is done by literature study, data collection (observation), documentation regarding natural dyes and batik techniques, and interviews. This research is carried out in an e-commerce application for Katura Batik Cirebon stores to facilitate the customer to see the product by accessing the website e-commerce store.

Keywords: website design, e-commerce, Katura Batik Cirebon

1 INTRODUCTION

Technological developments are now increasingly sophisticated. Technological transformation continues to develop from day to day, one of which is the internet. The internet is a means of information and communication that is easy, fast, and accurate. This has made many parties use the internet to facilitate various interests, one of which is business purposes. This starts from small entrepreneurs to large companies that take advantage of advances in internet technology as a medium that serves to promote products or advertisements via the internet. Apart from being used for promotional media, the internet can also be used as a medium for selling and purchasing products, services, and information with consumers, which is commonly referred to as e-commerce.

The growth of online shopping has also influenced the industry structure. E-commerce has revolutionized the way of transacting various businesses, such as bookstores and travel agencies. Generally, large companies can use economies of scale and offer lower prices. Individuals or business people involved in e-commerce, both buyers and sellers, rely on internet-based technology to carry out their transactions. E-commerce has the ability to allow transactions anytime and anywhere and makes it possible to increase the overall business value of a company. Therefore, it is very necessary to understand the characteristics and different types of e-commerce businesses. However, opportunity costs can occur, if the local strategy is not suitable for new markets, and the company can lose potential customers. (I. Ariyanti 2022).

By taking traditional forms from business processes and utilizing social networking via the internet, a business strategy can be successful if done right, which ultimately results in increased subscribers, brand awareness, and revenue. Customer purchasing decisions are influenced by perceptions, motivation, learning, attitudes, and beliefs. The cultural context shapes the use of communication technology and the usage patterns of social networking sites. The advantage obtained by using transactions via e-commerce is to increase revenue by using online sales which have lower costs as well as operational costs such as paper and catalog printing. "E-Commerce is the sale or purchase of goods and services, between companies, households, individuals, governments, and the public or other private organizations, which is done via computers on network media" (Ahmadi C. & Hermawan D. 2013).

Therefore, e-commerce is what makes the process of shopping, buying or trading, selling products and services and information electronically or direct selling using internet network facilities where there is a website that can provide get and delivery commerce services to consumers (get and send online) will change all marketing activities and at the same time save other operational costs for trading activities (Astuti P.D. 2013). Marketing activities and the sales process at Katura Batik Cirebon are still manual, where buyers have to come directly to the Katura Batik Studio to buy, learn to make batik, or just browse the products available at Katura Batik Cirebon. Promotional activities are still limited to word-of-mouth (A.O. Siagian 2020) or expect promotions from customers who have purchased products at Katura Batik Cirebon. The obstacles faced by the store at this time are that it is difficult to market or sell their products outside the city or region and can only market them in the Cirebon area, for example, by relying on the BT Batik Trusmi showroom as a supplier for Cirebon Batik craftsmen, due to the lack of product renewal/innovation. New in marketing, and so it has an impact on less-than-optimal store revenue. This problem must be minimized in order to make it easier for stores to market their products, provide the latest product information, and make prices more competitive because prices have been printed on product information so as to attract buyers from both inside and outside the Cirebon area, West Java (N. Istifadhoh 2022).

Based on the above background, the authors are interested in conducting research at the Katura Batik Studio Cirebon to create an e-commerce website that can later assist the sales process and the store's digital promotion.

2 RESEARCH METHODS

We designed this research first through problem identification, literature study, and observation with the participation of local Cirebon batik craftsmen and after that, we made reports and evaluated the results.

2.1 *Problem identification*

At this stage, the identification of problems in the Batik Katura Cirebon shop is carried out. It aims to determine the problems raised in research in the form of product information dissemination and to expand the range of product information dissemination to the public that is felt to be less than optimal.

2.2 *Literature studies*

At this stage, a search for theoretical foundations obtained from various books and sources from the internet is carried out to complement the vocabulary of concepts and theories regarding e-commerce-based online shops, so that they have a good and appropriate foundation and science. In the form of theories about design, system design, e-commerce, websites, system modeling tools, MySQL databases, PHP, and Adobe Dreamweaver CS5.

2.3 Data collection

At this stage, data collection is carried out to obtain the data needed for research. The research uses the method of observation or direct observation of the object of research and the method of interviewing parties related to Cirebon Katura Batik/research object.

3 RESULT AND DISCUSSION

3.1 Analysis of the current system

The Katura Batik Cirebon shop is located in the Cirebon area, West Java, and is engaged in selling various kinds of batik clothing typical of Cirebon. Promotion and sales activities still use conventional methods where the sales process is that every customer who wants to buy or order batik can go directly to the Cirebon Katura Batik Shop, or to a showroom/supplier of Katura Batik handicrafts. In the promotion process, we still go through stories from one person to another. Thus, the promotion only expects customers who have bought this shop's batik to promote Mirabella Batik Jambi shop to their acquaintances (Suyanto M. 2003).

It is referred to previous research from the author Nurul Istifadhoh with the title "Pemanfaatan Digital Marketing Pada Pelaku Usaha Batik Ecoprint" in 2022. In this previous study, there were marketing media used, namely using social media and not making platforms or web-based applications. The novelty in this research is in terms of developing small and medium enterprises that use e-commerce as an improvement program for Katura Batik Cirebon as new media in marketing.

3.2 Problem solving

Of the several weaknesses contained in the running system, the researcher wants to implement an e-commerce system at the Batik Katura Cirebon shop, where the researcher proposes an online sales system using an e-commerce website that has advantages such as being able to carry out massive promotions and be able to expand product marketing. Which covers all of Indonesia to foreign countries, can increase sales because customers can order products according to their wishes which includes all internet users (Wong J. 2013), more effectively in managing the product data offered clearly so that there are no errors in promotion or delivery of information about products that are offered.

3.3 Implementation

Implementation is the process of translating designs that have been made into application programs that can be used by users. At this stage, the writer implements the results of the software design that has been converted into an output, namely in the form of an e-commerce system application.

The implementation of the website-based e-commerce design program at the Batik Katura Cirebon shop is as follows:

- Login and registration page
  On the login page, users are required to enter the correct username and password to enter the main page. Then on the registration page, the user can register an account.
- Visitor main page display
  The main page of visitors is a display that is used by visitors to access other pages and there are menus to display other pages.

This study has identified the design of website-based e-commerce with business processes from e-marketplace and social commerce. E-marketplaces used to identify e-commerce business processes are *Tokopedia, Bukalapak, Shopee, Blibli.com,* and *Blanja.com.* Meanwhile, social commerce is the use of social media to promote their product and the use

Table 1. The e-commerce design process.

Business Process	Description
Making email	Making username and password email for online shop
E-commerce registration	Making username and password online shop in e-commerce
Making online shop profile	Digitalizing the physical data of the shop and arranging the online shop profile
Uploading product	Digitalizing the physical data of the product and arranging the product catalog of the online shop
Promotion	Promotion with various promotional features on e-commerce
Sales Management	Communicate with consumers, check order status, and record orders received
Revenue management	Ensure the revenue has been received into the online shop account
Shipping product	Packing and shipping products to consumers
Complaint management	Discuss, provide solutions, and record complaints from consumers

of instant messaging to make transactions with consumers. Social commerce used to identify e-commerce business processes are *Instagram, Facebook, Line*, and *WhatsApp*. Based on the business process of e-marketplace and social commerce, we conclude the business process of e-commerce as shown in Table 1.

Figure 1. Login page display and login page display for admin (Source: Private).

Figure 2. The main page for consumer/visitors and shopping page display for consumers (Source: Private).

Table 2. The e-commerce competency program for Katura Batik Cirebon.

Business Process	Competency
Making online shop profile	Able to make and display online shop rules and profile of the seller
Uploading product	Able to take, make, and upload product photos and descriptions
	Able to upload products at online shop
Promotion	Able to promote products on e-marketplace and social media
Sales Management	Able to make a list of frequently asked questions and answers
	Able to check and record the transaction status of product sales
Revenue management	Able to record data of online shop revenue and the invoice that has been paid to the online shop account
Shipping product	Able to record and track data of shipped product
Complaint management	Able to explain, provide, and record any complaint or solutions.

4 CONCLUSION

The promotion system at the Batik Katura Cirebon shop still relies on product promotion by word of mouth and Batik Cirebon suppliers, such as BT Batik Trusmi, do not have special media to promote their products. For this reason, the e-commerce application for the Cirebon Katura Batik shop produced in this study can facilitate business sales with the Business to Customer (B2C) e-commerce transaction type that directly connects customers to view products in detail and place orders without having to come. to the shop simply by accessing the e-commerce website of the Batik Katura Cirebon shop.

Therefore, this competency is used to plan an e-commerce competency improvement program for Katura Batik Cirebon, providing a conceptual overview of the types of e-commerce in Indonesia. For further research, it is necessary to test empirically both qualitatively and quantitatively, so that information is found such as what types of e-commerce are most effectively used in business, or what types are most widely used in Indonesia.

REFERENCES

Ahmadi, C. and Hermawan, D., 2013. E-business dan E-commerce. *Penerbit Andi. Jogyakarta*, pp.47–53.

Ariyanti I., Novita N., Khairunnisa D., and Ganiardi A., "Website Sebagai Sarana Promosi UKM Sopyan," *Aptekmas J. Pengabdi*, vol. 3, no. 3, 2022, doi: https://doi.org/10.36257/apts.v3i3.2125.

Astuti, P.D., 2013, March. Perancangan Sistem Informasi Penjualan Obat Pada Apotek Jati Farma Arjosari. In *Seruni-Seminar Riset Unggulan Nasional Inoformatika dan Komputer* (Vol. 2, No. 1).

Hidayat, T., 2008. *Panduan Membuat Toko Online Dengan OSCommerce*. MediaKita.

Istifadhoh N., Wardah I., and Stikoma T., "Pemanfaatan Digital Marketing Pada Pelaku Usaha Batik Ecoprint," *Aptekmas J. Pengabdi.*, vol. 5, no. 2, 2022, doi: https://dx.doi.org/10.36257/apts.vxix

Media K. 2018. *Gandeng "Marketplace" Pemerintah Ajak UMKM Indonesia Mulai Berjualan Online* [Online]. Available at: https://biz.kompas.com/read/2018/04/24/193853228/gandeng-marketplace-pemerintah-ajak-umkm-indonesia-mulai-berjualan-online (Accessed 1 June 2018)

Munawar, K., 2009. E-commerce. *Diakses July, 21*, p.2015.

Siagian A. O., Martiwi R., and Indra N., "Kemajuan Pemasaran Produk Dalam Memanfaatkan Media Sosial Di Era Digital," *J. Pemasar. Kompetitif*, vol. 3, no. 3, 2020, doi: 10.32493/jpkpk.v3i3.4497.

Suyanto, M., 2003. *Strategi Periklanan Pada E-commerce Perusahaan Top Dunia*. Penerbit Andi.

Wong, J., 2013. *Internet Marketing for Beginners*. Elex Media Komputindo.

Thrift shop-based business rebranding practice, Case study: Andys_store24

G.P.S. Basten & R. Febriani
Telkom University, Bandung, Indonesia

ORCID ID: 0000-0001-5113-6416

ABSTRACT: In this modern era, the rapid development of fashion has led Indonesian people to continue fulfill a quality lifestyle by following trends. Saving money or buying used items at low prices for personal use can now be a viable solution because they have a more affordable price and relatively high quality added value for clothes. In the second-hand goods sector, the fashion style in great demand is the vintage style due to the notion of classic and antique production clothing—past and present models. Facts in the field show that many fans of vintage clothing styles are behind the quantitative research method with a focus on business rebranding process research, aiming to increase sales profits and meet market needs.

Keywords: Branding, Thrift, Vintage Style

1 INTRODUCTION

In this modern era, many developments in various fields are rapidly and continuously updated, one of which is in the field of fashion. This phenomenon occurs because of buying clothes to follow a trend or lifestyle (Fatimah 2016). Buying used goods could be a solution because used goods are purchased at affordable prices and of good quality. And it can fulfill the great desire of the Indonesian people to follow fashion trends, as well as increases public awareness of the use of second-hand goods to reduce textile waste.

Fashion enthusiasts make some people, including researchers, choose to become business actors who sell decent used clothes by carrying the color block/multi-colored fashion style as a brand image in their business. Currently, researchers know the types of clothing that are in great demand by consumers from various circles, namely vintage models. The fact of this phenomenon became the inspiration to change the image from selling color-block jacket products to vintage-style jacket products, through a rebranding process. The rebranding process can help in analyzing and determining marketing strategies for something new for buyers (Argenti & Putri 2010), which aims to get more significant profits and has more potential to reduce clothing waste.

2 RESEARCH METHODS

This research was conducted using qualitative research methods. This method was selected as it is based on the philosophy of the positivism paradigm. The methodology applied was a case study approach, which aims to research and understand things related to actual conditions/facts, comprehensively and integratively, and using the principle of triangulation

(Hakim 2017). In addition, this type of research is classified as field research, namely analyzing and describing data obtained from research targets/respondents through data instruments such as interviews, observations, documentation, and so on (Nata 2000). This type of qualitative descriptive research is to obtain the right strategy for changing and introducing a new brand image of the researcher's business.

2.1 *Methods of data collecting*

The research data were obtained from primary and secondary sources. Data from primary sources are considered more accurate and detailed because they are received directly (without intermediaries) through the instruments used in primary data collection. For example, interviews are conducted to obtain information from more experienced informants. And observations are made to see and seek documentation of phenomena in accordance with field facts. Data from secondary sources are obtained indirectly because they come from existing data sources and have been collected. The instrument used in collecting primary data is a literature study obtained through books, journals, theses, and articles.

3 RESULT AND DISCUSSION

3.1 *Thrift vintage style study*

Thrift is a word of English origin. According to the OCBC NISP (2022) article, the word thrift is the activity of buying or using used goods for reuse. Many products are sought after during frugal activities, one of which is vintage products. Thrift products can be said to be vintage if the product is no longer produced, unique, and rare (Gafara 2019). According to the Oxford dictionary, the word "Vintage" means "old and very high quality". This attracts researchers (as business actors) to take advantage of the activities of used goods and vintage products as opportunities because vintage products can be used by all groups and in various situations, so that it creates opportunities for more users to get economical products. It also has an impact on the environment because it is realized that fashion products are the second largest source of accumulation of waste and pollution (Rukhaya *et al.* 2021).

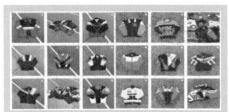 

Color-block style product   Vintage Style product

3.2 *Business re-branding study*

According to Muzellec and Lambkin (2006), rebranding creates elements for a brand that carries out new development, aiming to help sellers analyze prospective markets and determine marketing strategies reflecting the product to provide something new for buyers (Argenti & Putri 2010). Business development is carried out by researchers as business actors are product rebranding.

The choice of rebranding is a business development strategy because it looks at the facts on the ground, that vintage products are more in demand by the public because they can be used in all circles and various event situations. This is in accordance with the statement of Lomax and Mador (2002) that there is business development through rebranding which is motivated by internal factors (changes in company strategy), namely the experience and observations of sellers from facts on the ground.

Researchers do business rebranding through social media content (especially Instagram and Shopee applications) which are the media for marketing strategies and the introduction or promotion of the latest products to be launched. The content created can be in the form of setting up Instagram feeds (layout product photos, using models, uploading photos in stages as product transitions, making several events such as sales, and giveaways), promoting with other accounts, and optimally and consistently using the features on the Instagram website or mobile application (for example, live, insta-story, free shipping, etc.) by presenting vintage-style products as the main products sold. This strategy is expected to attract new customers without losing existing customers.

The success of this rebranding strategy can be seen from the increase in turnover and sales insight, as well as the many interactions with consumers about products through comments, saving product photos, active consumers providing feedback stories, and live sales in the application. And in this study, researchers succeeded in developing business through product rebranding, evidenced by an increase in turnover, insight, and active consumer feedback. The following are the differences in the insights obtained from interactions and promotions with consumers:

Vintage-style product introduction and promotion content on the Instagram app

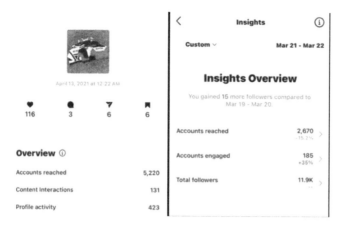

Color-block style product

insight results

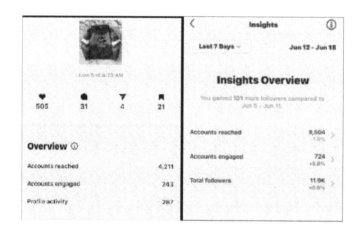

Vintage-style product

insight results

4 CONCLUSION

The right decision as a savings-based business actor to develop a business by rebranding and changing product collections from color-block style to vintage-style products is the right decision. The rebranding process can maximize the concept of thrifting goals and apply sustainable products to reduce fashion waste, increasing sales turnover, which means a high interest in using second-hand goods.

The choice of vintage style as a rebranding option is also considered appropriate because researchers as business people not only look at it from the seller's perspective but also consider how Indonesian people dress in all walks of life. Therefore, from the observations and facts from the field, the researcher decided to take a style that can be worn by all groups and follow the dress norms of the Indonesian people so that later it can increase sales turnover, which is vintage.

The process carried out by the author to carry out rebranding is by focusing on creating content in sales applications, by creating content following those that are viral or that are liked by consumers, such as giveaways, selling prices, and photos. Furthermore, the product is made as attractive as possible using models, and the layout of the Instagram feed is made as neat and reasonable as possible so that consumers are comfortable viewing visual accounts or bringing them online. The method of rebranding using content is considered more effective, as evidenced by the range; it can attract a large number of buyers and can increase sales turnover.

REFERENCES

Argenti, P. A., & Putri, A. I. (2010). *Komunikasi Korporat*. Salemba Humanika.

Fatimah, L. (2016). *Peran Sikap, Norma Subjektif & Perceived Behavioral Control Tergadap Intensi Membeli Pakaian Bekas*.

Gafara, G. (2019, April 19). A Brief History of Thrifting. *USS FEED*.

Hakim, A. (2017). *Metodologi Penelitian (Penelitian kualitatif, tindakan kelas, & studi kasus)*. http://repo.uinsatu.ac.id/11536/6/BAB%20III.pdf

Lomax, Wendy., Mador, Martha., Fitzhenry, Angelo., & Kingston Business School. (2002). *Corporate Rebranding: Learning from experience*. Kingston University.

Muzellec, L., & Lambkin, M. (2006). Corporate Rebranding: Destroying, Transferring or Creating Brand Equity? In *European Journal of Marketing* (Vol. 40, Issues 7–8, pp. 803–824). https://doi.org/10.1108/03090560610670007

Nata, A. (2000). *Metodologi Studi Islam*.

OCBC NISP With You. (2022, June 9). *Apa itu Thrift Shop? Ini Keuntungan dan Cara Memulainya*.

Rukhaya, S., Yadav, S., Rose, N. M., Grover, A., & Bisht, D. (2021). Sustainable Approach to Counter the Environmental Impact of Fast Fashion. ~ 517 ~ *The Pharma Innovation Journal*, 8, 517–523. http://www.thepharmajournal.com

The physical and philosophical values of the Mappalette Bola tradition and its potential in industrial development design earthquake resistant house foundation construction

S.D.S. Rusdy & F. Ciptandi
Telkom University, Bandung, Indonesia

ORCID ID: 0000-0003-2753-3475

ABSTRACT: The diversity of traditions in Indonesia is part of a culture that cannot be separated from development and presents a lot of potential to be used as inspiration. One of the areas in eastern Indonesia that has interesting traditions to discuss and has the potential to be used as inspiration is South Sulawesi. The mappalette bola tradition is the tradition of moving the house of the Bugis tribe which involves moving the house in its entirety using human assistance. This qualitative-interpretive research will discuss the physical and philosophical values of the mappalette bola tradition, then reinterpret it for implementation purposes in the development of the earthquake-resistant house foundation construction design industry. This design can provide benefits to inspire designers in designing something with deep meaning and values inspired by traditional culture.

Keywords: Bugis, Construction, Earthquake-Resistant House, Mappalette Bola, Traditions

1 INTRODUCTION

According to the Crisis Center of the Indonesian Ministry of Health, Indonesia is a country that has a fairly high level of earthquake vulnerability. This is because 80% of Indonesia's position is located at the confluence of three active plates that are continuously moving. The three active plates are the Indies-Australia, the Pacific and Eurasia. This makes Indonesia one of the countries that has a fairly high threat of earthquakes compared to other countries. So that the potential for damage to Infrastructure and losses due to the earthquake caused are very high in Indonesia. In addition, the potential for casualties caused by building structures that collapse and hit humans due to not being able to withstand earthquakes is very high (Maharani *et al.* 2020). To anticipate this, a building foundation construction design is developed that can absorb shocks during an earthquake.

The Mappalette Bola tradition itself is a tradition originating from the Bugis tribe in South Sulawesi Province, which moves the house in its entirety using human assistance and is a legacy tradition from the ancestors of the Bugis tribe (Musnur 2018). With the physical and philosophical values contained in the mappalette bola tradition, the author will explore the potential to develop these values and their potential in the industry. This development will be realized in the design of earthquake-resistant house foundation construction. With the earthquake-resistant house foundation construction industry, it continues to grow every year. So the construction design of the developed house foundation can compete in the earthquake-resistant house foundation construction industry with the physical and philosophical values of the mappalette bola tradition in it. This research will focus on exploring the physical and philosophical values contained in the mappalette bola tradition.

1.1 *Mappalette Bola*

The mappalette bola tradition is a moving house tradition that comes from the Bugis tribe. Moving house in the sense of moving the house as a whole. Usually the mappalette bola tradition is carried out if one of the people wants to move and sell their house but not the land. The house that was moved is a traditional house on stilts made of wood, typical of the people of Sulawesi. The framework of the house usually uses poles and beams that are assembled without using nails. As well as with the shape of a rectangular building that is made extending towards the back. In the mappalette bola tradition, there is a process that must be done, namely before the house is moved, the household furniture in the house must be removed from the house to avoid damage. Then the poles under the stilt house are attached with bamboo which is useful for lifting the house. Involving almost tens or even hundreds of residents. Before the procession begins, prayers are also said together so that it runs smoothly and according to expectations. This procession is only carried out by men. The process of appointing and moving houses is generally led by a traditional leader to give the signal and direct the residents (Musnur 2018). There are previous research that have similar research topics and are related to the mappalette bola tradition. The previous research was Irfandi Musnur (2018) in his research on the symbolization and implementation of pacce (solidarity) as an analogy for the representation of togetherness in Bugis society. In this research, it was stated that the mappalette bola tradition itself is an implementation of the Pacce philosophy of the Bugis tribe which can be seen as a sense of togetherness at work.

Figure 1. The Mappalette Bola tradition.

2 RESEARCH METHODS

This paper is qualitative-interpretative, choosing a qualitative research method, namely Online Ethnography. The choice of the research method itself is because qualitative research methods are often called descriptive research and tend to use analysis by collecting data from literature studies (Rosyidah & Ciptandi 2019). Literature study is done by reading literature studies in the form of books, research journals, etc. While the interpretative method is a method of reinterpreting, the process, method, the act of reinterpreting the meaning that already exists (Ji-Ae and Seung-Hwan 2007). In this paper, the author will describe in detail the facts about the mappalette bola tradition, then interpret the physical or philosophical values contained in the mappalette bola tradition. From the results of the interpretation of the physical or philosophical values of the mappalette bola tradition. Then it will look for physical or philosophical values that can be developed into designs that have industrial value.

In writing this research, there are several stages of research that must be carried out, namely determining the subject, collecting data, analyzing, interpreting, seeking potential, to designing design concepts. The initial stage in this research is to determine the research subject, in this stage the author looks for an interesting subject or topic from the internet, books, research journals, etc. related to the culture of South Sulawesi. After determining the subject or research topic, the author will collect research data, more data collection techniques are literature studies carried out by reading literature studies in the form of books, research journals, etc. After the research data is collected, data analysis will be carried out. The data analysis technique that will be used in this paper is the data analysis technique of the Miles and Huberman model. the stages of the Miles and Huberman model data analysis technique are data collection, data reduction/summarization, data presentation, and verification (Sugiyono 2020). After the data analysis stage, the next stage is the stage of reinterpreting the data that has been obtained and then looking for potentials of industrial value. The last stage is to design a design concept that is inspired by the subject or topic raised.

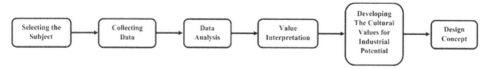

Figure 2. Research method (Research to designing scheme).

The design is done by processing and interpreting the values obtained and trying to see other forms of culture originating from Bugis customs. From the results of the analysis, these values were then developed using the mind-map method. From the results of the development, earthquake-resistant house foundation construction attributes are needed for the realization of the design concept for this potential.

3 RESULT AND DISCUSSION

Through the process and stages of collecting data, which is done by collecting information through journals, books, etc. Information data obtained that almost 70% of house damage is caused by the impact of earthquakes so that the use of earthquake-resistant house foundation construction is mandatory in Indonesia, but there are still many Indonesian people who do not understand earthquake-resistant buildings ('Teknologi Rumah Tahan Gempa' 2018). So that the development of earthquake-resistant building construction continues every year. The author in this research will explore the physical/philosophical meaning and interpret the values contained in the mappalette bola tradition. Then the author will develop this value through a design concept that will provide added value for the earthquake-resistant house foundation construction industry in the future.

3.1 *Analysis*

In analyzing the physical and philosophical values of the mappalette bola tradition, the author will try to see other forms of culture originating from Bugis customs. Other forms of culture originating from the Bugis tribe that are seen are Passapu and Kanna Bugis. Quoted from the writings of Dian Cahyadi and Karta Jayadi (2019), that Passapu is a traditional men's headband typical of South Sulawesi. Passapu was used by many people, ranging from soldiers, people who were hunting, to people in the royal circle. Passapu also symbolizes masculinity, resilience, strength, and agility. The mention of the name passapu itself is well known in the Makassarese/Bugis indigenous peoples, whose use is still sustainable in various traditional activities of the South Sulawesi people, especially the Bugis/Makassar tribes. As for the Bugis kanna itself, quoted from the Indonesian Ministry of Education and Culture, Bugis Kanna itself is one of the traditional weapons of South Sulawesi in the form of a shield. Unlike other traditional weapons of South Sulawesi, Kanna serves to protect themselves from attacks by weapons of the enemy. Kanna has been known since the heyday of the local kingdom among the Bugis people. This weapon is even mentioned in the folklore entitled Pau-Paunna Sawerigading.

From two other cultural forms of Bugis, the author explores the physical and philosophical values of the mappalette bola tradition and finds similarities or patterns in other forms of Bugis culture, namely patterns of resilience or protection. The following is an explanation of the pattern: In the mappalette bola tradition, the physical value obtained is from a traditional house on stilts made of wood, typical of the Sulawesi people, where the frame of the house usually uses poles and beams that are assembled without using nails so that the frame of the house itself is not too rigid. so that when there is an earthquake vibration the frame of the house can still withstand it. While the philosophical value of the mappalette bola tradition obtained is that the residence must be built in a balanced and sturdy manner so that it can be passed on to the next generation. while in passapu, which symbolizes a symbol of resilience and agility, the construction of the foundation of today's houses must be strong enough to withstand vibrations, especially vibrations from earthquakes. In Kanna Bugis, it also serves to protect oneself, which is where one of the goals of earthquake-resistant house foundation construction is usually to protect residents from earthquake vibrations. The author finds a pattern that is quite easy to see when examining other forms of culture belonging to the Bugis tribe. It can be concluded that there are values about patterns of resilience or protection that occur in the past, present, and future in a culture or tradition.

3.2 *Earthquake resistant house foundation construction*

Quoted from Zelly Rinaldi *et al.* (2015) regarding earthquake-resistant foundation construction, there are three main principles of earthquake-resistant construction, namely 1) A simple and symmetrical plan, a plan like this can withstand earthquake forces better because of the lack of torsional effects and its strength is more evenly distributed. 2) Building materials should be as light as possible, so that the building is not too heavy so that when an earthquake occurs it is not easy to collapse. 3) Adequate construction and foundation systems, The need for adequate load-bearing construction and foundation systems so that a building can withstand vibrations resulting from earthquakes. For the foundation itself usually uses a pedestal foundation and a local reinforced concrete foundation.

The Mappalette Bola tradition in relation to earthquake natural disasters can be linked to the development of the earthquake-resistant house foundation construction design industry. The earthquake-resistant house foundation construction design inspired by the mappalette bola tradition itself is expected to compete with the existing earthquake-resistant house foundation construction design. The development of earthquake-resistant foundation construction in residential houses aims to create a residential building that can prevent casualties, and minimize property losses when an earthquake occurs (Boen 2009). Also the use of earthquake-resistant foundation construction in residential houses in order to provide sufficient time for the occupants of the dwelling to save themselves before the residential building collapses due to the earthquake.

3.3 *Earthquake resistant house foundation construction design concept*

At the stage of developing the culture, there are stages that are carried out, including the search for values using the mind-map method. mind-map method will be chosen as a model in describing values that can be transformed into a design concept for earthquake-resistant house foundation construction. Values of resilience, balance and sturdiness are found in the mappalette bola tradition. In addition to the value of the mappalette bola tradition, the author also looks at other cultural forms originating from Bugis customs, namely Passapu and kanna bugis, where Passapu itself has the values of virility, resilience, masculinity, strength, and agility, while kanna bugis has the value of protection.

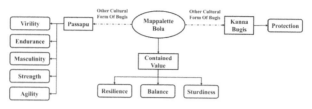

Figure 3. Mind-map (search for values).

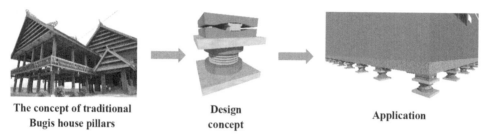

Figure 4. Design concept.

These values can be associated with residential buildings which will later become a source of inspiration for designing the concept of earthquake-resistant house foundation construction. The value of resilience, balance, and sturdiness symbolizes that the construction of earthquake-resistant house foundations has resilience in the face of vibrations from earthquakes and also has the properties of balance and is sturdy in the face of earthquakes. For the design concept of earthquake-resistant house foundation construction, it will take the concept of the traditional Bugis house pillars where the house poles are assembled without using nails and placed on a rock with balance.

4 CONCLUSION

Cultural forms contain physical and philosophical values in them that can be developed and have potential in today's industry. The mappalette bola tradition with the values of resilience, balance, and sturdiness found in it, has the potential to participate in the earthquake-resistant house foundation construction industry. It is hoped that the design of the earthquake-resistant house foundation construction concept itself can have potential in the industry and play a role like the role of tradition. The novelty in the design of this earthquake-resistant house foundation construction concept is that the house foundation construction design is assembled without using screws/locks so that the foundation construction of the house is not rigid so that it can dampen vibrations from earthquakes. In addition, there are many other potentials that can be explored from the physical and philosophical values contained in the mappalette bola tradition. The value of mutual help, the value of togetherness, etc. can be explored further for its potential other than as the construction of earthquake-resistant house foundations. The development of this potential can be in the form of activities such as events or logo/symbol designs inspired by these values.

REFERENCES

Boen, T. (2009) *Manual Bangunan Tahan Gempa*. Jakarta: World Seismic Safety Initiative.

Cahyadi, D. and Jayadi, K. (2019) 'Makassar Headdressed Passapu/Padompe'. *TANRA, Jurnal Desain Komunikasi Visual*, 6(1), pp. 100–117.

Ji-Ae, H. and Seung-Hwan, O. (2007) 'Reinterpretation Of Traditional Culutre By Developing A Real-Time Composer With Pattern And Sound -Based On Symbolism Of Korean Traditional Pattern'. *International Association of Societies of Design Research*, pp. 1–12.

Kementerian Pendidikan, K.R. dan T.R. (2010) *Senjata Tradisional Daerah Sulsel*. Available at: https://warisanbudaya.kemdikbud.go.id/?newdetail&detailCatat=562 (Accessed: 11 June 2022).

Maharani, N., Eka Kherismawati, N.P. and Widya Sari, N.L.P. (2020) 'Sosialisasi Dan Simulasi Gempa Bumi Di Smpn 3 Kuta Selatan Badung Bali'. *Jurnal Bakti Saraswati*, 9(1), pp. 31–37.

Musnur, I. (2018) 'Simbolisasi Dan Implementasi Pacce (Solidaritas) Sebagai Analogi Representasi Kebersamaan Dalam Masyarakat Bugis'. *NARADA, Jurnal Desain & Seni*, 5(2), pp. 77–97.

Pusat Krisis Kementerian Kesehatan RI. (2016) *Potensi Gempa Bumi Di Indonesia Masih Tinggi*. Available at: https://pusatkrisis.kemkes.go.id/potensi-gempa-bumi-di-indonesia-masih-tinggi (Accessed: 7 June 2022).

Rinaldi, Z., Purwantiasning, A.W. and Dewi Nur'aini, R. (2015) 'Analisa Konstruksi Tahan Gempa Rumah Tradisional Suku Besemah Di Kota Pagaralam Sumatera Selatan'. In *Seminar Nasional Sains Dan Teknologi*. Jakarta: Fakultas Teknik Universitas Muhammadiyah Jakarta, pp. 1–10.

Rosyidah, S. and Ciptandi, F. (2019) 'Pengembangan Kain Tenun Gedog Tuban Bertekstur Dengan Pewarna Alam Mahoni'. In *E-Proceeding of Art & Design*. pp. 2049–2057.

'Teknologi Rumah Tahan Gempa'. (2018) Indonesia: CNN Indonesia.

Sugiyono. (2020) *Metode Penelitian Kuantitatif, Kualitatif, Dan Kombinasi (Mixed Methods)*. 2nd Edition. Sutopo (ed.). Bandung: ALFABETA.

Marketing strategy of *Tipa-Tipa* (souvenirs from Lake Toba) with SWOT method

J.B.W. Jawak & I. Wirasari
Telkom University, Bandung, Indonesia

ORCID ID: 0000-0002-6074-0487

ABSTRACT: Tipa-Tipa is a typical food from Toba Regency designed as souvenir. One of the districts in Toba Regency, Sigumpar, is the home to 16 SMEs (in the form of Small to Medium Enterprises) which produce Tipa-Tipa. Data shows that Tipa-Tipa does not experience any development, both in serving and marketing methods. Thus, leaving an underdeveloped market segment for Tipa-Tipa. The Lake Toba Area is one of the National Tourism Strategic Areas (*Kawasan Strategis Pariwisata Nasional*, KSPN), which provides opportunities to develop local culture and attract tourists. Therefore, business owner should have a strategy to compete in the snacks and souvenirs market and harness the opportunities from being in the location of the KSPN. The methods involved in formulating the strategy recommendations are mixed techniques with Internal Factor Evaluation (IFE) matrix, External Factor Evaluation (EFE) and Strengths, Weaknesses, Opportunities, and Threats (SWOT). From the analysis, it is known that SMEs are in Grow and Build position so they need a new strategy, where there are 3 alternative strategies that can be used.

Keywords: Tipa-Tipa, strategy, marketing, SWOT

1 INTRODUCTION

Tipa-Tipa is a typical food that comes from one of the Lake Toba areas, namely Toba Regency, North Sumatra. This food is intended as a souvenir because of its very distinctive shape and taste. The manufacturing process itself starts from rice then soaked in water for approximately two nights. After that drained and roasted until cooked. In a hot state then crushed by pounding. The result is then called the Tipa-Tipa. At first this product was produced by each family after the harvest period, but this habit has been abandoned by the community so that the types are less well known to the public, especially teenagers and children (Simanjuntak 2018). In its development, types can be found at bus stops or public transportation along the Sumatran causeway. One of the sales locations for these products is at Simpang Silimbat, Jl. Siborong wholesale-Parapat. district. Sigumpar. Where, there are 16 business actors who sell along the road with the main product, namely Tipa-Tipa. Based on initial interviews with business actors, information was obtained that product sales did not experience development from the time it was opened until now. Then from initial observations, it is also seen that the marketing method is still quite simple, namely packaging that only uses clear plastic without any other identification.

The Lake Toba area itself has been designated by the government to be one of the 10 National Tourism Strategic Areas (*Kawasan Strategis Pariwisata Nasional*, KSPN), as stated in the Cabinet Secretariat Letter Number B-652/Seskab/Maritim/11/2015 regarding the presidential directive on tourism. Various improvements to facilities and resources were made to make it easier and attract the attention of tourists. Regional identity such as culture with all its elements is also the main attraction that will be presented when traveling. So that there are quite large opportunities for business actors with regional specialties to be able to develop their business by attracting the attention of visiting tourists.

Marketing strategy is a marketing logic in which companies seek to create customer value and achieve mutually beneficial relationships (Kotler & Keller 2007). One method that can be used in determining marketing strategy is the SWOT method which consists of Strengths, Weaknesses, Opportunities, and Treats. This method is needed in designing strategic developments for future businesses (Rahmallah & Wirasari 2022). In general, the stages of analysis carried out consist of input and matching stages. At the input stage, the Internal Factor Evaluation (IFE) and External Factor Evaluation (EFE) matrices are determined. Then, at the matching stage, the Internal External (IE) and SWOT matrix are determined. The advantage of this method is that the analysis is carried out on the internal and external environment so that the formulated strategy can be said to be more accurate because the analysis is carried out from two perspectives.

Based on the datas, the formulation of a marketing strategy using the SWOT method for Tipa-Tipa SMEs in Sigumpar District is one form of business that can be done to preserve regional specialties. In addition, the resulting strategy is also a solution for Tipa-Tipa product business actors in increasing their sales. A large market opportunity from visiting tourists can also be utilized from the formulation of this marketing strategy.

2 RESEARCH METHODS

In general, the research framework is as follows:

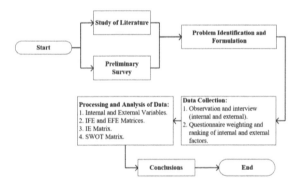

Figure 1. Research framework (Source:Personal data).

This study uses a mixed-method research approach which combines qualitative (obtaining internal and external factor variables) and quantitative (weighting and ranking) forms. The research is also limited by the scope, namely 16 Tipa-Tipa Sigumpar SMEs and 25 respondents consisting of customers or buyers of Tipa-Tipa products. The research period lasts for 4 months (October 2021 to January 2022). Primary data were obtained using several techniques, namely:

- Observation, by making direct observations of the location of SMEs to see operational activities and conditions in the field.
- Interviews, conversations with business actors and buyers familiar with Tipa-Tipa's stores and products. Questions related to management, marketing, finance and production to identify internal and external variables.
- Questionnaires, filling out questionnaires by 25 respondents from SMEs and customers or buyers related to internal and external factors assessment.

While secondary data is obtained from documentation in literature studies or internet sites related to the problem. Determination of internal and external variables using qualitative methods by looking for the overall relationship from the data collection results then formulated and interpreted. While the quantitative approach is used to obtain the value of weights, ratings and weight scores, which will be analysed in the form of IFE and EFE matrices, IE matrix, and SWOT matrix.

3 RESULT AND DISCUSSION

3.1 *Small and Medium Enterprises (SMEs) overview*

Sigumpar District is one of the locations for SMEs selling Tipa-Tipa products. The location is located at Simpang Silimbat, Jalan Lintas Sumatra. All business actors in this location, do not have a form or characteristic that distinguishes them. This can be seen from the product, price and sales facilities. The types of products sold are white rice and brown rice at a price of IDR 20,000/liter.

Figure 2. Tipa-Tipa SMEs products in Sigumpar DISTRICT (Source:Personal).

The production process starts from the process of checking for types of logs from suppliers then the packaging process by adjusting the size of the product contents in liters which are then sold at shops located at bus stops or public transportation along the road. The packaging used is quite simple, namely clear plastic without any product identity. The marketing process is still traditional, where the products are arranged on the trade table and marketed on the overhang of the store. Buyers of this product are classified into 2, namely domestic tourists and local people. Its use is as souvenirs or snacks for personal consumption. In addition, business actors also receive orders via cellular telephones, which then send their products using public transportation.

In general, the problem with the Tipa-Tipa SMEs in Sigumpar District is that there is no business development carried out to take advantage of existing opportunities. The opportunity is an increase in tourists to the Lake Toba area due to tourism development by the government. Based on the results of interviews, this is due to business actors who do not understand how to determine the right strategy to develop their business. It can be seen from the product packaging and sales methods which are relatively simple and less attractive to potential buyers. This is closely related to the marketing strategy that aims to attract potential buyers to the product. Utilization of opportunities can be done by formulating marketing strategies that are in accordance with current conditions.

3.2 *Internal Factor Evaluation (IFE)*

IFE The analysis consists of the strength and weakness variables along with their weights and ratings which are displayed in the following matrix form:

Based on the results of the IFE analysis in Table 1, it shows that the factor that becomes the main strength of SMEs is a strategic trading business location, which is located at a bus

Table 1. IFE matrix of Tipa-Tipa SMEs in Sigumpar District (Source:Personal).

	Internal Factors	Weight	Rating	Weight Score
Strenght	The quality of the product is good because it is produced without preservatives and quality checks are carried out	0,13	3	0,41
	A strategic trading business location, which is located at a bus stop or public transportation	0,14	3	0,43
	Have regular customers, namely bus passengers or public transportation	0,12	3	0,35
	Sub Total			1,19
Weakness	The product does not have flavor variants and only texture variants	0,12	2	0,29
	Sales are done traditionally only in one place	0,12	2	0,29
	Product packaging is less attractive with plain packaging without any labels or aesthetic enhancing ornaments	0,13	2	0,32
	Sales facilities are relatively simple and unattractive	0,12	3	0,35
	The product does not have a brand as a product identity	0,12	3	0,33
	Sub Total			1,58
	Total	1		2,78

stop or public transportation with a weight score of 0.41 while the main weakness is the sales facilities are relatively simple and unattractive with a weight score of 0.35. The results of the IFE analysis will be used to determine the position of the Tipa-Tipa SMEs in Sigumpar District on the Internal-External (IE) matrix.

3.3 *External Factor Evaluation (EFE)*

The analysis consists of opportunity and threat variables along with their weights and ratings, then displayed in the form of the following EFE matrix:

Based on the results of the EFE analysis in Table 2, the factor that becomes the main opportunity for SMEs is Tourism Development in the Lake Toba Region as KSPN with a

Table 2. EFE matrix of Tipa-Tipa SMEs in Sigumpar District (Source:Personal).

	External Factors	Weight	Rating	Weight Score
Opportunities	The product is known as a regional specialty food	0,17	3	0,56
	Product marketing can be done widely with technological developments	0,17	3	0,57
	Competitors and business actors do not have product differences	0,16	3	0,50
	Tourism development in the Lake Toba area as KSPN	0,17	3	0,59
	Sub Total			2,22
Treats	The increasing number of souvenir SMEs around the Toba Regency area	0,15	3	0,47
	The existence of modern food products that are an alternative to regional souvenirs	0,17	3	0,54
	Sub Total			1,58
	Total	1		2,78

weight score of 0.59. Meanwhile, the main threat is the existence of modern food products that become an alternative for regional souvenirs with a weight score of 0.54. The results of the EFE analysis will be used to determine the position of the Tipa-Tipa SMEs in Sigumpar District on the Internal-External (IE) matrix.

3.4 *Internal-External (IE) matrix*

The total weight scores of the IFE and EFE matrices of 2.78 and 3.23 are made in the IE matrix. The aim is to determine the position of the Tipa-Tipa SMEs in Sigumpar District. The following is the IE matrix:

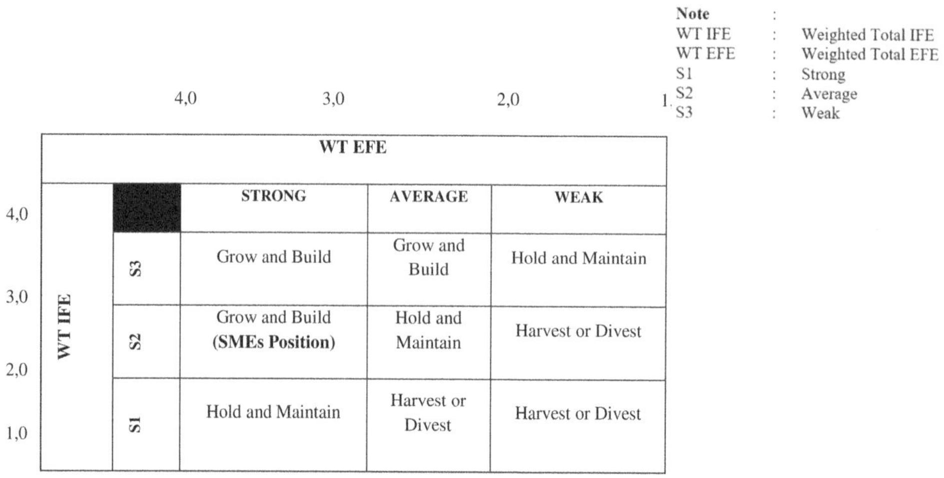

Note:
WT IFE : Weighted Total IFE
WT EFE : Weighted Total EFE
S1 : Strong
S2 : Average
S3 : Weak

	STRONG	AVERAGE	WEAK
S3	Grow and Build	Grow and Build	Hold and Maintain
S2	Grow and Build (SMEs Position)	Hold and Maintain	Harvest or Divest
S1	Hold and Maintain	Harvest or Divest	Harvest or Divest

Figure 3. IE matrix of Tipa-Tipa SMEs in Sigumpar District (Source:Personal).

Based on Figure 3, the position of the Tipa-Tipa SMEs in Sigumpar District is at Grow and Build. This means that Tipa-Tipa SMEs need a new marketing strategy to be able to develop their business. Analysis with the SWOT Matrix is needed to determine the strategy to be used in accordance with the position Tipa-Tipa SMEs in Sigumpar District. Where, the strategy is chosen by looking at the highest weight score when compared to other strategies.

3.5 *SWOT matrix*

The following is an alternative marketing strategy based on the SWOT matrix:
There are 4 strategy columns, namely SO (Strength-Opportunity), WO (Weakness-Opportunity), ST (Strength-Treats), and WT (Weakness-Treats) with each weighted score from the sum of the weights of the IFE and EFE matrices. Based on Table 3, the highest weighted score is the WO Strategy with 3,8. This means that this strategy is the most appropriate strategy for the current conditions of Tipa-Tipa SMEs in Sigumpar District.

Table 3. SWOT matrix of Tipa-Tipa SMEs in Sigumpar District (Source:Personal).

SO Strategy	WO Strategy
1. Utilization of technological developments (e-commerce and social media) to expand the product market 2. Carry out various promotions (offline and online) related to the types of souvenirs from the Lake Toba Region	1. Creating product innovations starting from adding flavor variants, more attractive packaging and brand strengthening. 2. Improve facilities to attract more tourists who come to the Lake Toba area. 3. Sales are carried out in various ways to target various market segments
Weight Score = 3,41	Weight Score = 3,8
ST Strategy	WT Strategy
1. Utilization of a strategic location for word of mouth promotion through regular consumers by highlighting messages of typical regional souvenirs	1. Seeing market trends that can be used as references in product development, sales and improvement of supporting facilities 2. Strengthening the Tipa-Tipa brand as local souvenirs through various promotional media
Weight Score = 2,2	Weight Score = 2,59

4 CONCLUSION

Based on the results of data processing and analysis that has been carried out, it can be concluded that the position of the Tipa-Tipa SMEs in Sigumpar District is Grow and Build position so that it requires a new strategy to develop. The alternative strategy that can be used is the WO Strategy, consisting of:

- Creating product innovations starting from adding flavor variants, more attractive packaging and brand strengthening.
- Improve facilities to attract more tourists who come to the Lake Toba area.
- Sales are carried out in various ways to target various market segments.

These results will be a strategic recommendation that can be used by Tipa-Tipa SMEs in Sigumpar District for Tipa Tipa products as souvenirs typical of the Lake Toba Region.

REFERENCES

Fadhilah, D. & Pratiwi, T., 2021. Strategi Pemasaran Produk UMKM Melalui Penerapan Digital Marketing. Jurnal Ilmiah Manajemen, pp. 17–22.

Kotler, P. & Keller, K. L., 2007. Manajemen Pemasaran. Jakarta: Erlangga.

Kotler, P. & Keller, L., 2014. Manajemen Pemasaran. Jakarta: Erlangga.

Munthe, A. & Simanjuntak, M., 2020. *Analisis Faktor-Faktor yang Mempengaruhi Minat Beli Wisatawan pada Kuliner Lokal yang ada di Kawasan Danau Toba (Jenis Makanan Ringan) Studi Kasus: Kabupaten Toba Samosir*. Bandung, IRWNS, pp. 1118–1124.

Rahmallah, Z. V. & Wirasari, I., 2022. Analisis Strategi Bisnis dan Penerapan Kebutuhan Desain Pada TUSHE di Masa Pandemi COVID-19. Jurnal Demandia: Desain Komunikasi Visual, Manajemen Desain dan Periklanan, Volume 07, pp. 141–160.

Relawati, R., Baroh, I. & Ariadi, B. Y., 2015. Analisis SWOT untuk Pengembangan Strategi Pemasaran Produk Olahan Apel di Malang Raya. SEPA, pp. 58–69.

Setyariningsih, E. & Utami, B., 2022. Analisis Strategi Pemasaran UMKM Tepung Bumbu ARIEN dengan Metode IFE, EFE, SWOT dan STP. Bisman, pp. 83–94.

Simanjuntak, M., 2018. Pemanfaatan Media Sosial Sebagai Sarana Pengembangan Strategi Bisnis Tipa-Tipa (Oleh-oleh Khas Toba Samosir). *Jurnaltio*.

Wijoyo, H. et al., 2020. Digitalisasi UMKM. Solok: Insan Cendekia Mandiri.

Designing Markaz Domba strategy business using design thinking approach

K.M. Syahid & I. Wirasari
Telkom University, Bandung, Indonesia

ABSTRACT: Markaz Domba is an animal husbandry business with a focus on selling products such as lamb and lamb processed meat. The Markaz Domba was founded in late 2019, Elan as the main founder of Markaz Domba utilizes human resources in the environment where the Markaz Domba was built, Pamoyanan, South Bogor. Utilization of the surrounding community aims to open a field there which is a job opportunity, so the profits generated by Markaz Domba will partly be used for empowerment of the Pamoyanan village itself. The big responsibility borne by Markaz Domba is not as dynamic as the profit makes. This phenomenon requires a new business strategy innovation so that the profits experienced by Markaz Domba can be achieved as expected. This study uses qualitative, collecting data through qualitative methods. This research uses the stages of the design thinking method, from the empathize stage to the define stage will eventually create a branding design solution for the Markaz Domba business to be more noticed by potential customers.

Keywords: Business, Branding, Design Thinking, Markaz Domba, Strategy

1 INTRODUCTION

Breeding is one of the businesses that can be relied upon to improve the lives of farmers because of its advantages. Animal husbandry is a promising business, because the target market is clear, namely humans. There are many farm animals from poultry such as chickens or quail to catfish, cows, sheep, and goats. Each farm animal has its own advantages and target market. Sheep is one of the common livestock and has a clear target market. Sheep have a market when there will be an Eid al-Adha celebration process or almost every day there are potential customers, namely sheep for the Aqiqah process.

Markaz Domba is one of the businesses that focuses on sheep. Sheep Markaz is located in the Pamoyanan area, South Bogor. Markaz Domba uses the surrounding community to actively care for the sheep, with this activity of the Pamoyanan village creates an empowered village, which uses the surrounding community to work, not only aiming to develop the village but also to create jobs where the majority of the people there are unemployed.

After less than 2 years, Markaz Domba is just waiting for potential consumers who want to buy the product without doing a significant promotion to a wider audience, this is a core problem experienced by Markaz Domba, so that from these problems it causes huge profits that far from the expectations of the Markaz Domba.

A company needs to implement the right business strategy to gain market share. Moreover, if the company does not develop significantly from its initial stand, then the products offered will experience difficulties and major challenges to gain market share from established competitors. Therefore, companies need to make promotions or attractive offers to consumers to change behavior and create brand awareness in potential consumers. The right balance is needed by the company to be able to gain advantage in the market, it must not only focus on meeting the core needs of the target market by providing products of equal quality to its competitors, but also must provide different value to consumers compared to its competitors. (Heriyadi 2018).

2 RESEARCH METHODS

This research uses qualitative methods by conducting direct observations and structured interviews as primary data collection methods. Secondary data uses literature studies related to the problems experienced by the object of research, namely promotion.

According to sugiono (2005) Qualitative research is suitable for the type of research that understands a social phenomenon from the participant's perspective. In simple terms, qualitative research is suitable when researching a condition or situation on the object of research.

This study uses the design thinking method, where at the empathize stage the process of observation, interviews, and supported by literacy related to promotions is carried out. Starting with the empathy stage, finding user needs by understanding the values and behaviors needed by Markaz Domba consumers, analyzing whether the promotional activities carried out by Markaz Domba are in accordance with consumer needs or not. Next, it will proceed to the define, ideate, prototype, and testing stages.

Research framework:

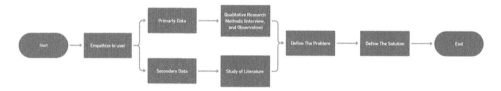

Figure 1. Research framework (Source: Personal).

3 RESULT AND DISCUSSION

One company that is good in marketing its product strategy is a company that will "survive" in market competition. The strategy of every business or business of each person must have their own characteristics or differences, but still with the same goal, namely to make their products known to a wide audience.

In today's modern marketing activities, company policies related to good product planning and development, pricing and distribution methods that are easily accessible to target consumers are marketing activities carried out within the company, or with partners in marketing. However, this is not complete, because the company must have the opportunity to communicate with audiences who may become customers, or have been customers before, so as to create an interaction between what will be determined by the company and what is needed by the company and desired by consumers.

According to Morissan (in Mutmainah dan Wirasari 2010:39) promotion has a purpose as:

(a) Introducing a company or product to the wider community.
(b) Educate consumers to be effective and understand in using the product.
(c) Changing the image of the company or product in the eyes of the public.

The results of interviews and observations that have been made to Markaz Domba shows that the promotions that have been carried out are not in accordance with the objectives of the promotion that should be. Markaz Domba only did two promotions, namely from online and offline media. Markaz Domba carries out online promotions through Instagram and WhatsApp content only, where this does not fulfill the purpose of the number one promotion point above, which is to introduce products to the wider community. The next offline promotion of Markaz Domba is words of mouth, where the promotion is less effective due to its very small reach. From these promotional activities, it is very common if the profits obtained are not in line with the expectations of the Markaz Domba.

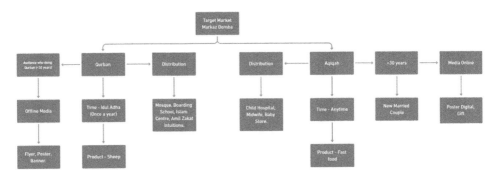

Figure 2. Target market (Source: Personal).

After analyzing the problem of Markaz Domba's promotion strategy, the next step is to map out the Markaz Domba's target market first, this needs to be done so that the resulting solution is right on target. Below is the target chart of the Markaz Domba:

From the picture above, it can be seen that Markaz Domba's target market is very broad due to the products it offers. The target market of the Markaz Domba is divided into two according to the slightly different products, namely qurban and aqiqah. Both targets do offer sheep, but for qurban, it tends to be live animals, while aqiqah tends to be processed sheep. From the two target markets, there is a significant difference in the age of the target market for qurban, which tends to be above 30 years, while the aqiqah target market is more towards 30 years and under. From the significant age difference, the promotion needs are also slightly different. If the target market is above 30 years, it tends to be conventional, while under 30 years it is the other way around.

When the analyzed of target market Markaz Domba is done, it can be concluded that the promotional solution that will be carried out by the Markaz Domba in the future is to carry out promotional activities that cover both parties of the target market. The promotion solution for Markaz Domba will be carried out through two media, namely online and offline. Online is more directed to the target market that tends to be modern, while offline is more directed to the conventional target market.

The first promotional solution is to optimize Instagram content to be more communicative, not only displaying content in the form of activities of the Markaz Domba, it must return to the original goal of Markaz Domba, which is a business where there is buying and selling of products, information about the products offered, information about the company, and creative content to make it look more dynamic. The purpose of optimizing social media content, especially Instagram, is to restore the usefulness that should be done, namely selling, because Markaz Domba is not a matter of social service, so content and products must match. The following is a social media content promotion solution:

Figure 3. Social media content educative (Source: Personal).

Figure 4. Social media content profile (Source: Personal).

Figure 5. Social media content hard-selling (Source: Personal).

Next is the offline promotion solution, for this solution the method used is the pick-up-ball promotion method. In simple terms, this strategy is carried out in contrast to the conventional system where it is usually the consumer who has to go to the seller of the product. However, with this system, sellers directly bring in potential customers openly or transparently. The advantage of this promotion method is that it makes it easier for consumers to get information and product offers, and will indirectly get a good impression when they do it well. The following is a promotion solution for the Sheep Markaz ball pick-up:

Figure 6. Coupon and gift (Source: Personal).

The distribution of this gift is only done on certain days, for example on Mother's Day or other special days. This gift will be given to prospective consumers who give birth on that day. The gift box contains baby needs or equipment, the two gift concepts above are equipped with a coupon given by the Markaz Domba. This activity was carried out with the aim of giving a good impression to the Sheep Markaz.

4 CONCLUSION

This research was conducted using qualitative research methods with the stages of research using the design thinking method. At the empathize stage, it was found that the promotion carried out by Markaz Domba was not optimal and not right on target, as evidenced by the results of interview data, direct observation, and analysis of the Markaz Domba target market. This requires a promotion strategy that is more optimal and on target with the suitability of the Markaz Domba target market that has been determined. The promotional strategy solution is in the form of optimizing Markaz Domba's social media content for Online, while for offline it is in the form of coupons and gifts which will create a good impression on Markaz Domba.

REFERENCES

Aldo R, Wirasari, Siti D (2015). *"Perancangan Promosi Danau Linow Di Sulawesi Utara"*.
Annisa N F (2018). *"Analisis Strategi Promosi pada UMKM Social Enterprise (Studi Kasus Pascorner Cafè and Gallery)"*.
Bayu A, Perwitasari F D (2019). *"Kajian Aspek Sosial dan Ekonomi Usaha Ternak Domba Secara Intensif di KTT Haur Kuning Desa Ciawigadjah"*.
Broto Wibow, S. Rusdiana, dan U. Adiati (2016). *"Pemasaran Ternak Domba di Pasar Hewan Palasari Kabupaten Indramayu"*.
Dominikus T (2012). *"Marketing Communication dan Brand Awarness"*.
Faqia F, Ira W (2019). *"Perancangan Promosi BCA Sakuku"*.
Razan F (2017). *"Perancangan Promosi Topeng Benjang Untuk Remaja Kota Bandung"*.
Ulfa Khaira, Indra Weni, Edi Saputra, Zainil Abidin, dan Yolla Noverina (2022). *"Penerapan Visual Branding Agro-Edu Wisata Desa Dataran Kempas Sebagai Strategi Penguatan Promosi Pariwisata"*.

The implementation of health protection-based design on Telkom University dormitory building

D. Murdowo*, V. Haristianti & R.H.W. Abdulhadi
Telkom University, Bandung, Indonesia

*ORCID ID: 0000-0001-6435-5108

ABSTRACT: The increase in Covid-19 patients is not being facilitated by the availability of patient care facilities. Data show bed occupancy rate (BOR) trend for treating COVID-19 patients in West Java has touched 90.32 percent in 2021, therefore efforts are needed to transform a place or space other than a hospital with a design that facilitates medical personnel in optimizing patient care and also as a form of healing environment. The result of the research is how the idea of adaptive reuse of buildings will be applied to Health Protection approach as design concept. This study used a mixed method with an R&D (Research and Development) approach. The purpose of this research was to make a design prototype to transform Telkom University Dormitory as an alternative to the Independent Isolation Building (ISOMAN) and could be a reference or model in designing a building that can be used as a location for self-isolation.

Keywords: dormitory room, covid-19, room design, self-isolate

1 INTRODUCTION

Pikobar website, West Java on 29th of June 2021, the total confirmed cases of COVID 19 in West Java reached 376,982 cases, recovered was 322,103, patients who were still under operational care 49,617. With so many patients who need to be treated at the hospital, the hospital's carrying capacity for handling COVID-19 patients is experiencing problems. The main problem with the West Java Government is that the capacity of the hospital, which is prioritized for cases with severe to critical symptoms, is very limited (Koh *et al*. 2020). Meanwhile, alternative places for Covid-19 patients without symptoms or with mild symptoms who are directed to be able to undergo Independent Isolation (ISOMAN) have not been mapped and prepared properly.

For this reason, it is necessary to increase the bed capacity in hospitals and the need for other alternatives (Al-Yafei *et al*. 2017), namely the provision of a place for patients with mild symptoms by self-isolation. The Daily Chair of the West Java COVID-19 Handling Task Force, Daud Achmad, revealed that in an effort to reduce home BOR, during the Emergency PPKM, the West Java Provincial government increased facilities by adding a building that functioned as a place for self-isolation for positive patients without symptoms. The problem that arises is to find a location / place / building as a problem center that meets the requirements and facilities.

Telkom University Dormitory that has good infrastructure (Murdowo 2018), can be an alternative as the ISOMAN Building, this needs to be studied and carried out in-depth be research. As a dormitory located in the Telkom University area of 50 hectares, with a green environment, around it there are still many rice fields and lakes so that the air still feels fresh. This study aims to make a design plan in order to change the function of the Telkom University Dormitory into an alternative place for the Independent Isolation (ISOMAN) Building. The design of the ISOMAN building is guided by the Procedure Guide for Establishing Quarantine Centers Outside Health Facilities (Kementerian Sosial Republik Indonesia 2020).

2 RESEARCH METHODS

This research is a pragmatic or combined research (mixed methods) with a Research and Development approach. This type of research was chosen because it allows for the collection and analysis of both qualitative and quantitative data in a study (Creswell 2003). In more detail the research was carried out in five major stages: (a) The Preliminary study was preceded by data collection, (b) Data Analysis Techniques using inductive techniques, (c) The design and alternative designs, (d) Model testing, (e) The publication phase is a form of the effectiveness of the socialization model to a wider audience.

2.1 *Methods of data collecting*

The data collection methods were conducted by field research (primary sources) that was carried namely observation, and interviews; and library research (secondary source) by literature study data collected from scientific journals, articles, and theories derived from literature books were used to find a theoretical basis In addition, a recapitulation of the results of the documentation of examples of self-isolation rooms that can be used as a reference is also collected.

2.1.1 *Observation*
At the observation stage, the researcher places himself as a non-participant (Sugiyono 2011), namely as an outside observer who observes the state of the participants/sampling by observing and recording events that took place during the observation stage. Things that exist include observation of the Telkom dorm room, in the form of photos and videos, reviewing the changes (if possible), and connecting the sample conditions with the theory that has been chosen.

2.1.2 *Interview*
An interview is a collection of data taken by way of interaction between one person and another, which can be one or more and has the purpose to get information from the interaction. Interviews were conducted using open interviews by making a list of questions that were expected to bring up information in the form of data and facts about the building and operations that occurred in the study that took place, namely the Telkom University dormitory building. The resource persons are policy makers, as well as the management of the dormitory building.

3 RESULT AND DISCUSSION

3.1 *Result*

The results of the research include two completions, namely the Monitoring System and Interior Design that support the dormitory as a place for handling covid 19. This can be explained as follows: IT-based Monitoring System that is applied in Adaptive Re-use as a General Idea for Independent Isolation Room Design. Adaptive re-use is a condition in which a building undergoes an overhaul in its spatial aspects to allow the building to experience a functional change in the classification of the building. This further results in the possibility of rezoning an area (Murdowo *et al.* 2021). The adaptive re-use of building space is one method that has been used by many developers and is considered very profitable from an economic point of view (Gewirtzman 2016).

The rationale for this design is what efforts are needed to be able to design a space that is not only an environmental healing facility for patients but can also facilitate medical personnel in optimizing patient care. This can then lead to the research objective, which is that it is hoped that both patients and medical personnel can speed up recovery efforts because they both get comfort in the dormitory room which has been converted into an independent isolation room. The realization of the approach is divided into categories namely (1) zoning and circulation flow, (2) RFID placement, (3) ventilation concept, (4) lighting concept, and (5) customize furniture intended for the approach. To achieve the objectives, here is a list of the main rooms (Table 1) that underwent functional changes and design adjustments:

Table 1. Room function change list.

No	The Function of the Plan in the ISOMAN Building	Initial Space Function	Location	Number of Spaces
1.	Receptionist and Lobby	Lobby	Ground floor	1
2.	Triage Room	Student bedroom	Ground floor	1
3.	PPE changing room	Student bedroom	Ground floor	2
4.	Nurse station	Communal room	GF – 3RD floor	4

Apart from the changes above, the function of the room remains the same, namely as a bedroom and toilet without changing the construction of the existing building. So that the capacity in a ISOMAN building is 91 patient rooms with details of 47 rooms for female patients, a capacity of 94 patients (located on the left wing of the building), and 44 rooms for male patients; capacity of 88 patients (located on the right wing of the building).

3.2 *Discussion*

In this section, we will explain in detail the design changes that occurred in the dormitory building in order to fulfill the requirements to become an ISOMAN building. The details of the changes can be discussed below:

(a) *Design variable: Zoning and blocking flow.*
Previously, the initial function of the building was as a female-only dormitory. Later The designation of this building was changed to a building that can be occupied by female and male patients. However, the separation is done by placing female patients in the left wing, and male patients in the right wing of the building area (Kucharski *et al.* 2020; Pratomo 2020); So, the circulation area in the building is divided into two parts, as shown in Figure 1 below:

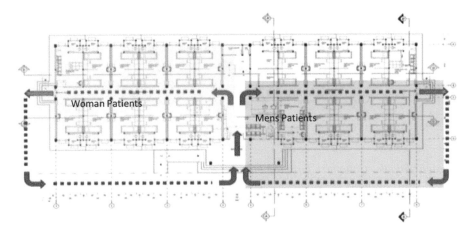

Figure 1. Circulation area.

(b) *Design variable: Ventilation concept.*

Before, Air circulations only relies on openings from room windows and doors (Figure 2). Therefore, improvements to the concept of ventilation were made because the existing air movement system did not meet the requirements and criteria in terms of the required air exchange, moreover a ventilation solution was needed without drastically changing the condition of the existing dormitory room. This problem can be overcome by adding an air purifier or air purifier equipped with a HEPA filter (Figure 3).

Figure 2. Existing air circulations.   Figure 3. Air purifier equipped with a HEPA filter.

(c) *Design variable: Lighting concept.*

In the existing building design, there is only general lighting from the ceiling. The solutions, the lighting can also be used to assist in creating a sterile isolation room. The use of short-wave UVC lamps is placed on the ceiling in the middle of the room to disinfect the room when the patient leaves the room, so that the room will always be sterile when used (Figure 4)

Figure 4. Short-wave UVC lamps.

(d) *Design variable: Customize furniture intended for the approach.*

Basically, furniture is adapted to the function of the building as a student dormitory with a function as a place for students to live while studying in the first year. The function and completeness of the room are in accordance with the requirements of the function of the building. So, it is necessary various customize furniture were made so that the change in the function of the dormitory building could meet the requirements to become an adequate ISOMAN building including. Before-after studies on the design of the ISOMAN building are:

1. Before, facade area without ramp and disinfectant booth (Figure 5). Afterward, added a disinfectant booth on the right and left stair areas of the building, Also adding a ramp for patient beds in the main entrance area (Figure 6).

Figure 5. Façade area without ramp.   Figure 6. Ramp for patients beds

2. Formerly, the bedroom capacity is used for four people with a bunk-bed system and there is no room divider provided (Figure 7). Subsequently, change the room design to suit the patient's needs, room capacity is used for two people with a bed divider (Figure 8).

Figure 7. Bedroom for 4 people.   Figure 8. Room for two people with a bed divider.

3. Beforehand, there is no signage inside the Dormitory Building. Next, added signage and a wayfinding system to facilitate the activities of health workers and patients.

4 CONCLUSION

Some further research is needed which is expected to be a reference for designing space designs with changes in function. The following are the minimum standards of room organization facilities for quarantined patients based on sources from "Procedures for Establishing Quarantine Centers Outside Health Facilities". The next research stage is to test the hypothesis through experimental research activities (quantitative). Research that reviews the function of the application of RFID technology and interior design needs to be tested for users so that later they can decide whether the hypothesis is accepted as the basis for this research.

REFERENCES

Al-Yafei, E., Ogunlana, S., Oyegoke, A. 2017. *Application of Value Engineering and Life Cycle Costing Techniques for Offshore Topside Facility Projects: Towards Sustainability*. Society of Petroleum Engineers – SPE Kuwait Oil and Gas Show and Conference 2017, (January).

Creswell, J. W. 2003. *Research Design: Qualitative, Quantitative, And Mixed Methods Approaches*. Second Edition, London: Sage Publications, International Education and Professional Publisher.

Gewirtzman, D.F. 2016. Adaptive Reuse Architecture Documentation Analysis. *Journal of Architectural Engineering Technology*, Volume 5 Issues 3.

Kementerian Sosial Republik Indonesia. 2020. *Panduan Penyiapan Fasilitas Shelter untuk Karantina dan Isolasi Terkait Covid 19 Berbasis Komunitas*.

Koh, W. C., Naing, L., & Wong, J. (2020). Estimating The Impact of Physical Distancing Measures in Containing COVID -19: An Empirical Analysis. *International Journal of Infectious Diseases*, 100, 42–49.

Kucharski, A. J., Klepac, P., Conlan, A. J. K., Kissler, S. M., Tang, M. L., Fry, H., … Edmunds, J. (2020). *Effectiveness of Isolation, Testing, Contact Tracing, and Physical Distancing on Reducing Transmission Of SARS-Cov-2 in Different Settings: A Mathematical Modelling Study*. https://doi.org/10.1016/S1473-3099 (20)30457-6

Murdowo, Djoko (2018). *Pendidikan Karakter Berbasis Asrma untuk Pembinaan Nilai Nilai Budaya Organisasi: Penelitian Grounded Theory Pada Universitas Telkom*. S3 thesis, Universitas Pendidikan Indonesia.

Murdowo, Djoko, Prameswari, N. S., & Meirissa, A. S. (2021). Engaging the Yin-Yang Concept to Produce Comfort and Spatial Experience: *An Interior Design for a Chinese Restaurant in Indonesia*. 8(2), 60–71.

Pratomo, H. (2020). From Social Distance to Physical Distance: A Challenge for Evaluating Public Health Interventions Against COVID-19. Kesmas: *Jurnal Kesehatan Masyarakat Nasional (National Public Health Journal)*, 0(0), 60–63.

Sugiyono. (2011). *Metode Penelitian Kuantitatif, Kualitatif, dan R & D*. Bandung: Alfabeta.

Redesign coffee packaging for sustainable local community development

W. Swasty
Telkom University, Bandung, Indonesia
Universiti Sains Malaysia, Pulau Pinang, Malaysia

A. Mustikawan & F.E. Naufalina
Telkom University, Bandung, Indonesia

ORCID ID: 0000-0001-6586-2697

ABSTRACT: Indonesia as a coffee producer and exporter occupies the fourth largest position globally. However, many local coffee farming communities, especially in Bandung Regency, embrace marketing problems, i.e., unattractive package design. Widjie Coffee as one of the local community products has a mission to build public awareness that cares about the importance of environmental sustainability and improves the welfare and social solidarity of the community. This study aims to improve local product marketing efforts by redesigning packaging which has a competitive advantage. Data collection was carried out by field observations and interviews with the owner. The creative design process was carried out by visual exploration. The design results recommend three alternative designs by highlighting environmental sustainability as one of this local community's missions. This study is expected to help developing sustainable local community through competitive packaging design, not only in offline retail but also in online marketplace.

Keywords: environmental sustainability, local community, package design

1 INTRODUCTION

According to the data from the Ministry of Industry, Indonesia as a coffee producer and a coffee exporter occupies the fourth largest global position, following Brazil, Vietnam, and Columbia (Kementerian Perindustrian RI 2017). This is of potential for redeveloping the coffee industry if productivity can be increased. However, according to empirical research, many local coffee farming communities, particularly in the Bandung Regency, embrace marketing challenges, such as the unattractive style of their product's package design. A similar thing is also found in other local products that are usually produced by small and medium enterprises (SMEs) that are still weak in package design (Istianah, 2020). The term "package" refers to a product that performs multiple tasks, including primary functions such as protection, transport, and information as well as secondary functions such as marketing, economic, and ecological functions (Ankiel and Grzybowska-Brzezińska 2020). In its development, packaging is not only to protect the product but can provide added value and competitiveness in the market by using attractive packaging techniques.

Numerous studies have attempted to redesign the coffee package (Darmanto *et al.* 2013; Yuliani *et al.* 2021; Zulkarnain *et al.* 2020). This paper discusses the redesigned coffee packaging of Widjie Coffee as one of the local community products to enrich the literature in the redesigned coffee package. This study aims to improve local product marketing efforts by redesigning packaging which has a competitive advantage. Widjie Coffee itself has a mission to build public awareness that cares about the importance of environmental sustainability, and improve the welfare and social solidarity of the community.

2 RESEARCH METHODS

This study is applied research with a qualitative approach. The object of the study is Widjie Coffee, one local community product from Bandung Regency. The object was selected due to its interesting mission which builds public awareness that cares about the importance of environmental sustainability. Data collection was conducted by field observation to gain product insight and the existing packaging design. The unstructured interview was performed with the owner as the key informant to obtain the design brief. Literature and design reviews were done to get design references. Afterward, the creative design process was carried out by visual exploration (such as mood board, sketching, and digitalise). Market analysis was done by identifying segmenting, targeting, and positioning of the product. This paper provides three alternative designs as a recommendation. Figure 1 illustrates the research framework, adapted from Zulkarnain *et al.* (2020).

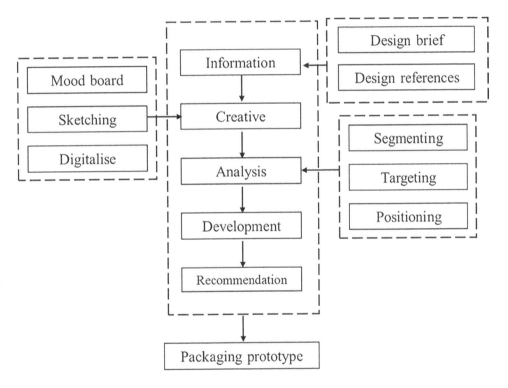

Figure 1. Research framework.

3 RESULT AND DISCUSSION

Widjie Coffee is taken from the Javanese language, namely wiji which means seeds. It also means Eternal or Savior in unisex language. It is aligned with their dream to be the savior of the environment and eternal hope until the end of this world. "Intrepid spirit and humble coffee roasters on a mission to conserve nature and finding the best coffee possible." In addition, to prevent the erosion of coffee plants, it is also economically valuable because Indonesian coffee is now worldwide, but the selling price is very high. This is one of the reasons why Widjie Coffee was founded in order to improve the situation of coffee production in Indonesia.

From an interview with the key informant (the owner of Widjie Coffee), some issues in the coffee industry have been revealed regarding the sustainable local community. Therefore, their mission is to build public awareness that cares about the importance of environmental sustainability, to improve the welfare and social solidarity of the community, as a mean to participate in the development of human resources, and to increase the knowledge and intelligence of the community through regular, planned, and sustainable efforts through coffee, conservation, and literacy. Utilization of biological natural resources needs to be done responsibly and wisely. This is to ensure that the supply of natural resources does not run out in a short time. Using it responsibly and wisely is called conservation. Their mission is aligned with the Sustainable Development Goals (SDGs) in responsible consumption and production which highlights transforming the relationship with nature is key to a sustainable future (United Nations 2019).

To redesign the coffee package, a market analysis has been performed. Table 1 shows the segmenting of Widjie Coffee. Whereas their target market is teenagers and adults ranging from the lower middle to the upper middle class, especially who are aware of the surrounding environment by utilizing natural resources through various methods of conservation with responsibility. The positioning statement is "For those who are aware of the surrounding environment, Widjie Coffee is the coffee that will consider the environmental sustainability so they can develop sustainable local community through coffee."

Table 1. Segmenting.

Segmenting	Item	Description
Geographic	Country	Indonesian
	Region	West Java
	City	Bandung and Surroundings
Demographic	Gender	Male & Female
	Age	Teenagers and Adults (17 Years and Over)
	Income	Standard economy
	Social Class	Middle-upper class
Psychographic	Lifestyle	Active, productive, nature and environment lover
	Characteristics	Have a taste for coffee and have an awareness of the environment
	Behavior	Have an interest in consuming local coffee

Following the market analysis and design development, finally, the recommendation is formulated in packaging prototypes. The recommendations provide three alternative designs by highlighting environmental sustainability as one of this local community's missions, as described in the following section.

1. *Design A*

The theme applied to this surface package brings the impression of "Fresh, Clean, Friendly". Fresh with a variety of colors and new concepts, raising social issues that are considered important and giving messages to buyers. This concept is still very fresh and novel among its competitors. Impression of clean is a simple layout style with a clear reading flow, supports fresh concepts. Friendly, overall design of this packaging still prioritizes the friendly element for the buyers.

This packaging uses a standing pouch made of brown paper, with screen printing on it. The label uses a sticker material, while the logo label above uses a sticker with a material that is easy to tear, making it easier when opening the package. Selected materials and techniques were aimed to facilitate and minimize the funds spent.

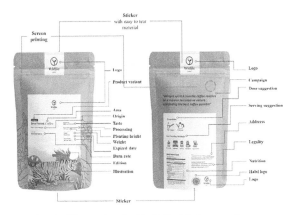

Figure 2. Design alternative A (illustration by Fadhilla Muhammad).

2. *Design B*

This package maintains the color style of the previous package design that did not have sufficient colors patterns. The previous packaging was black and white, and currently, it is white on the sticker label or product information. This new design only uses one color, namely gold by adding visual illustration elements to the packaging and compiling more detailed layouts and information using the illustration of coffee plants that are used on the front and back of the packaging surface as well as on the side to highlight the coffee product.

Figure 3. Design alternative B (illustration by Herdi Ramadhan).

3. *Design C*

The personality of this brand is also applied to surface packaging that has natural properties. This packaging design uses natural nuances with flat design illustrations. Additionally, new flavors are added by altering the color of the container, making the variations clear to the consumer. This packaging uses sealed foil packaging (packaging that is sealed with a layer of foil). Bags with triple-ply foil features are usually sealed as soon as the coffee beans are inserted to prevent air from enter in, but these packages also have valves to let carbon dioxide out. Until a few days after the roasting process, coffee beans will usually emit carbon dioxide as a natural process. This valve on the packaging is the way out for the carbon dioxide so in the end, it keeps the freshness of the coffee.

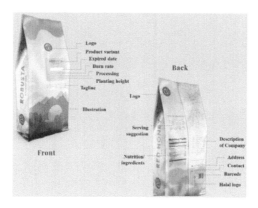

Figure 4. Design alternative C (illustration by Ratih Kurniati).

In contrast to an earlier study by Darmanto *et al.* (2013) that applied local/traditional ornament in the surface package, this present work utilizes environmental elements such as coffee plants, a lion, or a forest for the illustration on the surface package. It is highlighting its sustainable efforts through coffee, conservation, and literacy as one of its missions.

Sticker labels were used combined with the screen printing on the packaging paper. This recommendation was put forward considering the printing cost, where local SMEs usually do not have a lot of capital that can be allocated for packaging production. This is in line with previous research by Zulkarnain *et al.* (2020) who used sticker labels on their package design as they argue sticker labels can be ordered/printed on demand without a limit so that the SMEs are not too burdensome for their capital capacity.

Coffee packaging innovation is important to increase sales of coffee products. Innovation is a capital to generate renewal in every manner and aims for sustainable progress (Yuliani *et al.* 2021). They suggest the innovations in coffee packaging to be novelties that consist of information such as (1) brand of coffee; (2) variants (e.g., single-origin and espresso blend); (3) regions; (4) variety ; (5) roast level; (6) testing notes or cupping notes; (7) altitude; (8) coffee weight; (9) roast date; (10) seeds and degree of fineness. This present work recommends packaging prototypes with that kind of innovation which is in line with a prior work by Yuliani *et al.* (2021).

4 CONCLUSION

The purpose of this current study was to improve local product marketing efforts by redesigning packaging which has a competitive advantage. The design results provide three alternative designs by highlighting environmental sustainability as one of this local community's missions. This study is expected to help develop a sustainable local community through competitive packaging design, not only in offline retail but also in the online marketplace. By highlighting the relationship with nature in designing the illustration for packaging surface, it is expected that the value of sustainable development can emerge.

The findings will be of interest to the local community of coffee farmers as well as packaging designers. A limitation of this study is that it only addresses one brand as a case study as it is applied research that provides three alternatives design. A further study should be carried out to assess those alternatives design by doing consumer research and to recommend one design that can be applied by the company and accepted by the wider market.

ACKNOWLEDGEMENT

The authors would like to thank Fadhilla Muhammad, Herdi Ramadhan, and Ratih Kurniati for their assistance in the survey and 3D visualizing of the prototype.

REFERENCES

Ankiel, M. and Grzybowska-Brzezińska, M., 2020. Informative Value of Packaging as a Determinant of Food Purchase. *Marketing of Scientific and Research Organizations*, 36 (2), 31–44.

Darmanto, S.M., Ahmad, A., and Wijayanti S, A., 2013. Perancangan Corporate Identity Dan Kemasan Kopi Surya Kintamani Bali. *Jurnal DKV Adiwarna*, 1 (2), 1–12.

Istianah, R., 2020. Re-Design Label Kemasan Makanan Sistik Ubi Ungu 3A Saudara Baper. *Jurnal Seni dan Desain Serta Pembelajarannya*, 2 (1), 28–31.

Kementerian Perindustrian RI, 2017. *Peluang Usaha IKM Kopi*.

United Nations, 2019. *The Sustainable Development Goals Report 2022*. United Nations publication issued by the Department of Economic and Social Affairs (DESA).

Yuliani, D., Nursetiawan, I., and Taufiq, O.H., 2021. Inovasi Kemasan Kopi Robusta Kekinian Desa Sukamaju Berbasis Kearifan Lokal. *MALLOMO: Journal of Community Service*, 1 (2), 64–72.

Zulkarnain, Machfud, Marimin, Darmawati, E., and Sugiarto, 2020. Rancangan Model Purwarupa Kemasan Kopi Specialty. *Jurnal Teknologi Industri Pertanian*, 30 (1), 1–12.

The mosque rest area circulation system with a design thinking approach

D.P. Pangestu & D.W. Soewardikoen*
Telkom University, Bandung, Indonesia

*ORCID ID: 0000-0003-0823-7320

ABSTRACT: Rest areas are located every 30 kilometers as a temporary resting place for toll road users with various facilities, one of which is a mosque rest area with religious, economic, and social values. Al-Safar Mosque, which is located in the rest area of KM 88 B on the Cipularang toll road section, often experiences circulation problems because the application of circulation types, determination of circulation goals, and circulation patterns are not adjusted to the circulation activities of the mosque room users, especially before the holidays. The research question is whether the condition of the mosque's existing circulation system is to the circulation activities of mosque users. Empathize design thinking approach is used to find out the actual condition by conducting observations, interviews with users, and literature studies. It is known from the results of this study that the condition of the mosque's circulation system is not to the needs of mosque users' circulation activities.

Keywords: circulation system, design thinking, empathize, Mosque rest area

1 INTRODUCTION

The mosque rest area has several functions, including as a place of worship for Muslims, a place to support economic and social activities of the community around the mosque, a da'wah center for Islamic studies, as well as a supporting function of the existing facilities of the mosque, which can be used as a place for a short rest and toilet facilities can be used (Muslim 2004). Some of these functions cause circulation activities by mosque room users to the condition of the mosque's existing circulation system; under certain conditions of time it becomes a space circulation problem, so it is necessary to pay attention to aspects of the circulation system related to the type of circulation, circulation objectives and circulation configuration (Tofani 2011). Circulation is a traffic pattern or movement in an area or building that requires consideration of the value of flexibility, economy, and functionality for circulation users (Harris 1975). Including the mosque rest area, these three values can be considered to overcome circulation problems in space and time that are adjusted to the circulation activities of mosque visitors. Considerations regarding flexibility, economy, and functionality can be realized in several provisions of the circulation system, including determining the type of circulation intended for circulation of goods, circulation of vehicles, or circulation of people who pay attention to the area of circulation ergonomics, circulation floor patterns, clarity of circulation orientation and circulation lighting.

From several provisions regarding the space circulation system, this study aims to determine the perception of space users regarding the suitability between the mosque's existing circulation system and the needs of space user circulation activities. Furthermore, the purpose of circulation is divided into 2, namely having a direct nature with a predetermined circulation orientation with the optimal use of the duration, the next circulation goal with a recreational nature, namely the orientation of the circulation goal is not specifically determined with unlimited duration (Tofani 2011). Finally, there are six circulation configuration types: linear, radial, spiral, grid, network, and composite (Ching 2012).

2 RESEARCH METHODS

This study uses a qualitative paradigm with participatory observation instruments by being directly involved in the condition of the object of research, structured interviews with ten mosque visitors in the category of adolescent and adult males as the category that dominates mosque visitors and is supported by a literature study of the circulation system. The sample selection was purposively based on the criteria (Soewardikoen 2019), determined at the mosque located in a type A rest area with complete facilities, the widest rest area criteria, and a larger number of visitors when compared to type B and C rest areas. The sample was determined from the number of variables with the highest number and supported by literature studies related to space circulation. The object of research is Al-Safar Mosque KM 88 B Cipularang Toll Road. At this stage, the data collection process is carried out by conducting direct observations of the object of research, as well as the interview process with mosque room users to obtain more detailed and valid information and understand the needs and expectations of mosque circulation and supported by literacy related to space circulation. The process is human-centered or involves humans as the center in every stage of the design process (Wolniak 2017).

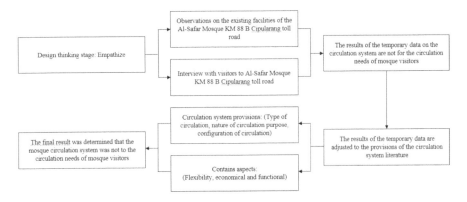

Figure 1. Research framework.

3 RESULT AND DISCUSSION

Collecting data through participatory observation and in-depth interviews on March 30, 2022, at the Al-Safar Mosque KM 88 B Cipularang Toll Road was carried out from 16.00 WIB to 18.30 WIB. During this time, observations were made (with direct participation) related to the circulation activities of mosque space users on the mosque's existing circulation system, and interviews were conducted with ten mosque room users.

Observations of the circulation system are carried out on the circulation of the corridor leading to the ablution area (the process of purification when going to pray) until the circulation of the corridor to the main prayer area; in the circulation area, the intensity of circulation problems will be prone to occur when compared to circulation in the main prayer area. According to (Panero 1979) in the book, Human Dimension and Interior Space, the standard circulation area of 4 lanes is 243.8 cm, the circulation area in the corridor leading to the ablution area has met the standard circulation area with an area of 270 cm until it continues until the main door of the main prayer area of the mosque. So that the consideration of the value of the flexibility of the space for movement in the area has been fulfilled; furthermore, the number of main doors of the mosque, there are only two doors to be used in and out of the main prayer area with an area of 200 cm for each circulation of the main door of the mosque. However, the circulation standard for two people is 122 cm. The circulation for wheelchair users for the disabled is 91.4 cm, so in certain time conditions, the circulation area of the main door of the mosque cannot meet the needs of the circulation of space users, with consideration that it is used by three categories of room users, namely men, women, and people with disabilities for circulation in and out of the main mosque prayer area, and can affect taharah ablution (maintaining the sanctity of ablution). So the consideration of the flexibility value of space users in the area has not been fulfilled.

Sacred limits to fulfill the flexibility value of space users. Furthermore, the floor pattern circulation system in the corridor circulation to the ablution area uses textured ceramic material with dark gray color supported by rubber carpets in several parts for non-slip space users to consider the functional value of the material that forms the floor pattern has been fulfilled. However, circulation problems occur because male and female room users are in the same circulation, starting from the corridor circulation to the ablution area to the corridor circulation to the main prayer area. As in the research on *Perancangan Masjid Manasik Haji dan Rest Area Berbasis Aspirasi Masyarakat di Desa Triharjo Kulon Progo* (Risfanda 2020), it was determined that the circulation design between men and women when entering the mosque area until leaving the mosque area must be separated for consideration of taharah ablution in the form of clarity of floor patterns. So that the value of the flexibility of the space in that area has not been fulfilled because men and women are in the same circulation.

The clarity of the circulation orientation has been supported by signage by the needs of the circulation orientation of space users, by containing information about the function of space textually, as well as information on the allocation of space users (men, women, and people with disabilities) in the form of symbols. However, in the observation process, there are still room users who have difficulty reaching the men's toilet area because the placement of the men's toilet signage is difficult to reach for sight, so the functional value of the signage has not been fulfilled. In the study of the *Tinjauan Perilaku Pengunjung Terhadap Pola Sirkulasi Majid Agung Jawa Tengah a* by (Nabilah et al. 2018) considering the optimization of the delivery of signage information related to the suitability of the size, placement, shape, and amount of signage so that it is easy to reach by sight with the aim of room users being easy to read, understood and trusted to understand the pattern of space circulation.

One of the functional and economic values in the lighting of the circulation system can be achieved by maximizing natural lighting from sunlight for the needs of high-intensity circulation activities from morning to evening. Disturbing (Wicaksono 2014). Furthermore, it is supported by artificial lighting as circulation lighting in the afternoon until the morning with excess lighting that can be adjusted according to needs and does not depend on time and weather (Budianto et al. 2019). It is known that the functional and economic values have been fulfilled because lighting can support activities in the circulation of the mosque so that the activities of room users when they want to pray until they finish praying in the context time can be optimal. The functional and economic value is supported by the combination of LED fluorescent lamps with electrical energy saving and openings for sunlight entering the prayer area to optimize activities in the main prayer area.

After knowing some elements of space that can support the value of flexibility, economy, and functionality of the circulation system, it is known based on the observation that the circulation purpose of the mosque circulation is the direct nature of circulation in the form of a linear circulation configuration so that the circulation pattern of space users can be traced hierarchically starting from coming into the mosque area to going to the ablution area, then going to the main prayer area and leaving. Leave the mosque area.

After explaining the participatory observation process, further in-depth interviews were conducted to obtain more detailed information about the perception of space users related to the existing mosque circulation system with the needs of mosque space users' circulation activities to 10 mosque visitors, male, adolescent, and adult categories. As a result, 6 out of 10 mosque visitors stated that the existing circulation system of the mosque, space users feel uncomfortable because the circulation path is still integrated with women. The number and placement of male toilet signage are not optimal, and the number and area of the main door of the prayer area are limited, which impacts the non-optimal circulation process. Exit and enter the main prayer area.

Figure 2. Circulation conditions to the ablution area.

Figure 3. Circulation conditions of the main door leading to the main prayer area of the mosque.

Figure 4. The condition of the main prayer area of the mosque.

4 CONCLUSION

The results of the study using a design thinking approach at the empathize stage showed that the condition of the existing circulation system of the Al-Safar Mosque KM 88 B Cipularang Toll Road is not by the needs of the circulation activities of mosque room users. It is proven from the data collection results by participatory observation methods and in-depth interviews. There is a problem that the condition of the circulation system in some mosque facilities still does not meet the flexibility, economic and functional aspects that are not by the provisions of the circulation system. For this reason, it is necessary to pay attention to the needs of circulation activities and the expectations of space users that are adjusted to the provisions of the circulation system to achieve an ideal circulation system by taking into account the aspects of flexibility, economy, and functionality.

REFERENCES

Budianto, C. A., Wardoyo, A., Kristianto, T. A., Rucitra, A. A., & Ardianto, O. P. S. (2019). Studi Sistem Pencahayaan Buatan Adaptif untuk Selasar Aktivitas Gedung Baru Departemen Desain Interior ITS. *Jurnal Desain Interior*, *4*(1), 71. https://doi.org/10.12962/j12345678.v4i1.5263

Ching, F. D. K. (2012). *Architecture Form, Space, and Order* (Third Edit). Wiley. https://www.google.co.id/books/edition/Architecture/GryqqV58cXcC?hl=en&gbpv=0

Harris. (1975). *Dictionary of Architecture and Construction* (C. M. Harris (ed.); (Fourth Edit). McGraw-Hill Education. https://www.google.co.id/books/edition/_/fM5aRKFSQvYC?hl=en&sa=X&ved=2ahUKEwjS9pbyvPj3AhWjS2wGHewCAwMQ7_IDegQIERAC

Muslim, A. (2004). Manajemen Pengelolaan Masjid. *Jurnal Aplikasi Llmu-Ilmu Agama*, *5*(2), 105–114. http://digilib.uin-suka.ac.id/8309/1/AZIZ Muslim Manajemen Pengelolaan Masjid.pdf

Nabilah, A., Pribadi, S. B., & Alfia riza, M. A. (2018). Tinjauan Perilaku Pengunjung Terhadap Pola Sirkulasi Masjid Agung Jawa Tengah. *Modul*, *18*(2), 54. https://doi.org/10.14710/mdl.18.2.2018.54-59

Panero, J. (1979). *Human Dimension & Interior Space* (J. Panero (ed.); 1st revise). Whitney Library of Design. https://www.google.co.id/books/edition/Human_Dimension_Interior_Space/iBBQAAAAMAAJ?hl=en&gbpv=0&bsq=human dimension %26 interior space&kptab=overview

Risfanda, M. (2020). *Perancangan Masjid Manasik Haji dan Rest Area Berbasis Aspirasi Masyarakat di Desa Triharjo Kulon Progo* [Universitas Islam Indonesia]. https://dspace.uii.ac.id/handle/123456789/28466

Soewardikoen, D. W. (2019). *Metodologi Penelitian Desain Komunikasi Visual* (B. A. F. Maharani (ed.)). PT Kanisius Yogyakarta. https://www.google.co.id/books/edition/Metodologi_Penelitian/-uQWEAAAQBAJ?hl=en&gbpv=0

Tofani, L. (2011). *Terminal Imbanagara Kabupaten Ciamis Clarity* [Universitas Komputer Indonesia]. https://repository.unikom.ac.id/18465/

Wicaksono, A. A. (2014). *Teori Interior*. GRIYA KREASI. https://www.google.co.id/books/edition/Teori_Interior/03rQBgAAQBAJ?hl=en&gbpv=0

Wolniak, R. (2017). The Design Thinking method and its stages. *Systemy Wspomagania w Inżynierii Produkcji: Support Systems in Production Engineering*, *6*(6), 247–255. http://yadda.icm.edu.pl/baztech/element/bwmeta1.element.baztech-81d700a1-e4ea-4257-87cf-d0b790873bc8

Personal values vs UTAUT factors on mobile gamers in Indonesia

R. Aulia
Telkom University, Bandung, Indonesia

A. Bismo & R. Amansyah
Bina Nusantara University, Jakarta, Indonesia

ABSTRACT: The mobile game industry is an industry that is growing rapidly in the world as well as in Indonesia itself in line with the development of current technology. The purpose of this study is to determine the influence of personal values and UTAUT model factors that can affect the behaviour of mobile game users, profiling factors of personal values and UTAUT in each long-playing segment of mobile game users in Indonesia, namely the duration of play is less than 1 hour, 1–3 hours, and more than 3 hours. This study uses the MANOVA analysis method and descriptive statistics with the questionnaire data collection method to 400 respondents. The results found that here were no differences in the average of 9 factors in the personal values and UTAUT factors. Suggestions for the mobile game industry are to create strategies that are tailored to the characteristics of game old segment.

Keywords: mobile games, personal values, user behavior, UTAUT

1 INTRODUCTION

In the modernization era, technology continues to develop rapidly, this makes all needs can be met with the help of technological tools such as smartphones to achieve the desired satisfaction by users. According to We Are Social, 2019 in conducting a survey of technological developments in Indonesia, the number of Mobile users which have reached 355.5 million. Based on these data, although playing games can be done on a console or PC platform, the percentage of survey data on the activities of mobile users by playing games shows that 83% of respondents have said so. This can be an opportunity for Indonesian companies and start-up developers who are engaged in gaming targeted at these mobile users. Market segment with the most game revenue is on the Smartphone platform with the percentage of the market segment is 35% with a total game revenue of $ 164.6 billion in 2018 and is predicted to increase until the year 2022 predicted to reach 41% with a total game revenue of $ 196 billion. Through the results, it is seen that there is a change in user behavior in playing a game from the Downloaded / Boxed PC platform and the Console moves to the Smartphone platform (Wijman 2018).

In fact based on CNBC Indonesia, it shows the infographics examined by local game developer company, stating that the market share owned by all companies Indonesian developers only have 0.4% with game revenue which can only be $ 2 million in a year. They also made a comparison of the market share of their respective local companies such as Japan, with a total share of 81% of the local developer industry, making the comparison with the total industry share of Indonesia's local companies significantly (Adharsyah 2019). Agate International 2019 also said that effective marketing methods for the target market that local developers want to achieve are still very limited so it is difficult for them to attract users in accordance with the group of users they want to go to and ultimately none of the marketing methods can drive emotional impulse from users in wanting to find out more about the game that will be launched later as foreign developers do in Indonesia. Then, the local game developer company Digital Happiness, has difficulty in achieving profits of less than 5% which is caused by the majority of gaming products in Indonesia that have been controlled by foreign games. In fact, they say even though there are a number of game products published with low quality can still be viral so there are factors that are difficult to guess in explaining what makes users prefer the game over games with better quality (Koran Sindo 2019).

Based on the results of pre-test interview, the responses given by these respondents indicate that each of these users has their respective motivations in wanting to play mobile games to achieve satisfaction in playing what they want and even encouraged to make purchases on the game. According Tseng, 2011 regarding online segmentation of gamers based on their motivation, researchers show that of the many online game users, their motivation to play is divided into two things: the motivation of the need to explore the mechanics and features of the game and the motivation to fulfill the desire for hatred in himself by committing violence against other gamers and winning self-esteem that shows himself better than other gamers. This happens because of the diversity of user behavior.

Another fact from the results of the pre-test interview that some of the respondents received offers to purchase games by charging fees for the game and there were offers given with games without charging fees for buying the game but there were in-game purchases. Dividing users who bought the game at a given cost, users who get the game at no charge, and users who get the game for free but make in-app purchases of the game. This proves perceptions about the behavior of game users cannot be equated and the need for motivation and personal value research of each game user and become one of the goals of all game developer companies in identifying large specific game user market groups that can be market groups that use and provide good value offers for the company's long-term (Ramírez-Correa et al. 2020). So, the purpose of this study is to determine the differences in the Personal Values UTAUT factor based on the group based on the length of time playing mobile game users in Indonesia and identifying each of the characteristics of the old group playing mobile game users in Indonesia.

2 RESEARCH METHODS

2.1 UTAUT

In several previous studies the application of UTAUT theory is closely related to the main topic that raises technology directed at mobile technology such as smartphones, tables and other gadgets (Ericska et al. 2021; Rahadi et al., 2022).

2.2 Sample

In this study, researchers used survey data collection methods by distributing questionnaires to 322 respondents. However, out of a total of 322 respondents, there were 5 respondents whose questionnaire was invalid due to duplicate filling. Thereby reducing the initial respondents to 317 respondents.

2.3 Data collecting

This data collection technique will be carried out using a questionnaire which will be distributed online through Google Form to mobile game users in Indonesia.

2.4 Data research methods

After determining research methods, data collections techniques and sampling techniques, planning methods are needed in conducting data analysis to simplify the data for easy reading and interpretation. The research techniques used in this study are MANOVA and Descriptive statistics using SPSS 26.

3 RESULT AND DISCUSSION

Table 1. Differences in personal values factors of user behavior in each old segment of mobile game users in Indonesia.

Testing	Result
Simultaneously	
UTAUT factor variable	Differences Occurred
Individually	
Performance Expectancy	No Differences
Effort Expectancy	No Differences
Social Influence	Differences Occurred
Facilitating Condition	Differences Occurred

The results of this study aim to find out whether there is an average difference between the UTAUT factors in mobile game users, based on the table obtained data processing results that show that the 4 UTAUT factor variables have a mean difference simultaneously. This means that there are differences between the old segments of mobile game users in Indonesia based on the UTAUT factor. While the factors in the table above can be a reference to be considered by mobile game developers in Indonesia with the target of mobile game users to find out patterns of usage behavior when playing mobile games that can increase use behavior. While the results of individual differences test showed that 2 of the 4 factors did not display differences namely the Performance Expectancy and Effort Expectancy variables. In other hand, 2 other factors that have average differences between segments are the Social Influence and Facilitating Condition variables.

Table 2. Comparison of each old play segment and personal values factor in mobile game users in Indonesia.

Order of *Personal Values* factor	Old Play segment		
	(1) Less than an hour	(2) 1–3 hours	(3) More than 3 hours
1	Stimulation	Stimulation	Stimulation
2	Self-Direction	Self-Direction	Self-Direction
3	Achievement	Achievement	Universalism
4	Power	Universalism	Achievement
	Universalism		
5	Benevolence	Power	Conformity
6	Conformity	Benevolence	Power
7	Face	Conformity	Benevolence
8		Face	Face

From the table above it can be seen that each segment has priority factors in different order. First segment (less than an hour) has the characters of Stimulation, Self-Direction, Achievement, Power and Universalism, Benevolence, Conformity, Face. Second segment (1–3 hours) has the characters of Stimulation, Self-Direction, Achievement, Universalism, Power, Benevolence, Conformity, Face. Third segment (more than 3 hours) has the characters of Stimulation, Self-Direction, Universalism, Achievement, Conformity, Power, Benevolence, and Face.

Table 3. Comparative results of each old play segment and UTAUT factor in mobile game users in Indonesia.

Order of UTAUT factor	Old Play segment		
	(1) less than an hour	(2) 1–3 hours	(3) More than 3 hours
1	**Effort Expectancy**	**Effort Expectancy**	**Effort Expectancy**
	Capability	*User interaction friendly*	*User interaction friendly*
2	**Facilitating Condition**	**Facilitating Condition**	**Facilitating Condition**
	Sufficient technology	*Sufficient technology*	*Accessories*
3	**Social Influence**	**Social Influence**	**Social Influence**
	Society behavior driven	*Respect opinion*	*Relative*
4	**Performance Expectancy**	**Performance Expectancy**	**Performance Expectancy**
	Useful	*Helpful*	*Performance*

From the table it can be seen that each segment has factors in the same order but has different priority indicators. First. segment (less than an hour) chooses Capability (Effort Expectancy), Sufficient technology (Facilitating Condition), Society behavior driven (Social Influence), Useful (Performance Expectancy). Second segment (1–3 hours) selects User interaction friendly (Effort Expectancy), Sufficient technology (Facilitating Condition), Respect opinion (Social Influence), Helpful (Performance Expectancy). Third Segment (More than 3 hours) selects User interaction friendly (Effort Expectancy), Accessories (Facilitating Condition), Society behavior driven (Social Influence), Performance (Performance Expectancy). Following are the results of in-depth profiling of the 3 main playing segments and personal values factor and UTAUT obtained.

1. *First segment (Achievement seeker gamer, less than an hour)*

In this segment that makes them distinct from others are Achievement can be one of the motivations for playing motivation which tends to be driven to be the best and likes the challenges they can achieve with a faster duration of play. This can be one of the needs of those who want to be fulfilled during their free time. On the one hand they will do everything they can to bring down others (Power) to be the best compared to others. However, on the other hand, the characteristics of Universalism have the same magnitude as the characteristics of Power where they also have the desire to get to know others who can also encourage segment 1 users to achieve their goals. This segment also has a relationship with the characteristics of Benevolence in having the drive to be directly involved in a community in the mobile game in the pattern of mobile game usage, first segment tends to choose to play a mobile game from Capability where they prefer a game that they are able to play quickly and technology on their smartphone that can adequately play the mobile game (Sufficient technology). Their behavior tends to be driven by those closest to them because there is a connection with their character in Universalism and Benevolence. However, first segment users tend to respond to Useful when playing mobile games but not significantly or low responses.

2. *Second segment (Co-op to achieve gamer, 1–3 hours)*

In this segment what makes them different from the others is found in the Achievement characteristics followed by the Universalism factor. This indicates that on the one hand they have a goal to achieve the goals they want to achieve in playing mobile games for their own sake. However, on the other hand they also have the desire to get to know others but users of second segment put the urge to win awards first. In the pattern of mobile game usage, it appears that second segment chooses on User interaction friendly and Sufficient technology where they look first at the interaction of the use of mobile games that are easy to understand and can be adequate on the system and devices on the smartphone they have. Even though users in second segment tend to be less concerned with mobile games that can help them achieve their goals, it is still needed by users in second segment who have the desire to get to know others.

3. *Third segment (Social gamer, more than 3 hours)*

In this segment that makes them different from the others lies in the characteristics of Universalism where they are more concerned with getting to know others than achieving what they want to achieve. So, they tend to meet social needs first rather than winning awards for themselves. This segment is also characterized by Conformity where they play mobile games because they do not want to hurt others. In the pattern of mobile game usage, it appears that third segment chooses on User interaction friendly where the interaction of use when playing a mobile game should be easy to understand and the Accessories usually as long as they play the mobile game also require additional help tools when they play the mobile game.

4 CONCLUSION

Through this research it can be seen that the factors of Self-Direction, Stimulation, Achievement, Power, Face, Conformity, Universalism, and Benevolence there are differences in the average between the segments playing for less than 1 hour, 1–3 hours, more than 3 hours through the MANOVA Test simultaneously. Then, on the submission of individual factors, the Self Direction, Stimulation, Achievement, Face, Conformity, Universalism, and Benevolence factors do not have an average difference between the segments playing for less than 1 hour, 1–3 hours, more than 3 hours. While the Power factor has an average difference between the old segments playing less than 1 hour, 1–3 hours, more than 3 hours. After knowing the characteristics of personal values factors in each old segment playing mobile game users, a profile of UTAUT characteristic.

Through this research it can be seen that the factors of Performance Expectancy, Effort Expectancy, Social Influence, and Facilitating Conditions there are average differences between the segments playing for less than 1 hour, 1–3 hours, more than 3 hours through the MANOVA test simultaneously. Then, on the submission of individual factors, the Performance Expectancy and Effort Expectancy factors do not have an average difference between the segments playing for less than 1 hour, 1–3 hours, more than 3 hours. Whereas the factors of Social Influence and Facilitating Condition have an average difference between the segments playing for less than 1 hour, 1–3 hours, more than 3 hours. After knowing the characteristics of the UTAUT factor in each old segment playing mobile game users, a profile of the UTAUT factor characteristics of mobile game users in Indonesia was formed. Therefore based on this study, mobile game industries can adjust their strategy models effectively to suit the Indonesian market to create strategies that are tailored to the characteristics of game old segment.

REFERENCES

Adharsyah, T. (2019) *BI Sebut Game Online Bisa Rugikan Negara, Ini Faktanya!, CNBC Indonesia*. Available at: https://bit.ly/3T8FnBi.

Agate International (2019) *Tantangan Industri Mobile Game di Indonesia*. Available at: https://bit.ly/3etg8KP.

Ericska, R. A., Nelloh, L. A. M. and Pratama, S. (2021) 'Purchase Intention and Behavioural use of Freemium Mobile Games During Covid-19 Outbreak in Indonesia', in *Procedia Computer Science*. Elsevier B.V., pp. 403–409. doi: 10.1016/j.procs.2021.12.156.

Koran Sindo (2019) *Pangsa Pasar Game di Indonesia Makin Besar, Harus Direbut, sindonews*. Available at: https://bit.ly/3g9AEAn.

Rahadi, R. A. et al. (2022) 'Towards a cashless society: Use of electronic payment devices among generation z', *International Journal of Data and Network Science*, 6(1), pp. 137–146. doi: 10.5267/J.IJDNS.2021.9.014.

Ramírez-Correa, P. E., Rondán-Cataluña, F. J. and Arenas-Gaitán, J. (2020) 'A Posteriori Segmentation of Personal Profiles of Online Video Games' Players', *Games and Culture*, 15(3), pp. 227–247. doi: 10.1177/1555412018766786.

Tseng, F. C. (2011) 'Segmenting online gamers by motivation', *Expert Systems with Applications*, 38(6), pp. 7693–7697. doi: 10.1016/j.eswa.2010.12.142.

We Are Social (2019) *Indonesia Digital 2019: Tinjauan Umum, Websindo*. Available at: https://websindo.com/indonesia-digital-2019-tinjauan-umum/.

Wijman, T. (2018) *Mobile Revenues Account for More Than 50% of the Global Games Market as It Reaches $137.9 Billion in 2018, Newzoo Global Games Market*. Available at: https://bit.ly/2J4lsn0.

The efforts to develop Forest Farmer Group (FFG) business through designing visual identity packaging of mangrove coffee products in Indramayu

D.M. Septiani & I. Wirasari
Telkom University, Bandung, Indonesia

ABSTRACT: The Forest Farmers Group (FFG) is an association of Indonesian farmers with a background in forestry business management inside and outside the forest. FFG Karangsong, Indramayu Regency is one of the groups that developed a coffee production business from mangrove seedlings because mangrove plants are easy to find in the coastal area of Karangsong. Types of coffee beans that are processed into coffee are mangrove kerandang. This business product is still relatively new, and not a lot of people know about it. Therefore, it is necessary to design a visual identity on the packaging of mangrove coffee as a business development strategy carried out by FFG. The method used is qualitative, with data collection through observation, interviews, and literature study, then strengths, weaknesses, opportunities, and threats (SWOT) are used as the basis for the analysis phase. The design is expected to generate positive value for consumers to be interested in trying and knowing more about these innovative products.

Keywords: business, coffee, mangrove, packaging, promotion

1 INTRODUCTION

Indramayu Regency is located on the coast of north Island Java and borders directly with Java sea. Most of the region is low plains and coastal areas. Part of the coast widely encountered mangrove plants that hold medium rate abrasion. One of the coastal areas in the Indramayu Regency who already been active since 2008 in mangrove planting is Pabean Udik Village that formed into Forest Farmer Groups (FFG) named Jaka Kencana and led by Abdul Latif, until this moment already developed into 70 hectares of area planted with 34 species of mangrove plant by carrying the "Mangrove Village" concept. The planting aims of mangroves, in addition to being an effort to preserve the environment, also become an effort to develop the processed industry as the public needs it. Considering a lot of generated potential from mangrove plants, Jaka Kencana's Forest Farmer Group (FFG) utilizes it into processed products of economic value, so formed a name brand "Jakie Gold". One of the successfully developed innovations is coffee with an ingredient base from mangrove seeds. However, not all types of mangroves can be made into processed coffee, but the Kerandang mangrove seed type can be an option.

As time goes by, the development of product in the processed food field continued developments in brand identity. The important value of brand identity is useful as an identification or differentiator from other brands and can stimulate the five senses of consumers so that a relationship arises between products and consumers (Swasty 2016). The important values built are labels and fables. Labels include all visual elements in the form of logos, packaging, and taglines, while fables are aspects related to customer experience, advertising, company trust, and customer relationships (Mootee 2009). As processed food industry also needed packaging that can accommodate and protect the product when purchasing. In addition to being ergonomic, packaging must pay attention to creative and informational aspects (Wahyudi & Satriyono, 2017). Other packaging functions as a supportive medium in business development. According to a study (Dharmeria & SAB 2014), Packaging design is a strategy to compete in the business world, as well as becoming a brand image that can attract consumers to make purchase decisions for products.

On processed mangrove coffee products produced by Jaka Kencana's Forest Farmer Group, it still has not experienced a change in visual identity on the packaging properly. Therefore, this study is a design step to create a visual identity on mangrove coffee packaging that is relevant and able to become an effort to develop business in the field of processed food industry produced by the Jaka Kencana's farmer group and at the same time, able to maintain the existence of the organization. Another goal is to provide the latest brand image of processed mangrove coffee so that it can be an attraction for consumers to buy and try mangrove coffee.

2 METHODS

This study uses a qualitative approach method that emphasizes reasoning more on a certain situation and context, as well as researching more things related to daily life. Research with qualitative methods requires observation techniques that are directly involved with the source of the object of study. Conduct interviews with the necessary speakers with structural or non-structural interview techniques (Rukin 2019).

The data collection method used was an observation, conducting interviews with the Jaka Kencana Forest Farmer Group, and literature reviews. Data collection was carried out by direct observation of the mangrove coffee-making place in Pabean Udik Karangsong Village, Indramayu Regency. In order to find out how the process of making coffee from the beginning to the packaging stage. The interview was conducted directly by the head of the Jaka Kencana Forest Farmers Group Named Mr. Abdul Latief (50 years old), the originator of products from mangroves under the brand name "Jakie Gold". Interviews were conducted to find more information about the products being marketed, especially in mangrove coffee. Meanwhile, the literature review used is based on quotations from books, scientific journals, to official websites.

The data analysis method used is strength, weakness, opportunity, and threat (SWOT) analysis. The analysis is carried out by creating a matrix table between the outer factor on the vertical side and the inner factor on the horizontal side, then determining the design strategy based on the selection of one of the boxes of the merging results from the matrix table that has been created (Soewardikoen 2019).

3 ANALYSIS AND DISCUSSION

3.1 *SWOT analysis*

The study was conducted using SWOT analysis to assess mangrove coffee products based on internal and external factors. Internal factors consist of strength and weakness, while external factors consist of opportunity and threat.

The conclusion obtained after conducting a SWOT analysis in Table 1 is that the matrix on weakness–threat (W-T) has a high score to be used as a basis for discussing design concepts and strategies.

Table 1. SWOT analysis.

SWOT analysis	Strength	Weakness
	1. The Original raw material for Kerandang mangrove seeds 2. No production time limit	Lack of attractive appearance on the mangrove coffee packaging
Opportunity	SO	WO
1. Selected and superior mangrove seed processed products 2. Can be consumed by all circles	Become an opportunity in business development in the field of processed native mangroves and become a superior product.	Coffee with selected mangrove beans is packaged on good packaging to maintain product quality. As well as using a supportive and informative visual identity
Threat	ST	W-T
1. Consumers are less familiar with mangrove coffee products 2. The emergence of new competitors who are struggling in the field of mangrove processing	It offers original products with selected mangrove seed bases.	Designing a visual identity on packaging that is attractive and informative to become a differentiator from competitors, choosing good packaging to maintain quality, and becoming the latest brand image of products that support business development efforts in the processed industry.

Table 2. Conclusion of the SWOT analysis.

Weakness–Threat (WT)
Designing a visual identity on the packaging that is attractive and informative to become a differentiator from competitors, choosing good packaging to maintain quality, and becoming the latest brand image of products that support business development efforts in the processed industry.

3.2 *Segmentation, targeting, positioning (STP)*

Table 3. STP analysis of mangrove coffee products.

Segmentation	Targeting	Positioning
Geographical: Indramayu Regency, The Entire Territory of Indonesia Age: 18–50 years Occupation: Student, laborer, civil worker, entrepreneur.	– Coffee consumers – Traveler – People who like local products	Superior quality products with selected mangrove seed base materials. The visual identity used becomes a characteristic and at the same time introduces the product as the latest preparation.

3.3 *Concept*

Message concept

The concept of the message is taken based on the analysis of the data that has been carried out. In order to know what kind of products are needed by the market. The results of the data analysis produced three keywords as limitations in the design, namely, mangroves, farmers, and homemade. The message carried by mangrove coffee is to enjoy processed products from mangroves as the latest experience to create a new atmosphere and foster flavors to love local Indonesian products more.

Creative concept
The creative concept carried out is in the form of designing a visual identity, which consists of a logo and tagline, and determining the shape of the packaging as a solution to attract the attention of consumers. The design was made based on field data analysis.

Visual concept
The logo is formed in the form of a logotype and logogram that are put together and based on three keywords, namely, mangroves, farmers, and homemade. The logo design process that is carried out is to collect data, then find references, and the process of making manual to digital sketches.

Figure 1. Logo of the brand.

Visual identity packaging is designed in the form of silhouettes of mangrove trees and sea waves. It is divided into two sides, the backside, and the front side. On the front side, there are product descriptions, logos, taglines, weight information, and production sites. While the back side has a logo, information on how to present, composition, customer service, and product barcodes. Both designs are formed with a size of 9.3 cm x 15.7 cm and a green color taken based on the implementation of the plant and beauty.

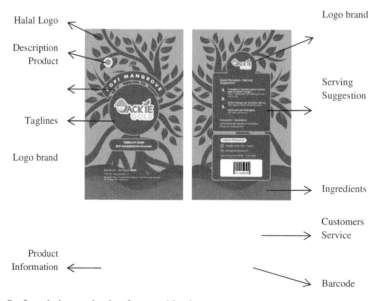

Figure 2. Surface design packaging front and back.

Media concept

The concept of the media used is a form of packaging that holds dry products, and it can protect the quality of the product for a long period of time. The specified packaging is in the form of a standing pouch that can accommodate products with a net weight of 100 grams. Made from brown cardboard on the outside and coated with aluminum foil as the inner layer of the package. The packaging cover uses a zip lock to be safe when the product is stored again. Visual identity packaging is made with printing techniques.

Figure 3. Packaging mockup weighing 100 grams.

4 CONCLUSION

The conclusion obtained is that the design of the visual identity of the packaging on mango coffee has an effect on building a brand as an effort to develop business in the processed food industry. A visual identity is designed based on research and direct observation, so that the visual identity on the packaging that is designed to make the product more different from other competitors can lift the brand image created by a small community of people, and maintain its existence in the long term. In addition, it can provide additional up-to-date information about the use of mangroves as products of economic value.

This research can provide added value to the latest information about the use of mangroves in products that have economic value. The drawback of this research is that there is no validation of the designs that have been made. Suggestions for further research to validate various related aspects in order to produce innovation and consistent design recognition.

REFERENCES

Dharmeria, V., & SAB. (2014). Analisis Pengaruh Keunikan Desain Kemasan Produk, Kondusivitas Store Environment, Kualitas Display Produk Terhadap Keputusan Pembelian Impulsif (Studi pada Pasaraya Sri Ratu Pemuda Semarang). *Jurnal Sains Pemasaran Indonesia*.

Maskun, V., Wirasari, I., & Ayafikarani, A. (2021). Perancangan Strategi Promosi UMKM Mochi Kaswari Lampion di Kota Sukabumi. *e-Proceeding of Art & Design*.

Mootee, I. (2009). *Brand Strategy*. United States: Idea Couture Inc.

Rukin, D. (2019). *Metodologi Penelitian Kualitatif*. Kabupaten Takalar: Yayasan Ahmad Cendekia Indonesia.

Swasty, W. (2016). *Branding (Memahami dan Merancang Strategi Merek)*. Bandung: PT. Remaja Rosdakarya.

Soewardikoen, D. W. (2019). *Metodologi Penelitian Desain Komunikasi Visual*. Daerah Istimewa Yogyakarta: Penerbit PT. Kanisius.

Wahyudi, N., & Satriyono, S. (2017). *Mantra Kemasan Juara*. Jakarta: PT. Elex Media Komputindo.

Winarno, F., & Andieta, O. (2020). *Bahan dan Kemasan Alami Perkembangan Kemasan Edible*. Jakarta: PT. Gramedia Pustaka Utama.

Exploring ideas of sustainable workplace through visual storytelling: A case study of #KomikCeritaIndah

D. Apsari
Telkom University, Bandung, Indonesia

P. Maharani
SOAS University of London, London, UK
ORCID ID: 0000-0001-8993-2987

ABSTRACT: The idea of sustainable workplace reflects a number of aspirations in the Sustainable Development Goals (SDGs), including gender equality (Goal 5), decent work (Goal 8), and partnership for the goals (Goal 17). Women Lead campaign, a collaboration between Indonesian feminist web magazine Magdalene and Investing in Women initiative by the Australian Government, is an example of such partnership. A part of the campaign was #KomikCeritaIndah (Comic of Indah's Story) webcomic series, published weekly on Magdalene's Instagram @magdaleneid. Using qualitative method, the authors investigated the application of visual storytelling theory to discover how the campaign's messages are delivered through digital content such as webcomic series. The research was conducted through literature study, observations, and in-depth interviews. The research later found that collaborative campaign content such as #KomikCeritaIndah was able to deliver key messages around equality and inclusivity at the workplace, while creating a sustainable work culture in order to achieve the SDGs.

Keywords: campaign, digital content, gender equality, visual storytelling, webcomic

1 INTRODUCTION

A supportive working environment for women employees can help countries achieve the Sustainable Development Goals (SDGs), some of which are gender equality and decent work (UNDP 2022). According to Ajmal *et al.* (2017), there is an increasing concern about incorporating social sustainability throughout their business operations, including equality, diversity, dan inclusivity in every aspect. A balance in the work environment will affect the cultural realm, works structures and environment, which will significantly impact social sustainability (Woodcraft 2015). One way to address such concern is through campaigns that aim to raise awareness on women's leadership at the workplace, such as the Women Lead campaign by Indonesian feminist web magazine Magdalene. Started in mid-2021, the campaign was supported by Investing in Women, an initiative by the Australian Government whose mission is to catalyze inclusive economic growth through the economic empowerment of women in Southeast Asia.

Initially, the campaign focused more on written pieces on women's endeavors to survive and thrive at the workplace, which were published on Magdalene's website. However, Magdalene's editorial team later decided to diversify their avenues, by including element of visual storytelling in the form of webcomic series titled #KomikCeritaIndah (Comic of Indah's Story), using their Instagram handle as main publication outlet of the series. This effort reflects the understanding that comics, as a form of visual storytelling, have become reliable because the campaigns can be more easily understood through visuals (Saputro et al. 2016). Moreover, the choice to utilize social media to publish #KomikCeritaIndah serves as an example, that one of the keys to a successful and effective campaign is that the campaign can reach out to public awareness (Ardha 2014). The development of Women Lead campaign in general and #KomikCeritaIndah in particular have sparked the authors' interest to further investigate the use of visual storytelling to raise awareness on gender equality at the workplace through social media. The International Labour Organization notes that women made up 39 percent of global employment in 2019, yet they are massively underrepresented in managerial positions at only 28 percent, and the stats are likely worsened due to COVID-19 pandemic (Karkee & Sodergren 2021). These facts speak volumes on the urgency to raise awareness on the importance of inclusive, sustainable workplaces, from which women could benefit.

Meanwhile, although there are already a number of studies done on campaigns in comic format, they are mostly focused on the novelty of the digital comic format, and none has yet to discuss digital comic specifically published on social media. For example, Santoso and Bezaleel (2018) use a 360-degree comic format combined with virtual reality technology. Thus, #KomikCeritaIndah was born.

2 RESEARCH METHODS

This study uses a qualitative approach, with data analysis obtained from visual observations and in-depth interviews. Observational data from this study were taken from the collaborative process held by the authors as the creative team and Magdalene's editorial team while developing the story for the webcomic from April to August 2021. The overarching theme of women at the workplace was later explored in various topics, such as traditional gender norms, perceptions of women's leadership in the workplace, gender equality at home, and harassment in the workplace.

Insights about stories and suitable characters were obtained from observation data, that were carried out on followers/readers of the Instagram account @magdaleneid, the majority of women aged 22 to 40 years. In the context of the visual storytelling process, observations are made by perceiving the reference that is relevant to the story making (Chute 2008), which is then recorded in the form of sketches or fieldnotes that are useful for making visuals related to the story. After that, to evaluate and compare the observation data that had been obtained from Magdalene's followers, in-depth interviews were also conducted with the creative team and editorial team of this #KomikCeritaIndah. All processes carried out by the authors is based on the stages of Digital Storytelling in creative process by DeNatale in Saputro et al. (2016), so that the authors can implement campaign objectives with key messages around equality and inclusivity through visual storytelling.

3 RESULT AND DISCUSSIONS

3.1 *Collaborate: #KomikCeritaIndah, Comic Series from Women Lead by Magdalene*

#KomikCeritaIndah is a weekly webcomic series published from August 2021 to January 2022, initiated by Magdalene, with visual storytelling implemented by Beginsat30. Beginsat30 is a collective work of Puji Maharani as the scriptwriter and Diani Apsari as the illustrator.

The comic series focuses on Indah, the main character, and her life as a working woman, mother to her baby daughter Salma, and wife of information technology (IT) freelancer Adit. Apart from Indah, there are also accompanying characters with interrelated character developments, as outlined in 24 episodes of the webcomic series. This collaboration process started from the breakdown of ideas from the Magdalene team and story building by the Beginsat30 team in the early period of this comic being made, in May 2021. During the story-building process, in-depth interviews were held with the creator to the editor team to gain more insight from

Magdalene's followers. After that, the Beginsat30 team began to create visual ideas in June-July 2021, accompanied by assistance from the editorial team, until it premiered in August 2021.

3.2 *Communicate: Campaign objectives through visual narratives*

The implementation of this visual narrative communication method can be divided into; campaign idea, creative concepts, and visual concepts.

A. Campaign idea through key messages

The aim of this campaign is portrayed in the form of webcomic series because it can convey concepts about equality and a sustainable workplace in a more visually comprehensive way than the conventional printed comics (Santoso & Bezaleel 2018). This complete picture of the world of living characters, in this context of #KomikCeritaIndah, is the idea of sustainable workplace, need to be remembered by the readers (Mcloud 2007). The story's content begins with the dynamics of the lives of women workers and the surrounding environment, which can invite discussion from readers and creators. A summary of the entire episodes and the represented themes can be described as follows:

The story begins with A, the main character, who decides to return to work after her maternity leave ends. The story focuses on anecdotal discourse from mothers who return to work. (Theme: Gender Norms Based On Traditions). At the office, A discussed with many other working mothers the problems that often arise in the lives of working mothers and their solutions. After coming home from work, A chats with B, the husband, —from their conversation; it was caught that the two of them were not only trying to survive in their daily lives but also had to support their families (Theme: Sandwich Generation). The next day, a family event was held at A's house. Unsolicited advice from family members remained unavoidable when they found out that A was back at work, even commenting on B working at home. (Theme: Gender Norms Based on Traditions and Religions)

The story continues with C, B's younger sister, an intern at a news agency. In the office, C faces many conflicts related to his boss, who is male, middle-aged and does a lot of mansplaining on him (Theme: Mansplaining). C's position as a young female intern sometimes puts her in a critical position because her work environment is quite sexist (Theme: Sexism). C also saw many of his colleagues/supervisors who were women experienced injustice in the work environment (Theme: Misperception of Gender Equality). In addition, C often experiences unpleasant actions threatening her while working in the field (theme: sexual harassment). What is annoying is that when C tried to complain, his colleagues even said, "Ah itu kan biasa!" (English: "That is how men are, you know that.") (Theme: Misconception of Gender Norms). C also came home from the office, where she was feeling sad and helpless. In this story, we plan to discuss the exchange of perceptions between a working woman (Theme: Office Politics). From colleagues A and C's stories about office politics, it turns out that many impacts the mental health of workers, especially women. Here, the story's focus shifts to B as a husband, who begins to learn to be involved in daily activities, including child care and household affairs (Theme: Mental Health, and How Men Can Help). From the conflicts that have been experienced by A, B, and C, the focus has shifted to A's child, a baby, who turns out to be a girl. The ending is planned to be an inspirational story about raising a daughter to be proud of herself as a woman–all about raising children to be leaders (Theme: How to Raise A Girl As A Leader).

B. Campaign media creative concept

This webcomic series is presented in 24 episodes, each of which episode contains 10 pages. The comic format is also adjusted to the appearance on Instagram, in 1:1 proportion, and the storytelling is displayed in a carousel format–post containing more than one photo or video, which users can view by swiping left on a post through the phone app.

C. Visual concept

Beginsat30 as a visual creator of this comic series, explores visuals in terms of character design, line design and colour keys that are adjusted to the former art style and colour keys of Magdalene. The explanation of this visual exploration will be explained in the following sub-chapter.

3.2.1 Communicate: Final visual implementation

A. Character design

The characters in this comic consist of the main character, a working mother, A (later named Indah); her husband B (later named Adit), a freelance worker and stay-at-home dad; and C (later named Dewi), Adit's sister, a fresh graduate who just started her first job as a journalist. Other than the main characters, there are also supporting characters consisting of the main character's family members, employers, and co-workers. The visualization of co-workers also explored various characters we can find in real life, from female leaders to disabled workers, covering the idea of equality and accessibility in the workplace.

The art style displayed is a simple black outline with two colour accents, adapted to Beginsat30's style as the artist combined with Magdalene's color palette.

Figure 1. Final visualization of #KomikCeritaIndah's comic characters.

B. Final visualization

The following is the final look of the #KomikCeritaIndah. As previously explained, the 24 episodes of this webcomic series show the dynamics and dilemmas of working women. The sample shown is from episode 9 entitled "Ah, Itu Biasa!" (Ah, That's Nothing Special!), an episode that describes the dynamics of the main characters' different work place but faces the same challenges.

Figure 2. Sample images from final visualization of #KomikCeritaIndah. (source: Magdalene's Instagram page: https://www.instagram.com/p/CUr7e-cAulf/)

3.3 *Connect: Circulating The #KomikCeritaIndah on Instagram*

The development of the webcomic series focuses on the use of Magdalene's Instagram and its followers. Besides that, social media can also be used as a forum for discussions from Magdalene as the initiator and the readers through the "comments" and "likes" features. Interactivity can also be achieved because social media, such as Instagram, can be a successful viral marketing media campaign that can comprise an engaging message that involves imagination, fun, intrigue, encourages the ease of use and visibility, targets credible sources, and leverages combinations of technology (Dobele *et al.* in Agam 2017).

This comic is published once a week, with 24 episodes, on Magdalene's Instagram account, @magdaleneid, from August 2021 to January 2022. The social media team of Magdalene publishes this comic series in each edition along with captions that explain the contents of the story. Supporting hashtags are also embedded in each caption, in order as a bookmark for posts that related to these topics, including #WomenLeadByMagdalene and #KomikCeritaIndah. The Beginsat30 team also released a teaser for this comic series on their Instagram account, @beginsat30, in the form of Instagram Post and Instagram Story. Each post that is circulated contains a page that links readers to the official Women in lead website, https://womenlead.magdalene.co.

4 CONCLUSION

From the creative process mentioned in the previous chapters, we can see that multi-stakeholder collaboration can promote the achievement of the sustainable development goals in the form of campaign visualization, which is circulated on social media. Of course, from a visual perspective, this media campaign must follow the prescribed social media format in order to obtain which visual form is the most effective. The continuity of the story with themes that can represent the dynamics of women in the workplace must also be considered so that readers can feel connected to the characters in the story, both visually and in terms of stories. These processes involve collaboration, communication, and connectivity between stakeholders, creators and readers. It is hoped that in the future, after the #KomikCeritaIndah, there will be more innovations related to campaigns which aim to build awareness for employers who need to consider the needs of their workers to build a sustainable workplace for women.

REFERENCES

Agam, D. N. L. A. (2017) 'The Impact of Viral Marketing Through Instagram', *Journal of Business Management*, 4(September), p. 7.

Ajmal, M. M. *et al.* (2017) 'Conceptualizing Social Sustainability in the Business Operations', *Proceedings of the International Conference on Industrial Engineering and Operations Management*, pp. 2027–2040.

Ardha, B. (2014) 'Social Media Sebagai Media Kampanye Partai ...', *Sosial Media Sebagai Media Kampanye Partai Politik*, 13(1), pp. 105–120.

Chute, H. (2008) 'Comics as Literature? Reading Graphic Narrative', *PMLA*, 123(2), pp. 452–465. Available at: http://www.jstor.org/stable/25501865.

Karkee, V. and Sodergren, M. (2021) *'How Women are Being Left Behind in the Quest for Decent Work for All'*, Available Online on https://ilostat.ilo.org/how-women-are-being-left-behind-in-the-quest-for-decent-work-for-all/, accessed August 8, 2022.

Mcloud, S. (2007) *Making Comics: Storytelling Secrets of Comics, Manga, and Graphic Novels*. HarperCollins Publisher.

Santoso, B. A. and Bezaleel, M. (2018) 'Perancangan Komik 360 sebagai Media Informasi tentang Pelecehan Seksual Cat Calling', *ANDHARUPA: Jurnal Desain Komunikasi Visual & Multimedia*, 4(01), pp. 14–24. doi: 10.33633/andharupa.v4i01.1544.

Saputro, G. E., Haryadi, T. and Yanuarsari, D. H. (2016) 'Perancangan Purwarupa Komik Interaktif Safety Riding Berkonsep Digital Storytelling', *ANDHARUPA: Jurnal Desain Komunikasi Visual & Multimedia*, 2 (02), pp. 195–206. doi: 10.33633/andharupa.v2i02.1207.

UNDP (2022) *Sustainable Development Goals_United Nations Development Programme, UNDP*. Available at: https://www.undp.org/sustainable-development-goals?utm_source=EN&utm_medium=GSR&utm_content= US_UNDP_PaidSearch_Brand_English&utm_campaign=CENTRAL&c_src=CENTRAL&c_src2=GSR& gclid=Cj0KCQjw2MWVBhCQARIsAIjbwoPTxGhR3nGTx4B_NF1e_X-_IMj-MrmkbXFEu4sbAFmd-G9UKVfF-9 (Accessed: 22 June 2022).

Woodcraft, S. (2015) 'Understanding and measuring social sustainability', *Journal of Urban Regeneration and Renewal*, 8, pp. 133–144.

How SME-scaled branding agencies creates a high-quality yet affordable brand identity design for SMEs in Indonesia

R.A. Siswanto*
Telkom University, Bandung, Indonesia
Universiti Sains Malaysia, Penang, Malaysia

J.B. Dolah
Universiti Sains Malaysia, Penang, Malaysia

*ORCID ID: 0000-0002-8085-0739

ABSTRACT: Small-medium enterprises are considered the backbone of the Indonesian economy. They have an increasingly important role in maintaining economic stability and as one of the solutions for the revival of the Indonesian economy after the pandemic. The need for SMEs to market their products online is becoming increasingly important during the pandemic. Therefore, SMEs need an effective way to build their brand through existing digital channels. However, the ability of SMEs in designing brand identities is minimal, so recently, a phenomenon has emerged where many small-scale design agencies provide identity creation services and unique branding strategies for SMEs in Indonesia. Therefore, this study seeks to map these small design agencies that offer special services for SMEs, both in the design process to their strategy in creating a proper identity yet still affordable for SMEs.

Keywords: Branding, Brand Identity, Logo Design, Packaging Design, SME

1 INTRODUCTION

In this digital age, small and medium-sized enterprises (SMEs) are compelled to adapt, since the digital world may be both an excellent chance for growth and a hindrance for SMEs. Because without preparation, it would be more difficult for SME enterprises to further develop. This is aggravated by their brand and logo design difficulties. (Himawan 2019. The key reasons why SMEs may finally update devices with strong product branding are proficiency with software and the capability to do so. They have a crucial role in sustaining economic resilience and as a remedy for Indonesia's post-pandemic economic rebound. During the pandemic, it is imperative that SMBs sell their goods online. (Amankwah-Amoah J. 2021) Consequently, SMEs need an efficient method for building their brand using current digital platforms. Nonetheless, the capacity of SMEs to build brand identities is quite limited, and as a result, a phenomenon has lately evolved in which several small-scale design businesses provide distinctive identity creation services and branding strategies to SMEs in Indonesia. Uniquely, the service providers for SME entrepreneurs are small and medium-sized organizations.

Despite their restrictions, these tiny agencies are able to meet the demands and problems of SMBs. Suyanto (2019) mentioned in his article detailing the difficulties SMEs face. obtaining quality design and branding is very expensive. While the ideal creative process is expensive and time-consuming. Therefore, this study is motivated by encouragement and curiosity about how these tiny agencies that handle SMEs thrive and can exist based on SME entrepreneurs' capabilities. then what are the sacrifices? Therefore, the purpose of this research is to develop a small agency that provides specialized services for small and medium-sized enterprises (SMEs), from the design process to their strategy for generating the proper invention while remaining inexpensive for SMEs.

2 RESEARCH METHODS

This research aims to comprehend the performance and design process used by brand agencies for SME customers. In contrast to the first section, which examines design work, the second section examines the design process and technique used by branding businesses. This is significant since numerous limits and conditions may undoubtedly impact the organization's workflow and decision-making processes. According to Creswell, J.W., and Creswell, J.D., (2017), qualitative research may be conducted to examine a phenomenon. By conducting in-depth interviews with purposefully selected samples (Sutopo H.B. 2006), we will be able to investigate this phenomenon's mechanism and process. To gather samples and case studies relevant to the study aims, it is necessary to develop criteria for purposive sampling.

In the first step, social media and linked websites are scanned at random to identify agencies. This was done due to the digital world's potential as a habitat for selecting this study sample. Agencies must be able to sell their services using digital channels that are generally accessible. Following the collection of many samples, the total should be decreased using the following criteria:

1. The sample must apply to businesses and not individuals: (freelance). This should be criteria since this study should also investigate workflow management and agency cooperation.
2. The sample was then compacted using the minimal criteria that had existed for the last three years. The first three years are the most crucial for the survival of small enterprises, thus it is crucial to pick a sample with a greater degree of experience in order to have better comprehension.
3. Because most Instagram search results encourage search portfolios with the hashtags #logoukm #brandingukm #kemasanukm, the sample search was done using Instagram observations.

Then, 28 agencies were chosen and evaluated using the second criteria, which included a focus on branding services and products, logo design, and social media, so that up to 14 agencies could be further screened. The fourteen agencies were then re-evaluated based on the criterion that they had been in operation for three years and that they had to operate together rather than independently. Then we recruit up to six agencies to conduct in-depth interviews (sugiyono 2015)

The Agencies are:

- **Markaz Design** – Sidoarjo, East Java Province
- **Dakocan Brand Strategiy Partners** – Malang, East Java Province
- **Jarjas Design** – Surabaya, East Java Province
- **Diji Media** – Semarang, Central Java Province
- **Jurangan Kreatif** – lamogan, East Jave Province
- **Jagoan Branding** – Cilacap, Central Java Province

The data from the interviews were collected first, and then the analytical process was carried out with the help of Atlas.Ti software using the following codes: Discover, Research, Brainstorm, Design, Sketch Present, and Deliver. This method was based on Slade-Brooking's theory of branding (2016).

3 RESULT AND DISCUSSION

During the stage of discovery as demonstrated on the Table 1, each organization used a method that was unique to itself. Agencies like Markaz, Dakocan, and DijiMedia are the ones responsible for conducting the more traditional face-to-face interviews. This is not an interview; rather, it is a dialogue between two people. The provision of organizations with data that is both more accurate and more narrowly targeted is one of the benefits that this strategy offers. Because they do not want their customers to feel like they are being interrogated in any manner, shape, or form. Jarjas and Juragan Kreatif use online forms to collect customer information. This approach has the advantage of reducing both financial and time commitments. The branding winner has the most unique approach out of the other five organizations, which helps them stand out from the competition. Under the guise of "Bang Jabran," Bravo Branding was the one who carried out the digital interview. Customers who have their own mascot are less likely to be sceptical, which increases their acceptance of the designer and makes the creative process easier to accomplish. The comparative study concludes that logo-focused organizations care less about strategy and depend more on SMS-based mass communication, avoiding face-to-face encounters. Numerous businesses use project management software to enhance administrative and customer interactions. Brands with a higher price tag are more strategic and demand a more tailored strategy.

At this point, it is feasible to assert that the stages are substantial and that their outcomes may have a considerable impact. Communication prioritization will very probably have a direct effect on outcomes. Several institutes that exerted considerable effort into this method produced a more robust notion. Goals and aspirations, as well as impediments to the brand selection, must be well understood throughout the creative process. Alternatively, this strategy is strongly tied to the agency's personal branding. The position of the owner as a "savior" who knows the industry and is innovative is crucial in establishing trust among SMBs. Each of the six surveyed universities has a distinct research focus. For instance, firms such as Markaz, Dakocan, Dijimedia, and Jagoan branding focus on establishing a unique selling proposition, which is the most crucial element. In addition, the owner's intentions and objectives are taken into account, since brand identity must also be in vain. They are also seeking brand souls to add vitality to the identities they construct. In the meanwhile, Jarjas and Juragan Kreatif focus on logo design. They often want to establish a brand identity as rapidly as feasible. Create the highest quality logo in the least amount of time. Once a strong brand identity has been developed, they are aware of the potential ramifications of terrible ideas that may emerge. Typically, agencies such as Markaz and Dakocan DijiMedia engage their clients in the brainstorming process; they always encourage brand owners to participate since they believe that consumers know more about the brand than agency employees do. It just helps to a clever strategy. This is done to increase the brand owner's feeling of ownership; enhancing engagement requires displaying the customer's interest in the created ideas. Customers will only get what the agency creates, contend the other three corporations. Although participation is restricted, under some circumstances this approach might increase consumer confidence and agency professionalism.

There are certain government entities that have moved away from the more traditional pen-and-paper method of drawing. The consensus of those in attendance was that the streamlining of this strategy will help them save both time and money. The design process is carried out by a team at each and every agency. Companies like Jarjas and juragankreatif.com make use of remote workers for their personnel. They never spoke to one another in person. They are able to cooperate with the assistance of technological advancements, which may result in a reduction in operational expenditures. In the meanwhile, several companies have actual studios and offices, which allow their employees to communicate more clearly and do work in a more timely and productive manner.

Table 1. Comparative analysis table of branding prosess amongst the design agencies.

	Case 1 Markaz Design	Case 2 Dakocan Brand Development Partner	Case 3 Jarjas Branding	Case 4 Diji Creative Media Agency	Case 5 Juragankreatif.com	Case 6 Jagoan Branding
Discover	In-Depth Inteview	In-Depth Inteview	Form Filling	In-Depth Inteview	Form Filling	In-Depth Inteview via Admin
Research	Searching for: USP Brand's owner Dream Brand Soul	Searching for: USP Brand's owner Dream Brand Soul	Searching for: Brand Owner's Design Direction Logo Reference	Searching for: USP Brand's owner Dream Brand Soul	Searching for: Brand Owner's Design Direction Logo Reference	Searching for: USP Brand's owner Dream Brand Soul
Brainstorm	Working together with Brand Owner	Working together with Brand Owner	Providing Ideas for Brand Owner	Working together with Brand Owner	Providing Ideas for Brand Owner	Providing Ideas for Brand Owner
Design	Digital Sketch	Digital Sketch	Digital Sketch	Digital Sketch	Digital Sketch	Digital Sketch
Sketch	Number of alternative acording to project's budget	Number of alternative acording to project's budget	Number of alternative acording to project's budget	Number of alternative acording to project's budget	Number of alternative acording to project's budget	Number of alternative acording to project's budget
Present	Presenting Logo also Rough Business Model	Presenting Logo also Rough Business Model	Presenting Logo	Presenting Logo also Rough Business Model	Presenting Logo	Presenting Logo also Rough Business Model
Deliver	Focus on Packaging design	Only providing design	Providing Social Media Admin if requested	Only providing design	Focus on Packaging design	Focus on Packaging design

While Markaz, Dakocan Dijimedia, and Jagoan Branding carried out the presentation process directly by providing a variety of alternative ideas and solutions, Jarjas and JuraganKreatif.com merely offered the customer-required master designs and logos. Jarjas and JuraganKreatif.com offered the customer-required master designs and logos. Markaz is a one-stop shop that caters to the needs of small companies by offering solutions for branding, design, and packaging. Only. Juragankreatif.com offers a wide variety of media design and development alternatives that are appropriate for businesses of any size or kind. Jarjas is a renowned authority on the branding of SMEs via the use of digital media. Additionally, they provide administrative services for the social media profiles maintained by companies. In return for an additional fee

4 CONCLUSIONS

Each of the six firms acknowledges and encounters similar obstacles when communicating with their customers. However, when it comes to providing services to their clientele, each of the six groups operates differently and employs distinct approaches. These conversations illustrate the high level of service given by these organizations. Organizations that do in-person and in-depth interviews are more focused on discovering the USP, brand advantages, and brand owners' objectives and wants. Other organizations get brand information by simply filling out a form, indicating that the agency's primary objective is logo creation. The prices and charges are proportional to the quality of service. Those who concentrated on the logo alone were charged less than those who performed in-depth interviews.

Six samples revealed differences between the two different approaches, most notably Jarjas, which offers a bigger-scale service and gathers customer information through online forms. However, they also provide other services, including social media administration services for small businesses with administrative needs. Because of the comparative study presented above, we can infer that logo agencies are less concerned with strategy and depend on SMS-based mass communication while avoiding face-to-face meetings. Some firms utilize project management systems to facilitate client-administrator interactions. Premium brands are more strategic, but not overbearing, and use a more personalized approach.

Thus, it can be inferred that the adaption of these agencies reduces the drawing process, which is regarded to be a rather costly and time-consuming activity. This is significant if it relates to Kodrat's (2009) assertion that this method may provide "low budget, big impact" results. In addition, the discovery phase, which is based on a more natural and relaxed interview procedure, is the primary and distinguishing method when working with SME-scale businesses.

ACKNOWLEDGEMENT

This research was carried out as a part of the first writer's Ph.D. research, and we would like to extend our sincere gratitude to everyone who participated in it, specifically Mr. Hendrik Bayu and the Markaz Design Team, Mr. Akbar and the Diji Media Team, Mr. Erick Ashof and the Jarjas Team, Mr. Romy and the Jagoan Branding team, and Mr. Tatag and the Dakocan Branding team. The proprietors of Juragan Branding, Mr. Nuril and Mrs. Luna.

REFERENCES

Amankwah-Amoah, J., Khan, Z., Wood, G. and Knight, G., (2021). COVID-19 and Digitalization: The Great Acceleration. *Journal of Business Research*, 136, pp. 602–611.

Creswell, J.W. and Creswell, J.D., 2017. *Research Design: Qualitative, Quantitative, and Mixed Methods Approach*. Sage publications.

Himawan, A. F. I. (2019). Digital Marketing: Peningkatan Kapasitas dan Brand Awareness Usaha Kecil Menengah. *Jurnal Analisis Bisnis Ekonomi*, 17(2), 85–103.

Kodrat, D. S. (2009). Membangun Strategi "Low Budget High Impact" di Era NeW wave Marketing. *Jurnal Manajemen Bisnis*, 2 No. 1, 59–86.

Slade-Brooking, C., 2016. *Creating a Brand Identity*: A Guide for Designers. Hachette UK.

Suyanto, B., Sugihartati, R., Hidayat, M. and Subiakto, H., 2019. Global vs. Local: Lifestyle and Consumption Behaviour Among the Urban Middle Class in East Java, Indonesia. *South East Asia Research*, 27(4), pp. 398–417.

Sugiyono, P., 2015. Metode Penelitian Kombinasi (Mixed Methods). *Bandung: Alfabeta*,

Sutopo, H.B., 2006. Penelitian Kualitatif: *Dasar Teori dan Terapannya Dalam Penelitian*.

Philosophical potential of *Ngalokat Cai* ritual for Cimahi city branding design based on sustainable development goals

D.S. Pratiwi & M.I.P. Koesoemadinata
Telkom University, Bandung, Indonesia

ORCID ID: 0000-0002-7134-1078

ABSTRACT: The image of Cimahi is known as a military city, but this image is not able to be represented fundamentally and comprehensively. The existing image only represents the values of most certain parties, so a city branding effort is needed to strengthen the city's identity. Various resources can be used as a basis to improve the city's image to strengthen the city identity. The *Ngalokat Cai* ritual is a traditional ceremony that aims to glorify water and build awareness of all elements of society about the importance of keeping water sustainable. This ritual has the same vision as the city's mission and Sustainable Development Goals (SDGs) in terms of protecting the environment so that the city's identity becomes stronger which has an impact on increasing the city's reputation, regional economic growth, and population welfare. The study was examined using interpretive qualitative case study methods through literature studies and observations to reveal the philosophical values of the ritual. These values are then interpreted as alternative images through the proposed design concept.

Keywords: City Branding, Design, Local Wisdom, Sustainable Development Goals (SDGs)

1 INTRODUCTION

In recent years, most of the cities in Indonesia's West Java Province have competed with each other to develop strategies to sell their cities. This happens because of globalization's influence, which causes every country to take strategic steps to increase global public awareness of its image among all stakeholders (Hon Lee 2015). The main motive behind this attempt is to create an identifiable image. A positive city image is born from a strong city identity, so it impacts increasing the city's reputation, regional economic growth, and the welfare of the people. However, a strong city identity is not only shaped by a positive city image but also built from an image that can represent the city's image fundamentally and comprehensively that comes from the values of the city's local wisdom. Cimahi is known as a military city but the image only represents the values of the majority of certain parties, so a city branding effort is needed that can represent the values that are recognized by all levels of society so that a sense of belonging can be realized.

City branding efforts can be carried out through a variety of creative ideas such as combining the values of local wisdom with global values which are the common goals of the entire nation. By adopting the concept of "think globally, act locally" which is widely practiced by the ancestors of the Southeast Asian nation (Koesoemadinata 2013), city branding efforts can be carried out not only based on the values of local wisdom in the city but also based on one of the pillars of SDGs in this case, the environmental concern as a global agreement is capable of being the basic, authentic and comprehensive of city branding efforts. According to (Sameh *et al.* 2018) there are three main approaches to promoting the city, there are the city's core values, characteristics, and aspirations. The approach used in this research is to improve the city's core values and characteristics through the restoration and promotion of heritage. One of the heritages of Cimahi is the *Ngalokat Cai* ritual where there are potential philosophical values that can be used in city branding efforts. So far, no research has been conducted on *the Ngalokat Cai* ritual in the context of design and SDGs. This paper aims to explore the potential of the philosophical values contained in the ritual that is in line with the vision and mission of Cimahi and the SDGs on the environmental aspect to then could be interpreted into a city media design concept as one of the city branding efforts.

2 METHODOLOGY

The study was examined using interpretive qualitative case study methods through literature studies and observations to reveal the philosophical values of the ritual. These values are then interpreted as alternative images through the proposed city media design concept.

Figure 1. Research to design scheme.

As shown in Figure 1, the methodology begins with selecting the object as a media for an alternative solution by searching for a city's local wisdom that has the same values as the values of the SDG's environmental pillar and the vision-mission of the city. The interpretation process is carried out through an adaptation process by understanding the intrinsic and extrinsic values of the selected object which is the *Ngalokat Cai* ritual. The results of the interpretation of these values are adapted for the city branding of Cimahi through the city media design concept. This method is carried out because Cimahi does not yet have a city identity based on local wisdom values that are in line with the city's vision mission and SDGs on environmental aspects.

2.1 Method of collecting data

Data collection was carried out through literature studies and observations to reveal the philosophical values of the ritual. A literature study is done by looking for references that are relevant to the problem and research objectives. References were obtained from various online journal papers with the same theme generated from previous studies and some online articles. The observation is done by paying good attention to city iconic objects to then be used as a comparison reference for the media design concept that will be proposed.

3 RESULT AND DISCUSSION

At this stage, the author will explore the meaning contained in the *Ngalokat Cai* ritual and then be interpreted it into media design for city branding efforts.

3.1 Analysis

One of the missions initiated by the Mayor of Cimahi for the 2017–2022 period is to realize environmentally sustainable development by improving the quality of life for people with justice. This is in line with the SDGs on the environmental pillar. While the vision is to be achieved by making the city advanced in religion and culture. As mentioned before Cimahi has the image of a military city but the image is only representing the values of the majority of certain parties. In that sense, a city branding effort is needed to achieve the city's vision and mission, SDGs in environmental aspect objective and represent the values recognized by all levels of society so that a sense of belonging can be realized. All of these considerations can strengthen the identity of Cimahi and the efforts were carried out by extracting philosophical values of the *Ngalokat Cai* ritual.

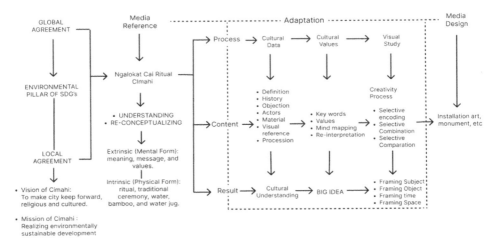

Figure 2. The adaptation process of *Ngalokat Cai* into a media design concept.

To have a deep understanding of *Ngalokat Cai* ritual, the author explores the main factors of cultural heritage there are intrinsic and extrinsic (Ummu H *et al.* 2012) as shown in Figure 2. The intrinsic factor of the *Ngalokat Cai* ritual is the procession in the form of a carnival where water obtained from several springs in Cimahi is symbolically carried by 8 dancers using *lodong* media there is a large tube made of bamboo trees. The actors of this carnival are the dancers, traditional Sundanese music players, and *bangbarongan*, or a kind of traditional clown as a symbol of the bad behavior of people who don't care about the environment. After all the carnival participants arrived at the *Ngalokat Cai* procession, they all pray as the expressions of gratitude were made by all participants in the traditional ceremony and then continued with a traditional dance performance. The next process is the core procession where the village heads from 15 urban villages in Cimahi take turns entering and mixing water from water sources in their respective villages into large jars. After all the water has been collected, the officials of the Cimahi government: the Mayor, Deputy Mayor, and Regional Secretary of Cimahi are invited to open the three lids for the water jars used for washing hands, washing faces, and watering plants around the water jars. While extrinsic factors can be extracted from philosophical values related to the meanings, values, and messages contained in the *Ngalokat Cai* ritual. The philosophical values are defined as a symbol of national unity through togetherness between the government and the community in building Cimahi by unifying determination, words, and steps; as a symbol of togetherness in each region in Cimahi in protecting the environment and the availability of clean water; as a medium of communication, a forum for gathering between the community and policymakers in expressing their thoughts and feelings; as a ritual of expressing gratitude to God who has given fertility to the soil and sufficient clean water.

The *ngalokat cai* ritual is one of the local wisdom of Cimahi which serves as an educational medium to increase public awareness and policymakers to work together to unite to protect the environment, especially water from pollution and scarcity of clean water. The word "*Ngalokat*" in Sundanese means to guard, preserve, maintain, and clean. Water in the context of this ritual serves as a medium for unifying thoughts and feelings in terms of protecting water, land, and culture. The origin of the words "Cimahi" or "Cai" and "Mahi" means the word "sufficient water" in Sundanese where in the North Cimahi area there are many springs that influence the culture of the people. Originally, the culture came from the ritual called *Ngabungbang*, which is a ritual related to the water where people wash utensils that are considered heirlooms and take a midnight bath at seven springs. Another form of culture is the ritual of cleaning rivers and springs and performing a thanksgiving ritual by displaying a variety of Sundanese arts at night. With the development of the times and changes in the function of land from agriculture to settlements, the culture related to water is gradually fading. The ritual disappeared along with the disappearance of dozens of springs in the North Cimahi area. With these conditions, people no longer carry out activities related to water. The impact of the disappearance of dozens of springs is the reduced availability of clean water and frequent flooding. For this reason, efforts to overcome the negative impact are needed that can be carried out through a cultural approach.

From the process of analyzing the intrinsic and extrinsic factors, a pattern is obtained that is in line with the problem and research objectives. This is done through a process of creative and critical thinking by utilizing cultural data, cultural values and visual study as Birgili (2015) said that creative thinking is an individual ability that uses imagination and thinking skills to generate ideas that are used as alternative solutions to solve problems. While critical thinking is a kind of ability to think by zooming in zooming out, paying attention to details, analyzing, and making connections to produce conclusions (Thurman 2009). Through the creative and critical thinking process, the main patterns were obtained: protecting the availability of springs in the long term, protecting the environment, and togetherness which was adapted as the basis for the city media design concept. In this study, the writer chooses installation art as a media design that comes from the adaptation process of Ngalokat Cai as Najmul K *et al.* (2016) said in their works that installation art can be encompassed traditional media such as readymade found objects by using natural materials, ie bamboos.

3.2 *City identity*

The pattern obtained from the process of encoding the philosophical values of the *Ngalokat Cai* ritual is used in the media design process to strengthen the identity of Cimahi. The image of the city through the design of monumental and iconic buildings has a very large contribution to influencing the satisfaction and well-being of citizens and visitors (M. Riza *et al.* 2012). The pattern obtained is then realized in media design that can increase the identity of Cimahi.

3.3 *City branding through media design concept*

As mentioned before that the way to do city branding could be developed through city media design which comes from the process of interpretation of the philosophical values of *Ngalokat Cai*. There are several icons of the city that can be used as a reference for comparison with the proposed media design concept shown in Table 1 below.

Table 1. Icon media of Cimahi.

No	Icon media	Meaning	Source
1.	Kujang statue	As a symbol of West Java's traditional weapon that serves to protect the city from all harm.	Files of Disperkim Cimahi
2.	R.A. Kartini statue	As a medium for remembering national heroes.	Files of Disperkim Cimahi
3.	Lighting installation	In a hope that Cimahi always shining in the good works that can build the city.	Files of Disperkim Cimahi
4.	Tank war	As a symbol of military force that always guards the city	Files of Disperkim Cimahi
5.	Nol kilometer statue	As a city that has a zero point for measurement to other places.	Files of Disperkim Cimahi

*Disperkim: Dinas Permukiman dan Perumahan.

Here is the media design concept based on the analytical process which has been made as shown in Figure 3 below:

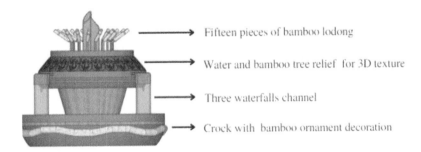

Figure 3. Media design concept based on philosophical values of *Ngalokat Cai*.

The fifteen bamboo *lodongs* as a symbol of national unity through togetherness between government and the community in building Cimahi and also as adoption of fifteen springs in fifteen Cimahi villages; the water and bamboo tree relief for 3D texture as a symbol of the key point that found in the *Ngalokat Cai*, namely protecting the planet by saving water and nature; the three of waterfall channel as a symbol of togetherness in each region in Cimahi and as a representation of three springs in three sub-districts of Cimahi; The crock with bamboo ornament decoration stands for as an expression of gratitude to God who has given fertility to the soil and sufficient clean water. With all of these values that are implied in the proposed design media concept, it gave a contribution to the city branding effort to strengthen the city identity, in achieving the city vision and mission, and also to support SDG's environmental concern.

4 CONCLUSION

The influence of globalization causes the city government to focus more on values that are modern and contemporary to improve the image of the city without using the authentic local culture that has more meaningful values for the residents and people who visit the place. The potential values of local wisdom combined with the values contained in one of the pillars of the SDGs and the vision mission of the city can produce a better and authentic city branding effort. Increasing people's awareness of the city's identity by inserting authentic local cultural values can enlarge the value of the city with outcomes the reputation of the city will increase as well as the economic growth, and also the level of welfare of people will escalate. There are still many cultural potentials from the local wisdom of Cimahi that have not been disclosed to use as the basis for efforts to preserve culture and the environment for the realization of SDGs as the ideals of all nations in the world to be achieved by 2030.

REFERENCES

Birgili. B. (2015). Creative and Critical Thinking Skills in Problem-based Learning Environments. *Journal of Gifted Education and Creativity*: 2(2) 71–80. Turkey.

Hoon Lee, K. (2015). The Conceptualization of Country Attractiveness: A Review of Research. *International Reviews of Administrative Sciences*: 0(0) 1–20. Soul, South Korea.

Koesoemadinata. M. I. P. (2013). Wayang Kulit Cirebon: Warisan Diplomasi Seni Budaya Nusantara. *J. Vis. Art & Des*: 4(2) 142–154. Bandung, Indonesia.

Riza. M, Doratli. N, Fasli. M. (2012). City Branding and Identity. *Procedia-Social and Behavioural Sciences*: 35 (2012) 293–300. Famagusta, North Cyprus.

Sameh. H, El-Aziz. H. M. A, Hefnawy. N. H. (2018). Building a Successful City Branding, Case Study: Dubai. *Journal of Al Azhar University Engineering Sector*: 13(48) 1058–1065. Cairo.

Thurman, A. B. (2009). Teaching Critical Thinking Skills in the English Content Area in South Dakota Public High Schools and Colleges. *Doctor of Philosophy Dissertation*. The USA.

Ummu Hanni *et al.* (2012). Preserving Cultural Heritage Through the Creative Industry: A Lesson from Saung Angklung Udjo. *Procedia Economic and Finance* 4 (2012) 193–200.

Najmul K et al. (2016). The Traditional and Cultural Practice of Installation Art: A Contextual Study. IOSR *Journal of Humanities and Social Science*. Vol 21, Issue 3, Ver. IV.

Files Documentation of Icon Media Cimahi by Dinas Permukiman dan Perumahan Cimahi, 2022.

Zine on sharia investment basics for college students

S. Desintha, A.T.Z. Abdalloh & S. Hidayat
Telkom University, Bandung, Indonesia

ORCID ID: 0000-0003-0569-9044

ABSTRACT: Investment is a good benchmark for Indonesia's economic development. However, at the same time, it is used by several parties who give a wrong understanding of sharia investment, such as giving birth to various misconceptions among general audiences interested in sharia investment. These misconceptions about sharia investment can result in people, especially college students who are new to sharia investment, turning their interest to other forms of investment and even retracting their assets in sharia investment. Therefore, a form of media that can increase literacy about sharia investment for college students to avoid misconceptions is needed. The research method used for this design is qualitative, while the data analysis method is a comparison matrix and SWOT analysis. So, it is hoped that this research can help students to get accurate information and foster a more serious interest in sharia investment.

Keywords: college student, education, sharia investment

1 INTRODUCTION

The high interest of the younger generation, especially students, in sharia investment is a good benchmark for Indonesia's economic development. According to data taken by Kustodian Sentral Efek Indonesia (KSEI) in June 2021, the number of sharia securities ownership based on single investor identification (SID) reached around 991.000 SID or a 36,48 percent increase in just about six months. Even though the number of sharia SIDs is increasing, there are still misconceptions and misunderstandings about sharia investment, most of which come from people under 40 years old. It shows that interest in investment without sufficient financial literacy is not enough, especially among college students and beginners in Bandung, West Java. It was proven that the number of news and crimes around illegal investment has been increasing these past few years. The importance of sharia investment and securities is even more apparent amidst this news because they offer an extra layer of security and certainty that other types of investments do not have. Sharia investment talks about making profits or bearing losses under sharia law (Abdalloh 34:2020). Profit and losses in sharia investment are obtained through buying and selling sharia stocks. Sharia stocks are investment products that have already fulfilled Islamic principles and can be traded in the Islamic Capital Market (Abdalloh 80:2018). To better reach the younger generation, a media alternative that is easy to understand and suitable for the younger audiences is required to improve people understanding of sharia investments.

The conventional media for sharia investments are usually dull and formal, especially for beginners and the younger generation. Even though modern media are starting to emerge like social media and website articles, there are still some problems with those media, especially with their accessibility and practicality. One of the more contemporary and modern media often used by the younger generation is a zine. A zine is a media made and published by an indie group or an individual for a specific group of people and preferences (Bartel

1:2004). For this research, the zine was chosen as an alternative media to break the existing images of conventional sharia investment media that is usually formal, dull, full of text and charts, and generally not beginners and younger generation friendly.

Figure 1. Zine example. Source: Auli Tamma Zhillan Abdalloh, 2021.

2 RESEARCH METHODS

The research for this design has been done through a qualitative method that consists of interviews, observations, surveys, and archives. The data taken from those methods will be used to prove the validity of the problems and the media relevance for college students in Bandung interested in sharia investment. After all the required data was collected, it was analyzed using a comparison matrix and SWOT method.

2.1 *Methods of data collecting*

The first method of data collecting in this research is the interview method. The focus of this interview is to know the perspective of sharia investment and the usage of a zine as a media for sharia investment education from experts such as Mr. Irwan Abdalloh, project giver representative Mr. Derry Yustria Surya Dharma, and sharia investment community Investor Syariah Pemula Bandung (ISP Bandung). The second method was observing the behavior of the research target audiences and the media used. The third data was collected from surveys distributed to a population sample in Bandung. The specification of the target in this survey is active college students, 18–25 years old, and interested in sharia investment. Surveys are used to ensure that the problem and the media used as the solution are acceptable from the perspective of the target audiences. The last method is researching past archives related to sharia investment and zines like journals, books, and website article.

2.2 *Methods of data analysis*

Soewardikoen (2019), in his book Metodologi Penelitian Desain Komunikasi Visual, explained that a comparison matrix is an analysis method used to compare two or more subjects of a similar topic to reach a conclusion that is helpful for the research. SWOT (Strength, Weakness Opportunity, Threat) analysis method is done by making a table of an external factor on the vertical side and an internal factor on the flat side, then selecting one box with a merged result to be used as a strategy for the research.

3 RESULTS AND DISCUSSION

From the data analyzed, the most common problem around sharia investment is that there is still much unawareness that causes negative misconceptions. It got even worse from the lack of accessible and modern media that talks about the basics of sharia investment for beginners. There is a need for media alternatives that can better reach younger audiences, especially the college students in Bandung. One of the characteristics of the younger generation is that they quickly feel bored, so reading long paragraphs about something they do not understand can be a little bit hard. Before going into details, they always need a brief explanation of what they will get. The problem with most conventional sharia investment media is that they do not offer an easy-to-understand-info. However, the information is also presented in rather dull and formal imagery; some of it is even inaccessible to most people. Zine as a media is already a trend amongst college students. A zine can offer an experience and explanation for people who like to read and those who do not. The variety of forms and content in zine can provide a different reading experience that everyone can find interesting. A zine can break the formality and rigidity in sharia investment, making it more suitable for college students in Bandung.

3.1 *Media and content concept*

The main content for this media will be about all the basics and standard information found in sharia investment. The content will be divided into three chapters; chapter one will be about sharia investment, chapter two will be about the Islamic capital market, and the last chapter will be about sharia investors. The core message in this zine is that sharia investment needs to be easy to understand through fun media, giving the impression that sharia investment is something everyone can comfortably start and learn. The visual graphic in the zine was inspired by arabesque art, especially one from The Damascus Room. The Damascus Room was chosen as inspiration because it has some similarities to this zine contextually. The content in this zine will also incorporate Sundanese language and other local characteristics so the target audiences can connect to it better.

The media used for this design consists of primary and secondary media. The primary media in this project is a printed zine titled "*Gak Kerja, Tapi Kaya: 100% Halal*". The provocative nature of the title can draw more interest and attention from the target audiences. Inside, there are also smaller interactive and other minor content pages to make the zine feel more approachable and easier to read. The secondary media is an e-zine, poster, display set, sticker, lanyard, and paper bag. All secondary media was chosen as a part of promotion and to help ease the rigid and formal image of sharia investment. AISAS (Attention, Interest, Search, Action, Share) will be used as a communication strategy to distribute the primary and secondary media.

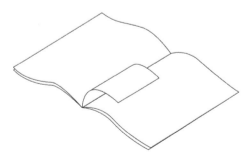

Figure 2. Isometric image of the zine. Source: Auli Tamma Zhillan Abdalloh, 2022.

Figure 3. *Gak Kerja, tapi kaya*: *100%* Zine Cover. Source: Auli Tamma Zhillan Abdalloh, 2022.

4 CONCLUSION

The basics of sharia investment can be daunting to learn at first, especially when conventional media presents a lot of complex language and information that is hard to process by a beginner. It's also possible to pack the basics of sharia investment in a more unconventional media that is easier and fun to learn. If more people could understand the basics of sharia investment, many of the negative misconceptions in Bandung could be reduced and even eradicated. A zine can be a unique and accessible option for the college student in Bandung to read. Sharia investment can be explained more relaxedly without losing many contexts through images, illustrations, and graphic elements inside the zine. In the long term, fun and unique media could ease off the formal and stiff imagery about sharia investment and show people that sharia investment is something anybody can start and learn.

REFERENCES

Abdalloh, Irwan. 2020. *Kaya Harta, Kaya Amal*. Jakarta: PT Elex Media Komputindo.
Abdalloh, Irwan. 2018. *Pasar Modal Syariah*. Jakarta: PT Elex Media Komputindo.
Bartel, Julie. 2004. *From a to Zine: Building a Winning Zine Collection in Your Library*. Chicago: ALA Editions.
Soewardikoen, Didit Widiatmoko. 2019. *Metodologi Penelitian Desain Komunikasi Visual*. Yogyakarta: PT Kanisius.

Fish consumption culture in Manado as a consideration to a web design

R.B. Adji, D.W. Soewardikoen & Ilhamsyah
Telkom Univeristy, Bandung, Indonesia

ORCID ID: 0000-0002-1272-7026

ABSTRACT: There is a culture that has become a strong part of the local's daily life in manado, it's about consuming fish as a daily dish. This culture has given birth to a slogan that invites people to eat fish. This expression is also supported by cultural aspects that shown a strong involvement in consuming fish. The problem that occurs today is that there are still many people in Manado City who don't know the detailed information about the cultural value of consuming fish from Minahasa's ancestors. Through qualitative research methods that will elaborate aspects of culture and communication and to accommodate systems that create the culture, the researcher aims to answer the problem about how to preserve the local's culture using the participatory design methods from the people of Manado. The result of this research is in a form of web design that works as a guide for cultural information regarding fish consumption in Manado and can be distributed to the public, domestic and foreign tourists.

Keywords: Fish Consumption, Manado, Participatory Design, Web Design

1 INTRODUCTION

In the city of Manado, serving fish as a daily dish has become part of the life of the local community. According to the people of Manado City, This culture is not limited only to serve a marine or freshwater fishes. Related to the local culture, the activity of consuming fish dishes is part of the culture of the locals, where every event from a birthday to a mourning of a loved ones is always associated with consuming fish as dish. The habits of the people of Manado City, known as "kumpul-kumpul makan ikan" which translated to gather and eat fish also strengthen the existing phenomenon. Fish-consumption in Manado City, North Sulawesi is also supported from the cultural aspect, there is no culture without the involvement of food in it. The environment, earth, nature change, and civilization develop from time to time, which is followed by aspects of local food culture (Bardosono et al. 2018). One aspect of food culture that has been known so far is the culinary specialties of the region. With the development of science related to food, the City of Manado, which is also a coastal city, utilizes fish as a tourism attraction for the community, especially those in Manado City and in general for domestic and foreign tourists. Six of the ten typical foods of Manado City, use fish as the main raw material for the manufacturing process, including Cakalang Fufu, Mujair Fish, Woku Fish, Fish Soup, Roa Fish, and Nike Fish. Seeing the potential of culinary products in Manado City, known as Nyiur Melambai, the percentage of local natural resources is strengthened by the existence of basic cultural elements. In traditional events and modern events that exist in the Minahasa tribe, various types of fish food are usually found. The event, which still seems traditional, is always in contact with aspects of culinary culture by consuming fish combined with several other dishes. The cuisine of the Minahasa community is indeed closely related to the identity or habits of previous customs.

Consuming Minahasa cuisine, especially fish, is like repeating the memory of the previous community's life (Pamantung, n.d.). At this time, with the existence of various types of fish in the waters of Manado City, people have not been able to identify the types that can be used as guidelines for choosing the type of fish to be consumed.

With the phenomena that occur, a structured and comprehensive information guide is needed so that the community, especially those in the city of Manado, can find out specifically the types of fish that can be consumed, starting from the nutritional content, management, distribution and economic aspects of the fish species. This will certainly have a direct impact on the sustainability of the culture of consuming fish in the city of Manado. By looking at the massive potential of fans of Minahasa cuisine, especially fish dishes, it must also be strengthened by knowledge in terms of culture, nutritional content, and health aspects for people who consume these foods which are often encountered at traditional community events in Manado City. This research aims to improve the culture of consuming fish for the community, especially those in the city of Manado and using web design as a medium of information that functions specifically to identify aspects related to consuming fish. Similar research that has been done previously is about the relationship between the eating habits of the Minahasa ethnic food in North Sulawesi Province. The theoretical framework that will be implemented in this study uses two theories that can help relate aspects related to the culture of consuming fish in Manado City, namely the theory of cultural elements as an institutional system in culture, which explains that culture is a system framework because it has a role as the main element in interact with members of the culture. This theory explains the three levels of culture as a system. The first level is history, identity, belief, values, and worldview. Furthermore, the second level explains about, rules, customs, and communication systems. At the third level describe some aspects of economic institutions, family, politics, health, and religion. The next theory is related to web design, the criteria for a functional website can be seen from several things including structure, visual design, content, process, compatibility, functionality, and interactivity (Ekarini 2017). There are several variables that can be used in designing a website including defining a good design, which discusses the point of view on function, presenting information effectively and efficiently. Anatomy of a web page, which divides several parts of the study including logo, navigation, content, footer, and spacing. Next up is grid theory, which explains the third rule of proportion. Balance, a visual harmony that is in harmony with physical form. Unity, which describes unity as the different guidelines of composition that interact with each other. Bread and butter layouts, talk about the layout used in the appearance of the website. Free trends, identify the latest trend ideas so that they can be designed according to the newness that occurs.

2 RESEARCH METHODS

Methodology is a study of regulations in the scientific method (DW Soewardikoen 2019). Research on the culture of consuming fish in Manado City will be carried out in Manado City and will use a qualitative approach based on phenomena that occur in the research area. The stages of research include the people of Manado City as the main instrument, observations, and interviews conducted with parties related to the object of research, literature studies about local community activities that are directly related to the culture of consuming fish in Manado City, and analyzing data collected inductively. In this study, there are several indicators that will be studied, including the relationship between fish and the meaning of life for the local community, ancestral values, community habits, investment generated from fish production, findings of experts especially related to fish potential, regulations from the local government, nutritional content on fish and typical cuisine of Manado City community.

3 RESULT AND DISCUSSION

The interpretation of cultural values on the culture of consuming fish in Manado City is based on the theory of cultural elements as a system which includes three stages and can be applied in identifying the culture of consuming fish in Manado City. In the first stage, explaining the history of fish in Manado City, which started from Minahasa folklore showing the emergence of the word sera or fish obtained from the river. At that time, a Minahasa figure named Tiow Mararatu was born, who liked to hunt and fish. The river is also one of the sources of food procurement in the Minahasa tribe.

The Minahasa community is also actively involved in discussing the management of fish to become dishes served for Minahasa traditional events. (Pamantung, R.P 2019). The next section deals with identity, that goldfish or in Minahasa language is pongkor fish is a fish that has a higher value than other types of fish and food found in Minahasa in the past.

Goldfish are often presented during events or traditional activities of the Minahasa community, including the inauguration of a new house for the local community, weddings carried out by the Minahasa community, and the birthday of the Minahasa community. The concept of Minahasa culture that can be considered in the sustainability of civilization in Minahasa is that the symbol of Minahasa food and drink is a mediator between humans and the creator of the universe so that on the other hand the Minahasa community makes it a symbol of blessing or gratitude (Weichart 2014). Furthermore, regarding the aspects of belief related to the culture that consumes fish in Manado City which is related to the Mapalus Minahasa philosophy which translated to as the act of helping the other person as a community to ease the progress of any kind of event that the other person undergo. The social interactions that are presented in the Minahasa community, even though they come from various regions, different cultural, religious, and occupational backgrounds, but the social interaction runs smoothly. The indication is the participation of the surrounding community in social activities such as community service, maintaining environmental security, unity of joy and sorrow, and social gathering. Close identification with the value aspect generated by consuming fish in Manado City, North Sulawesi, that most of the food and culinary products produced by the people of Manado City are the result of local cultural natural resources related to cultural aspects. In traditional events in the Minahasa tribe itself, various types of fish food are usually found, one of which is woku blanga fish. Meanwhile, at traditional events that still seem traditional in rural areas, woku blanga fish is processed with banana leaves roasted over a fire. Minahasa cuisine is indeed related to the characteristics or habits of ancient times because of this, consuming Minahasa cuisine is defined as an iterative process with the lives of Minahasa figures. In terms of the world view of the impact of fish on the people of Manado City, it is one of the gateways for domestic and foreign tourists who travel to Manado City. This is in accordance with the findings in 1997 by Mark Erdmann, a researcher at the International Conservation. Mark Erdmann when he received research funding from National Geographic to examine the existence of Coelecanth Fish in North Sulawesi, precisely on Manado Tua Island. The process of observation and research is also carried out in collaboration with fisheries and environmental conservation communities in the city of Manado. Based on these findings, the community and abroad know the potential in terms of fisheries and marine in the city of Manado. In the aspect of educational institutions, in terms of the approach chosen to be realized in strengthening the culture of consuming fish in Manado City, in fact little by little the culture began to enter the city of Manado. In terms of the role of educational institutions, the application of the curriculum regarding aspects of fish culture and the identification of problems that occur at this time the management of the fish industry in Manado City is important to realize. Educational institutions have a vital role in developing a culture of consuming fish in Manado City. Meanwhile, in terms of health institutions, judging from the benefits of consuming fish for the community in general, fish has an adequate nutritional essence from the aspect of public health. Fish is included in one of the categories with a high level of protein absorption if identified from various other animal species. In terms of policies related to the economy and the work system of the local government, in this case, the Government of North Sulawesi, the cultural aspects are contained in the export policy of direct fish shipments from Manado City to Tokyo, Japan. Fishery commodities in North Sulawesi are worth 66,000 US dollars. The fishery product that is the material for export is Tuna, which is also a High Migratory fish and is the mainstay, especially for markets in Tokyo, Japan. In terms of belief, with the culture of consuming fish in Manado City, when talking to local culture, consuming fish is one part of the cultural character of Manado City, where activities that are born until the day of death, are the basis for the people of Manado City. In order for the local community to recognize the term "Kumpul nda kumpul tetap makan ikan" which means to gather or not to gather, we must consume fish. In terms of government policies related to fish consumption culture in Manado City, with the declaration of Indonesia as the world's maritime axis on November 13, 2014 at the 9th East Asia Summit. The speech delivered by the President of the Republic of Indonesia, Joko Widodo about Indonesia as the world's maritime axis is the latest development method.

There are five pillars that are the goals of Indonesia's cooperation with the world related to the maritime axis, among others, rebuilding the maritime cultural ecosystem, maintaining and managing marine resources by focusing on designing seafood sovereignty through improving the fishing industry by placing fishermen as the main reference.

The third pillar is improvement in the aspects of maritime infrastructure and connectivity by designing sea highways, deep seaports, logistics, shipping industry, maritime tourism, and increasing maritime diplomacy by eliminating sources of conflict at sea. And the fifth pillar is designing maritime defense forces in Indonesia. In an interview conducted through the Manado City Tourism Office (Pelealu 2021), said that fish has a significant impact on fishermen in the coastal city of Manado, as a regulator the local government has prepared regional regulations and Mayor regulations regarding the development of the fishing industry in Manado City. By looking at the continuity between the cultural elements that have been discussed, the role of the community, especially those in the city of Manado, is to maintain and protect the culture of fish. Through the web design process, it is hoped that the culture of consuming fish in Manado City can have a positive impact on the welfare of the local community. The web design describes the history of the city of Manado, which has various post-production impacts on fish through investments made every year by involving various distributors. Second, explain about the various types of fish and the nutritional aspects contained in each type of fish. Third, explain the various historical stories of the characters in Minahasa land who do fish hunting activities. Fourth, describe related to the types of fish found in the waters of Manado City.

Figure 1. Image of web design of fish consumsing culture in Manado.

4 CONCLUSION

The findings of research on the culture of consuming fish in Manado City are an inseparable phenomenon in the lives of local people. The existence of foreign cultures that enter the city of Manado makes people get the choice to consume various variants of food that have an impact on survival. However, through the process of classifying cultural models that are expressed starting from aspects of history, health, belief, education, local government policies and investment, it provides direct reinforcement of the culture of consuming fish in Manado City. The web design is expected to help the aspects of developing the culture of consuming fish. The advantages of this research are in the description of various aspects of the cultural model applied by the community and those who are directly involved with the culture. And there are shortcomings in this study in the form of the number of correspondents who were asked questions about the culture of consuming fish in the city of Manado. It is hoped that further research recommendations can be carried out by designing different objects and activities and having a direct impact on the target community.

REFERENCES

Bardosono, S., Kapoyos, S. S. J. S. R. C., Basrowi, R. W., & Rares, V. U. A. S. M. T. (2018). *Aspek Budaya Nutrisi dan Kesehatan.*

Ekarini, F. (2017). Analisis Desain Website Bni, Bukopin, J.Co Donuts Dan Mcdonals Menurut Buku "the Principles of Beautiful Web Design." *Elinvo (Electronics, Informatics, and Vocational Education)*, 2(1), 8–20. https://doi.org/10.21831/elinvo.v2i1.14489

Pamantung, R. P. (2019). *Tradisi Minahasa Terkait Dengan Makanan Tradisional.* Pascasarjana Program Studi Linguistik Universitas Sam Ratulangi.

Soewardikoen D. W. (2019). *Metodologi Penelitian Desain Komunikasi Visual.* Kanisius.

Soewardikoen D. W., Prabawa B. (2020). *Digital Promotion Media For Small Medium Enterprise Proceedings, Contemporary Research on Business and Managemen*: Proceedings of the International Seminar of Contemporary Research on Business and Management (ISCRBM 2019).

Soewardikoen D. W., Prabawa B. (2020). *Visual Identity and Promotion Media of Small and Medium Enterprishe, Convash 2019*: Procheedings of the First Confrence of Visual Art, Design, and Social Humanities, Convash, 2 November 2019, Surakarta, Central Java, Indonesia.

Weichart, G. (2014). Identitas Minahasa: Sebuah Praktik Kuliner. *Antropologi Indonesia*, 0(74), 59–80. https://doi.org/10.7454/ai.v0i74.3510

Study of digital branding as media strategic toward metaverse

G.A. Prahara
Telkom University, Bandung, Indonesia

J. Adler
Universitas Komputer Indonesia, Bandung, Indonesia
ORCID ID: 0000-0002-7055-715X

ABSTRACT: Metaverse will be a testing ground for producers to face the changing dynamics of marketing and digital markets in the future. Rapid changes in consumer behavior affect the company's policy on promotion, especially on media strategy. Digital media provides unlimited access to all platforms flexibly. Several major brands have established themselves on the platform metaverse. This research aims to study brand strategies, especially in the digital era in positioning brands in the digital market through the metaverse platform. The AISAS model is also used in analyzing the activity of a brand. Observations are made through analysis of brand content based on the media space contained in the metaverse. A comparative study of several big brands as a sample of objective data on the metaverse platform is conducted. Brands have a great effect and involvement as a strategic form, especially in adapting to market trends swarmed by consumers. Although it has not yet become a form of consumer behavior, the existence of the metaverse platform will be one of the brand's strategies in anticipating changes in information technology today.

Keywords: brand, digital media, metaverse, strategic media

1 INTRODUCTION

The presence of the metaverse as a space for social interaction has surprised internet users around the world. Zuckerberg provides innovative ideas as well as controversy. Internet users today are very dependent and cannot let go of the internet in their lives. The presence of the metaverse becomes a new alternative to behaving with the internet. Although users must register and equip themselves with virtual reality (VR) tools and prepare cryptocurrency to be able to enter activities in the virtual dimension of the metaverse. Several brands are currently trying to enter and start their business activities in the metaverse world.

The metaverse is a post-reality world that combines a virtual reality environment with a digital world (Mystakidis 2022). Very much metaverse research has been done concerning its relationship with other sciences. The existence of a pandemic then became one of the reasons people carried out activities from home and Metaverse became a solution as a second world to answer problems (Article & Narin, n.d.). Digital technology that is increasingly converging makes the basis for the formation of a metaverse that requires other platforms to be able to immerse themselves in it (Lee 2021). The current development of the metaverse is part of the evolution of traditional monologue media into new media that are peer-to-peer dialogue (Timothy C Cunningham 2009).

More and more manufacturers are entering the metaverse platform, appealing to internet 3.0 even more. Sooner or later, the technology that continues to run will make it easier for people to get involved. Problems arise when the decision to join the metaverse is only used as a trend or an unwanted form of fatigue. The tendency is toward a metaverse that combines work and leisure activities in one place as the digital world gets older and more developed (DI_Digital-Media-Trends, n.d.). Digital advertising makes the metaverse an era of development that will become a new form in the world of digital advertising (Taylor 2022). In general, people are still waiting for the development of public awareness of the presence of the metaverse during the development of the digital world today.

This virtual approach based on the metaverse platform based on 3 Dimensions visually shows much more immersion in the real world. The use of avatars is an interactive aspect of the representation of the individuals in it. This makes the virtual world a simulation that facilitates interaction in real time regardless of location (Dionisio *et al.* 2013).

Global brands have entered the world of the metaverse with products that were previously known in the real world. Ease of transacting is now a step forward with digitization changing the way consumers buy products. The desire to buy is influenced by the need or influenced by the strength or drive for the service of a particular brand or brand (Shen *et al.* 2021). Advertising is a representation of the brand that is conveyed informatively and persuasively so that it is embedded in the minds of consumers. The problem is how brands adapt to today's digital media developments that encourage them to engage in the metaverse. Brand problems in the media strategy are used as part of the penetration of the development of the new digital world in the future.

Research related to the metaverse is generally still related to technical matters in the realm of information technology science while research related to the relationship between brands and the metaverse is currently more concerned with the presence of the metaverse in the new marketing world (Hollensen, Kotler, Opresnik 2022), metaverse in jewelry products (Rim Kang 2022), and the metaverse in advertising (Kim 2021). The problem of brand and strategy in the metaverse has not been widely discussed.

The objective of this research is the motivation of brand strategy in entering new media in the metaverse. Brand adaptation in following the direction of the development of people's behavior in the use of the internet after internet 2.0. The media brand strategy in a virtual world immersed in consumer activities in gathering and working as well as other virtual social activities. Several big local brands in Indonesia have placed their brands in the metaverse ecosystem, including **BRI** and **BNI**, from the banking sector. Alfamart, and Kalbe Farma from the retail sector and pharmaceutical products. The response of both consumers and customers is one measure of brand value after the brand joins the metaverse.

2 RESEARCH METHODS

This research puts forward a literature study based on the latest theory obtained from journal sources that focus on the development of the metaverse. Analyzing problems through a digital branding theory approach is a cutting-edge strategy. Metaverse is a virtual world that is becoming a highly anticipated new media, especially in the development of the entire digital sector that is connected with augmented reality and virtual reality technology. Digital technology requires users to be able to own and use tools that make them more integrated with the world of reality. AISAS (i.e., Attention, Interest, Search, Action, and Share) which is an indicator variable model related to advertising that is also used to determine brand awareness is carried out to determine the role of the metaverse in the company's branding strategy.

Digital branding is developing along with the increasingly widespread familiarity of people in using social media which elevates themselves amid other communities which is ultimately known as personal digital branding (Kleppinger & Cain 2015). Digital branding is currently very much needed in the digital era which has an impact on the progress of the company's industry (David *et al*., n.d.).

Figure 1. Consumer buying decisions (Edelman 2010).

Consumer behavior in today's digital era has changed compared to the previous era. The theory of consumer buying decisions for digital brands can make a jump in consumer purchases of products directly as a result of evaluation and advocacy through digital media, social media, or other two-way commercial media. This makes buying decisions faster without the need to consider or choose (Edelman 2010).

The case studies in this research were conducted on several local brands that have registered themselves as metaverse mediums through collaboration with the WIR group company that designed their placement in the metaverse world. The search was carried out on local brands of BRI, BNI, Kalbefarma, and Alfamart banks. Brand loyalty survey and brand engagement were carried out on the community as consumers in the digital era as many as 240 respondents filled out a limited questionnaire through an online survey. Interviews related to brands and consumer proximity to brands to determine the value of consumer preferences for brands.

3 RESULT AND DISCUSSION

The development of the metaverse, which began to be widely discussed after Mark Zuckerberg changed the Facebook brand name and logo to Meta, identifies a complete change in the face of changes in information technology. Companies are trying to find out more by starting to invest in digital platforms using cryptocurrencies and NFTs. Big brands have penetrated the market for a long time since its inception, so that the brand is very well known and already has many consumers and customers from these brands.

In strengthening the brand in the digital era, especially internet marketing, it is necessary to pay attention to the four basic pillars: marketing communication, customer understanding, interactivity, and content (Lipiäinen & Karjaluoto 2015). This is to further strengthen the brand equity owned by the company in the minds of consumers. In the search, each brand has a very good position and supports brand equity. Through digital marketing with various application platforms that are used to build digital equity. The company's website contains information and all of the company's digital marketing activities. Social media such as Instagram, Facebook, and Twitter allow relationships and engagement with digital consumers to be built. Chat media such as WhatsApp and telegram make the interactivity of consumer and brand relationships directly connected. The use of digital media in the current AISAS model is carried out in stages as a whole as part of the brand strategy, from the attention stage to share using digital commercial media.

Table 1. Internet digital marketing Strategy with AISAS.

Brand (AISAS)	Digital marketing (Interest, Search)	Customer Engagement (followers) (Attention, Share)	Interactivity (Attention, Share)	Content (Interest)
BNI	Official Website, Mobile Application	*Instagram 1.1M, Facebook 394.4K Twitter 15.7K.*	Whatsapp, Telegram, Tiktok 364 M View.	Brand, Product Information, Promotion, contact
BRI	Official Website, Mobile Application	*Instagram 830K, Facebook 850.3K, Twitter 206K.*	WhatsApp, Telegram, Tiktok 159.3M View	Brand, Product Information, Promotion, contact
Kalbe farma	Official Website, Mobile Application	*Instagram 24K, Facebook 15K, Twitter 1,768* (Attention, Share)	Tiktok 1.2 M View	Brand, Product Information, Promotion, contact
Alfamart	Official Website, Mobile Application	*Instagram 3M, Facebook 1,9M, Twitter 346.9K*	WhatsApp, Tiktok 1.1 M Like	Brand, Product Information, Promotion, contact

The current series of digital media strategies by these brands shows that the company's internet marketing fundamentals are very active and aggressive. The ability to reach digital consumers is carried out on all digital media and applications available on the internet. The company is trying to strengthen its brand equity to keep up with the flow of information technology developments. The motivation of these brands in entering the metaverse dimension is an opening for directions in the context of exploring early anticipation for the next 4 to 5 years to see the development of global internet trends. The analysis of media strategy on internet 2.0, as shown in Table 1, shows that these local brands do not want to be left behind in anticipating the development of future technological trends. Even though the current metaverse world is still very premature to say that it is ready and still measuring the tendency of internet users, the majority of whom still use internet devices in the form of cellphones or computers, either PCs or laptops.

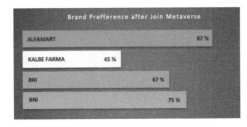

Brand Awareness after brand join Metaverse

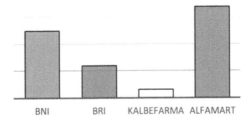

Figure 2. Brand awareness and preference brand after join metaverse.

The response from consumers as awareness of the incorporation of the above brands in the metaverse shows that they do not/have not paid attention to brand activity. This is a natural response because there are still limited opportunities for people to join due to Oculus AR/VR devices and the use of crypto money to be able to be present in the metaverse. However, the brand preference indirectly joins the brand with the metaverse and gets consumer appreciation, thereby increasing the level of equity of the brands. Consumers and all internet users are still waiting for the development of the metaverse as the development of internet 3.0. The development of metaverse retailing provides a much more mature consumer experience to more pervasively assess a brand or product (Bourlakis *et al.* 2009).

The media brand strategy to join the metaverse is the company's choice to join now or later. Business communication via the internet requires speed and accuracy in determining business decisions. The rapid and comprehensive development of information technology is possible to facilitate the penetration of the metaverse to all its users. Year 2028 is expected to be the year that metaverse users start to increase worldwide (Kaur & Gupta n.d.). Careful planning from the start will reduce various obstacles that will arise in the future. The metaverse infrastructure is gradually being developed and makes it easier for internet users to access and use it, especially cryptocurrencies and asset management through the blockchain.

4 CONCLUSION

The development of digital brands is in line with the development of digital technology. Internet 2.0 which became the main infrastructure began to evolve with the introduction of internet 3.0. Metaverse is a new multi-reality digital space to approach consumers in the future. The life activities of the digital community will begin to shift toward a pervasive experience. Brand activities will assist consumers to go through cyberspace so that the company's strategy to join the metaverse is the right choice. The development of retail consumers who prioritize experience as a determinant of customer value must be a concern for all companies to have strength in brand equity. The media branding strategy in the metaverse is carried out as consumer habits also evolve. The company's anticipation from a long time ago is an adaptive form that aims to strengthen product and company brand equity. Further research can focus on how brand equity can be maintained when the metaverse has become a habit in future experiences.

REFERENCES

Article, R., & Narin, N. G. (n.d.). *Journal of Metaverse A Content Analysis of the Metaverse Articles*. www.secondlife.com

Bourlakis, M., Papagiannidis, S., & Li, F. (2009). Retail Spatial Evolution: Paving the way from Traditional to Metaverse Retailing. *Electronic Commerce Research*, 9(1–2), 135–148. https://doi.org/10.1007/s10660-009-9030-8

David, A., Sudhakar, B. D., Nagarjuna, K., Arokia Raj, D., & Professor, A. (n.d.). *Marketing Assistance and Digital Branding-an Insight for Technology up-Gradation for Msmes*. 5, 2017. http://www.aguijmsr.com

DI_Digital-media-trends. (n.d.).

Dionisio, J. D. N., Burns, W. G., & Gilbert, R. (2013). 3D virtual worlds and the metaverse: Current status and future possibilities. *ACM Computing Surveys*, 45(3). https://doi.org/10.1145/2480741.2480751

Edelman, D. C. (2010). *S Potlight on SOcial MEdia and the New R ules of Branding Branding in the Digital Age You're Spending Your Money in All the Wrong Places*. www.hbr.org

Kaur, M., & Gupta, B. B. (n.d.). *Metaverse Technology and the Current Market*.

Kleppinger, C. A., & Cain, J. (2015). *Statement Personal Digital Branding as a Professional Asset in the Digital Age*. http://www.ajpe.org

Lee, K.-A. (2021). A Study on Immersive Media Technology in the Metaverse World. 한국컴퓨터정보학회논문지 *Journal of The Korea Society of Computer and Information*, *26*(9), 73–79. https://doi.org/10.9708/jksci.2021.26.09.073

Lipiäinen, H. S. M., & Karjaluoto, H. (2015). Industrial branding in the digital age. *Journal of Business and Industrial Marketing*, *30*(6), 733–741. https://doi.org/10.1108/JBIM-04-2013-0089

Mystakidis, S. (2022). Metaverse. *Encyclopedia*, 2(1), 486–497. https://doi.org/10.3390/encyclopedia2010031

Shen, B., Tan, W., Guo, J., Zhao, L., & Qin, P. (2021). How to Promote user Purchase in Metaverse? A Systematic Literature Review on Consumer Behavior Research and Virtual Commerce Application Design. In *Applied Sciences (Switzerland)* (Vol. 11, Issue 23). MDPI. https://doi.org/10.3390/app112311087

Taylor, C. R. (2022). Research on Advertising in the Metaverse: a Call to Action. In *International Journal of Advertising* (Vol. 41, Issue 3, pp. 383–384). Taylor and Francis Ltd. https://doi.org/10.1080/02650487.2022.2058786

Timothy C Cunningham, by M. (2009). *Marching Toward the Metaverse: Strategic Communication Through the New Media A Monograph*.

Identification of building around the Bogor market as forming the identity of the Chinatown area in Bogor City

H. Anwar, R. Hambali & T. Mulya Raja
Telkom University, Bandung, Indonesia

ORCID ID: 0000-0002-0099-7846

ABSTRACT: A city is a place in which it has the identity of the city itself. The city of Bogor is a city in which there is an integration of various ethnic groups in it, one of which is the Arab tribe with an Arab village in it and the Chinese tribe with the Chinatown area in Bogor. In the long journey of the city, there is a history of social integration of the Chinese community in Bogor, as we often find in Chinatown areas in other cities where the existence of Chinatowns is very closely related to very exclusive trade economic interests by creating settlements that unite functions with economic activities. Their trade, which we can generally recognize from the ethnic Chinese area or Chinatown is the emergence of a collection of shophouse buildings (ruko). As in the Chinatown area of other cities, the social portrait of the Chinese settlements in Bogor also forms a similar thing. Based on this background, it is interesting that we discuss the Chinatown area in the city of Bogor, where the Bogor market which was the center of economic activity from the past was even the forerunner to the formation of the city of Bogor. The very long history of the Chinatown area around the Bogor market which gives a very large cultural significance value to the city of Bogor. This research will identify the characteristics of Chinatown architecture in buildings around Bogor Market. This identification is carried out by conducting a study starting from the front view of the building to detailing in the buildings around the Bogor market so that it can be concluded that the Chinatown character in the buildings around the Bogor Market is the hallmark of forming regional identity in the city of Bogor. This identification will also be taken into consideration whether the buildings around the Bogor Market can be categorized as forming a conservation area in Bogor City.

Keywords: Architecture, Bogor, Chinatown

1 INTRODUCTION

It is interesting that we discuss the Chinatown area in the city of Bogor, where the Bogor market which was the center of economic activity from the past was even the forerunner to the formation of the city of Bogor. The very long history of the Chinatown area around the Bogor market which gives a very large cultural significance value to the city of Bogor. This Bogor Market, which since the Dutch colonial era has been called Pasar Bogor, is located in a village area that is very close to the Bogor palace or better known during the Colonial period as the Buitenzorg Palace and very close to the Ciliwung river. At first the Bogor Market was only opened once a week, but even though it only operates once a week, this Bogor Market can generate considerable profits, especially from the market rental profits given to the government of the Governor General of the Dutch East Indies at that time in 1777. This condition made it attractive for business. at Bogor Market, one of which is ethnic Chinese.

The increasing number of ethnic Chinese doing business around Bogor Market has led to the formation of a community that lives and does business with ethnic Chinese around Bogor Market and on the main access road to Chinatown around Bogor Market, namely Jalan Suryakencana. This was reaffirmed In 1845, at that time the Dutch East Indies colonial government made a policy regarding ethnic Chinese who had to be on the postal highway, this was intended so that the necessities of life that were sold would be easier to obtain by the residents, making the ethnic Chinese increasingly mushrooming. inhabit the area around Bogor Market and Jalan Suryakencana Bogor for living and doing business. From the above background, it is necessary to identify the buildings around the Bogor Market as the identity of the Chinatown area in the city of Bogor, while some of the studies are:

a) What building elements are the identity of the ethnic Chinese buildings found in Buildings around the Bogor market?
b) Get the Chinatown area in Bogor The area becomes a reserve area culture?

This research focuses on the buildings around the Bogor market, namely on Jalan Suryakencana Bogor which is a Chinatown area in the city of Bogor. It is hoped that it can enrich knowledge about the architecture of Chinese ethnic buildings in Indonesia and is expected to be a design guide for the Chinatown area and can become knowledge about the criteria for cultural heritage areas.

2 RESEARCH METHODS

Chinatown. Chinatown area is very synonymous with Chinese architecture. Quoted from the source "Chinese Architecture in The Straits Settlements and Western Malaya", the work of the author Kohl Hadinoto who wrote that the characteristics of Chinese buildings, especially in the Southeast Asian region, are:

a) *Courtyard*
   The courtyard that is often found in the Chinese house is an open space that is more private in nature. This courtyard is often found in North Chinese-style houses, which form an open space in the center of the building which is quite wide and sometimes more than one. We can find different types of Courtyards in the southern part of China, where in southern China there are many ethnic Chinese in Indonesia, the courtyards in the North China area are relatively smaller because the width of the house lots is not too big.
b) *Typical roof shape*
   Many types of roofs are typical of Chinese buildings and their details, but there are only a few types of roofs that we get most often in Chinese buildings in Indonesia. Some of them are the types of roofs Hsuan shan, Ngang shan, Tsuan tsien, Wu tien.
c) *Expose the roof structure and the combination of decoration on the structure*
   We often encounter many kinds of decoration in Chinese buildings, one of which is carvings exposed to wood construction, either for columns or for roofs. Decorative details that we commonly see in Chinese buildings in Indonesia such as roof construction or the meeting between columns and beams.
d) *The use of distinctive colors*
   Color in Chinese architecture has a fairly important role, even has a symbolic meaning. Many colors are often used in Chinese buildings, from several distinctive colors, there are red and golden yellow colors which are dominantly found in Chinese architecture in Indonesia, especially in the color of the columns as well as in the interior decoration of buildings and some specific elements in Chinese buildings.

Cultural Conservation Criteria. To determine the criteria for cultural heritage, according to UNESCO, cultural heritage is divided into Cultural heritage and Natural heritage. The two heritage criteria are as follows

a) to represent a work of genius from human creative work.
b) to demonstrate the importance of the exchange of human values, over a span of time or within a cultural area of the world, in the development of architecture or technology, monumental art, urban planning or landscape design.
c) to contain outstanding testimony/history or at least to a living/existing or lost cultural tradition or civilization.
d) to be an outstanding example of the type/style of building, architecture or technology or landscape that describes significant stages in human history.
e) to be an example of an attractive/unique traditional human settlement, land use, or marine use, which is representative of a culture (or cultures), or human interaction with the environment, especially when the cultural heritage is vulnerable to the impact of change.
f) Directly or markedly associated with events or living traditions, with ideas/way of life, or with beliefs, with universally significant works of art and literature. (The Committee considers that this criterion should be used in conjunction with other criteria).
g) contains natural phenomena that are superlative or the beauty of a natural area that is extraordinary / attractive and has aesthetic value.
h) to be an outstanding example representing the stages of Earth's history, including records of life, significant geological processes in the ongoing development of landforms, or geomorphic or physiographic features.
i) to be an outstanding example that represents the ongoing ecological and biological processes in the evolution and development of terrestrial, freshwater, coastal and marine ecosystems and assemblages of plants and animals
j) contains natural habitats of greatest importance and significance for the conservation of biological diversity, including threatened species of extraordinary universal value from a scientific or conservation point of view.

This research is an analytical research on architectural elements in the Chinatown area in Bogor City. The data collection method used is a combination of qualitative methods to obtain data related to building elements. Collecting data through direct observation in the Chinatown area of Bogor City regarding the identification of building typologies. There are a series of stages carried out in this research which include:

a. Determination of data sources from the phenomena found based on the results of previous surveys and observations;
b. Formulation of research methods refers to the expected research outputs;
c. Collecting data in accordance with the specified output needs;
d. Processing data from field surveys and literature studies;

3 RESULT AND DISCUSSION

Jon Lang identifies several aspects which are then assessed as aspects that can determine whether the area needs to be conserved or not. Based on several aspects that we have observed and analyzed directly, the Bogor Market Chinatown has fulfilled several of these aspects, some of which are as follows:

1) *Aesthetic value*

The aesthetic value of historic buildings in the Bogor market area can be seen from the details on the buildings and their typology. Unique or distinctive details are located on the roof / roof ridge. Besides that, the aesthetic value can also be seen from the configuration of the buildings that are lined up. The sleek, lined configuration is a hallmark of Chinese settlement.

2) *Value for architectural diversity*

The existence of buildings in the Bogor market, Jalan Surya Kencana shows the architectural diversity in the area. Where the variety of existing architecture represents the architectural style of the Colonial period and Chinese architecture. Some of these buildings are in good condition and some have been destroyed.

3) *Value for environmental diversity*

The Bogor market and the Jalan Surya Kencana area are the forerunners of the city of Bogor. Initially, it was a residential area for the Chinese ethnic community and was also the economic center of the city of Bogor. The current condition in the Bogor market area and the Suryakencana road corridor still maintains regional differences from other areas. Namely as an economic area of the city of Bogor that contributes to the economic development of the city of Bogor and also tourism.

4) *Value for functional divercity*

The value of the diversity of functions in the Bogor market area and Jalan Suryakencana which in the Dutch colonial era was a function of settlement and also as a function of trade. The current conditions are different from the conditions in the Colonial era. The function that still survives in the Bogor market which still functions as a trading function and the Dhanagun Vihara which still functions as a place of worship. Meanwhile, the buildings on Jalan Surya Kencana which have two floors only function as a place for trading.

5) *Resource value*

The resource value of buildings in the Suryakencana area and Bogor market can still be utilized from the original materials from these buildings. Most of the building materials are permanent buildings that use brick, concrete and wood walls. However, the condition of the roofing materials in the form of zinc and roof tiles has been partially damaged. But without dismantling these buildings can reduce the use of new materials in construction.

6) *Value for continuity of cultural memory and heritage value*

The cultural and heritage values in the Bogor market area and the buildings along Jalan Suryakencana are historical relics of the past, both for the memory/history of the Chinese ethnic community and the values contained in colonial buildings. Suryakencana buildings and roads record the history of life, activities that are of more value than other areas.

7) *Economic and commercial value*

The existence of the Bogor market and the buildings in the Suryakencana road corridor to this day still have economic and commercial value. Economic and commercial activities continue to occur in the area, and even become the center of the busiest economic and commercial activity in the city of Bogor. Thus these values can be maintained and very influential.

4 CONCLUSION

Based on the discussion above, that the feasibility values of the Bogor market area and the Surya Kencana road corridor can be concluded that the area is suitable for conservation, because it meets the conservation values of the area. research, as well as recommendations for further research.

Value Aspect	Yes / No
Aesthetic value	Yes
Value for architectural diversity	Yes
Value for environmental diversity	Yes
Value for functional divercity	Yes
Resource value	Yes
Value for continuity of cultural memory and heritage value	Yes
Economic and commercial value	Yes

Table conclusion

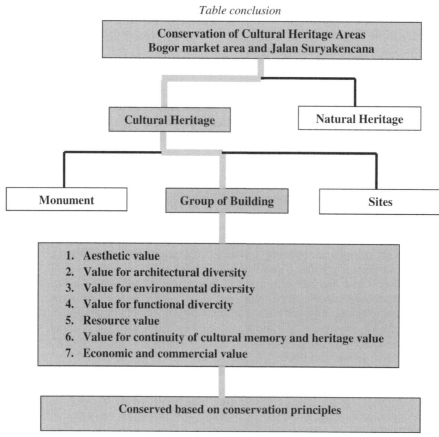

Table Conclusion

REFERENCES

Abidin, Y. (2016). *Tionghoa, Dakwah dan KeIndonesiaan*. Mimbar Pustaka.
Anwar, H., N. H. A. (2013). *Rumah Etnik Sunda*. Griya Kreasi.
Knapp, R. G. (2012). *The Peranakan Chinese Home*. Tuttle Publishing.
Poerwadarminta, W. J. S. (1976). *Kamus Umum Bahasa Indonesia*. Balai Pustaka.
Tan, S. (2014). *Chinese Auspicious Culture*. Kompas Gramedia.
Triharto, W., Hidayat, R., & Firman, R. (2018). *Kajian Perencanaan Penataan Kota Pusaka Bogor*. Http://Sejarah. Kompasiana.Com/2012/02/29/Menapaki-Jejak-Sang-Naga-Di-Bogor-443203.Html Http://Didisadili.Blogspot. Com Http://Sekarnegari.Wordpress.Com Https://Repository.Ipb.Ac.Id Www.Dspace.Uii.Ac.Id.

Tjap Go Meh animation process as a form of acculturation of Indonesian culture through edutainment media

I. Wirasari, D. Aditya, A.E. Adi, N.D. Nugaraha & S. Salayanti
Telkom University, Bandung, Indonesia

S. Anis
Multimedia University, Cyberjaya, Malaysia

ORCID ID: 0000-0003-3595-435X

ABSTRACT: This study aims to design learning animations for children. This Tjap Go Meh animation tells about the acculturation of Indonesian and Chinese cultures. This is necessary because today's young generation does not understand cultural acculturation properly and appropriately. This animation is also expected to be used as a learning media that can be used in informal education in Indonesia and Malaysia. By using qualitative methods with data collection methods in the form of observations and interviews, the animation design in the form of a story board is obtained for the prologue of this Tjap Go Meh animation. Characters and backgrounds have been generated from previous research.

Keywords: Animation, Tjap Go Meh, Acculturation, Culture, Edutainment media

1 INTRODUCTION

In this research, we will create a 10–15 minute animation that tells the story of a family of Chinese descent living in Semarang, the storyboard will be made lightly but contains very significant cultural values. So that an understanding of cultural acculturation in Indonesia for the younger generation, especially children, will be easily accepted and will lead to mutual respect and mutual respect among them. The research problems of the study is: " How is the process of making Tjap Go Meh animation that represents acculturation of Indonesian and Chinese culture, so that it can be used as a learning tool for the community?"

This animation will describe the acculturation of Indonesian and Malaysian cultures, which are intended as learning media for children in Indonesia and Malaysia. The creative process of character building and visual assets will be used in this animation. This cultural acculturation occurs between two different cultures, then meet and merge into harmony and peace. The combination of these two cultures can give birth to a new culture. Although there is an acculturation of two different cultures, it will not cause the old cultural elements to be lost. The original cultural elements or the old elements will still exist, so we don't have to worry if there is cultural acculturation.

2 RESEARCH METHODS

The method used in this research is a research and development approach. According to Sugiyono (2010: 407) what is meant by Research and Development consists of two words, namely Research & Development. Thus the Research and Development research method can be interpreted as research by producing certain products or developing existing products.

2.1 Culture acculturation theory

According to Berry (2005), acculturation is a process of cultural and psychological change that takes place as a result of contact between two or more cultural groups and their members. Every human being gets a cultural process. The process of socialization and cultural education that is instilled into behavior and personality that is already attached to the nervous system in each individual. With this learning process, humans must interact and communicate with each other

2.2 Animation theory

Animation can attract attention, and be able to convey something message well. The opinions of experts regarding animation are as follows: Animation is a collection of images arranged in sequence. When the series of images is displayed at sufficient speed, then the series of images will look moving (Hidayatullah et al. 2011:63).

3 RESULT AND DISCUSSION

Based on the results of the analysis using data collection methods in the form of observations and interviews, it was obtained a mapping of the cultural acculturation process that occurred in the Tjap Go Meh animation design process. The acculturation stage begins with the enculturation and deculturation stages of culture in society. If the process can occur properly, then the process of acculturation and cultural assimilation can be created properly. The process of merging Indonesian and Chinese culture can be seen from the animated characters produced, there are original Indonesian characters and there are also peranakan characters, accompanied by lion dance characters to strengthen the identity of the acculturation of the culture that occurred.

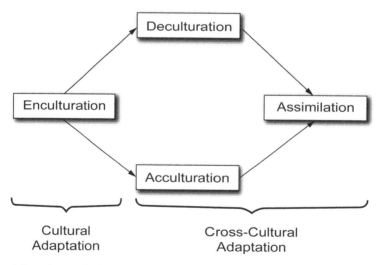

Picture 1. Adaptation method (source: researcher).

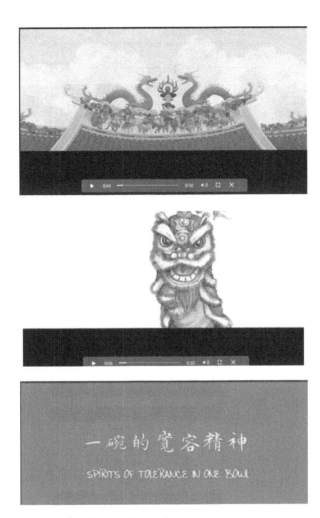

Picture 2. Capture of animation. Source: Researcher.

Based on the results of the analysis and observations above, we designed the Tjap Go Meh animation design as follows:

Scenes	Visual
1.	

(*continued*)

Continued

Scenes	Visual
2.	
3.	
4.	
5.	
6	
7	

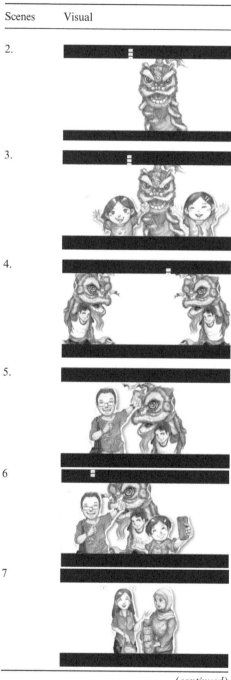

(*continued*)

Continued

Scenes	Visual
8	
9.	
10.	

The narration: the opening sequences will fill with title and credits, including the institution's visual identity. Sound: [Chinese Folk Song] due to the copyright issue the project use the free and open source common songs. Based on the description above, the design of the Tjap Go Meh animation for the prologue section for learning media has been completed. It is hoped that the animation design can represent the acculturation of Indonesian and Chinese cultures.

4 CONCLUSION

By designing the Tjap Go Meh animation, it is hoped that it can be a learning medium for more creative children. This animation can also be used in giving lessons to the students by using this animation. Cultural acculturation is something that must be understood by all people, so that it can increase the spirit of tolerance and togetherness in society, both nationally and globally.

REFERENCES

Agari, Priska, Budiman. (2019). Perancangan Storyboard Untuk Film Animasi 2D Lontong Cap Gomeh. *e-Proceeding Art Des*, vol. 6 no. 3, p. 4050.
A. S. R. Ghifari. (2019). Perancangan Background Untuk Film Animasi Pendek 2d Lontong Cap Gomeh. *Universitas Telkom*.
Beng, T. C. (2013). Chinese Food and Foodways in Southeast Asia and Beyond. *In Journal of social Issues in Southeast Asia*, NUS Press Vol. 28, Issue 3. https://doi.org/10.1355/sj28-3j

Berry, J. W., Poortinga, Y. H., Segall, M. H., & Dasen, P. R. (1999). Psikologi Lintas-Budaya. Riset dan Aplikasi. *Jakarta Penerbit PT Gramedia Pustaka Utama.*

Bromokusumo, A. Chen. (2013). Peranakan Tionghoa dalan Kuliner Nusantara. In Kompas Media Nusantara. *Kompas Media Nusantara.*

Hidayatullah, Priyanto, Amarullah Akbar dan Zaky Rahim. (2011). *Animasi Pendidikan Menggunakan Flash, Bandung, Informatika Bandung.*

Kim, Young Yun. (2001). *Becoming Intercultural: An Integrative Communication Theory and Cross-Cultural Adaptation.* USA: Sage Publication.

McCloud, S. (1994). *Understanding Comics*: The Invisible Art. In Understanding Comics.

Nopianty, Firda Ayu, Budiman. (2019). Perancangan Karakter Untuk Film Animasi Pendek 2d Lontong Cap Go Meh. *e-Proceeding Art Des.*, vol. 6, no. 3, p. 3835.

Sugiyono. (2010). Metode Penelitian Pendidikan. *Bandung, Alfabet.*

Preliminary study into the Genshin impact's aesthetics: The sustainability of visual culture through the character design

L. Agung & L.R. Wiwaha*
Telkom University, Bandung, Indonesia

*ORCID ID: 0000-0001-5627-5941

ABSTRACT: Genshin Impact became the most popular game in 2021. The game's character design is one of the reasons why the game became so popular. Besides the 'cuteish' anime character style, each character represents the aesthetics of each existing culture that we know. And this matter attracts our senses. Thus, in this study, we aim to find out how Genshin Impact adapts and represents the aesthetic value of existing culture in character designs and the most important question is how it influences the sustainability of the visual pattern of the culture it represents. In this research, we decide to choose Raiden Shogun because of her popularity and the intriguing visual of Japanese aesthetics. To do so, in this research, we collect the data qualitatively and to see the aesthetics of Raiden Shogun we use the visual elements analysis by Johnson Museum and a representation theory in the game proposed by Carr. As for the results, this study found that Raiden Shogun's character revolves around Japanese culture especially the Shinto religion and mythology yet the gravity of this game's impact on cultural sustainability is still unclear. It's safe to say that it aided players to recognize a similar culture in other media.

Keywords: Genshin Impact, Character Design, Aesthetic, and Representation

1 INTRODUCTION

Role-playing games (RPGs) are game genres with many fans. Agung (2021) stated an interesting finding in his research, an he stated that video games are basically a form of culture that relies heavily on visuals (Agung *et al.* 2021.). Therefore, we can categorize RPGs as cultural phenomenon because it acts as a medium to reflect our current cultural dynamics.

One of the RPG games that are currently popular is Genshin Impact. The game has won many players and is ranked number one in the game category in both the AppStore and Playstore, the instant game hit managed to accumulate more than 85 million downloads to date with monthly purchase revenues of 82 million USD according to statistics made by Clement (2022). The fact that this game is free to play is one of the factors that cause many players to be interested in downloading this game. However, what keeps players playing and having a sustained player base in this game is none other than the new characters that are released within a certain period.

Players are willing to play every day to collect in-game currency so they can get the characters they want or are even willing to spend a lot of money to receive the characters. According to Thon (2019), aesthetics contributes to the overall essence of the character and can help immerse the player even deeper into the game narrative. In other words, the character design in the Genshin Impact can attract and keep players attention.

Genshin Impact adapts the aesthetic value of the existing culture into character designs such as Chinese, Japanese, and Western aesthetics. Characters such as Raiden Shogun adapted many Japanese aesthetics and managed to reach 33 million USD in sales revenue and this sales figure is only regional sales in China not added to worldwide revenue. As of June 2022, Genshin Impact has reached a revenue figure of more than 500 million dollars and this figure is increasing every month as new characters are released.

Game characters not only serve as a form of representation for the player itself but also store game store certain cultural values. In this study, our most basic question is, do the characters in this game represent that particular culture in its entirety in its aesthetic context? Thus, this research is important to see how the aesthetic represented the existing culture in its current form through the latest media.

This study aims to find out how the Genshin Impact game adapts and represents the aesthetic value of existing culture in Raiden Shogun's character designs and its influence on the sustainability of the visual pattern of the culture it represents. Raiden Shogun is chosen to be analyzed because of her appeal and popularity considering it still holds the highest sales to date. We will be analyzing the character's promotional splash art that went viral on the character's promotional campaign last year. The reason that we analyzed the promotional splash art as Miller (2020) stated that virtual design is crucial to sustainable co-existence not only for the environmental issues but also for cultural discourse. Furthermore, this research is qualitative in nature because we collect the data by playing and observing Raiden Shogun, and also collect the data from other similar research.

2 RESEARCH METHODS

The method uses in this research is qualitative. According to Berg & Lune (2017.), qualitative research with meanings, concepts, definitions, characteristics, metaphors, symbols, and descriptions of things. We conducted data in this research as a player by playing the character and observing the character, we also read plenty of similar research. We used virtual analysis by Johnson Musem (2020), which is a method of understanding art that focuses on an artwork's visual elements. As a text, the character design can be analyzed by identifying their elements which are nothing but part of the cultural phenomenon itself. To do so, we analyzed the virtual elements of the character which is Raiden Shogun, by taking the splash art or the promotional art of Raiden Shogun and breaking down the visual element aspect: conceptual design, shape and clothing, and background design. By breaking down the visual elements of her character and linking the connection between it and the existing culture, as proposed by (Carr 2017), we used it to adapt. He proposed that the analysis of representations in games is complicated by play, player prerogative, and context. Thus, he makes the approach to games analysis that was influenced by games, by game studies literature. later he stated that the approach involved playing a game, fragmenting it, and then considering these fragments through three overlapping lenses: structural, textual, and intertextual. We also observed how players/fans reacted to her as a game character on social media.

3 RESULT AND DISCUSSION

Genshin Impact's popularity can be contributed to the character design, according to Genshinlab (2022), Raiden Shogun still held the record for the highest banner sales on the first run on the iOS platform. Raiden Shogun's character design uses many visual elements that adapt from Japanese cultural aesthetics. Players were first introduced to Raiden Shogun character design when Mihoyo now Hoyoverse released the character's splash art and causing the word Raiden Shogun to go trending.

Figure 1. Raiden Shogun "Plane of Euthymia" splash art.

Conceptual design

- The character Raiden Shogun of 'Reign of Serenity', where the word *raiden* means The God of Thunder and Shogun, is a military general and a Head of State, meaning she's a God of Thunder who rules and dictates a state. She adapts the aesthetics of an elegant and powerful Japanese woman showing her graceful figure with slender and lengthy legs, her pose looked positioned higher to show her asserting power and dominance. She's an ideal image of a mature and powerful woman in ideas.
- Her character design overall resembles a human although, in an anime style, this aspect is often called 'Moe-anthropomorphism'. Moe-anthropomorphism is a form of anthropomorphism which usually seen in media such as manga and anime where non-human beings are given the quality of moe.

Shape and clothing

- She wears a sakura with a little violet fan hair accessories on her rear head near the right ear.
- The clothing she wears is a kimono with long, draping sleeves adorned with the Mitsudomoe emblem of Inazuma (the nation which she rules) with violet, purple, and lilac as its color, underneath the kimono she wears a dark purple bodysuit that covers her arms and neck. A red and gold ribbon was tied at her back. She also wears a singular pauldron on her left shoulder which also has the same Mitsudomoe emblem and a dark purple thigh-high stocking with a pair of golden heels.
- She bears a calf-length ponytail with a purple-violet gradient and a hair bans that covers her forehead.

Background design

- Raiden Shogun is pulling a sword made from purple lighting which bears the same mitsudomoe on the handle.
- On the background, there's a Japanese style big building that looks like a castle, there are a few smaller building that resembles housing below the castle.
- Surrounding Raiden Shogun, there are Sakura trees and scattering Sakura petals.

Raiden Shogun is a character from Inazuma thus having many of the Inazuman aesthetic qualities. Inazuma itself was based on 16th-century Japan based on the lore surrounding the storyline. Based on this, Raiden Shogun adapts many Japanese aesthetics into her character. The name itself already indicates that the character is of Japanese origin, the word Raiden is a Japanese lightning God, and the word Shogun means commander-in-chief.

The clothing she wears although modified to be more fitting to the overall game aesthetics still resembles a traditional Yukata although shorter and the motives within the Yukata still bear an art style of a Japanese traditional painting. The background on her splash art also supports the Japanese aesthetic influence over her character, the Japanese castles or fortress (*shiro* or *jō*) erected behind her and a smaller Japanese housing below the castle. The Sakura Trees which being one of the Japanese unique characteristics are also shown on the character's splash art background.

Raiden Shogun's character design has a consistent, repeated pattern of Mitsutomoe symbols or three Tomoe symbols, which are associated with Shinto shrines, particularly to Hachiman, the Japanese God of War and archery. This design decision is befitting of Raiden Shogun's character, who not only serves as a God who rules a nation but also as a battle-hardened God who, according to the lore, has gone through countless wars.

When we approach this character the same way as the Carr methodology, structurally Raiden Shogun character serves as both NPC and playable character, she serves as NPC in a certain storyline and quest and can become playable for the players if obtained from the Gacha mechanic. Textually, Raiden Shogun is both a deity and ruler of Inazuma region and her aesthetic style revolves around her roles in the lore and Inazuma's aesthetic itself, and her visual elements use the same element as Inazuma's visual elements. While intertextually, she is inspired by Japanese culture aesthetics and Shinto religion.

In fact, she is made to be one of the strongest DPS yet easy to build and use in the game also became a factor of consideration for players to get her character. When combined with her stunning design, it is understandable why she became the most popular character in the game. On social media, many players/fans are desperate to get her character by draining their in-game currency to pull on her banner and flexing on social media if they acquire that character. Moreover, Japanese culture has already gained a place in popular culture through anime and manga media since the 1980s. This fact is why we think her popularity helped visual culture sustainability in Japanese culture.

4 CONCLUSION

Genshin Impact is adapting and implementing the visual aspect of an existing culture into game characters' designs, moreover on a highly popular game with a huge player base, which helps to sustain a culture's aesthetic through visual means. The players are exposed to the visual pattern of a certain culture and will recognize it if they were to see the same visual pattern in other media because the player is already familiar with the visual elements to a certain level. This is in accordance with with a study (Schröter & Thon 2019.) that game characters as 'communicatively constructed artifacts', as intersubjective constructs "based on normative abstractions about ideal character-imaginations" (Eder 2008), allow for sophisticated description of characters as compared to their 'external' medial representation.

Raiden Shogun's character style is Moe-anthropomorphism and her character design revolves around Japanese culture, especially the Shinto religion. By implementing Japanese and Shinto to the core of her character. The visual imagery shown on her character resembles a close visual pattern to the Japanese aesthetics though not only Raiden Shogun character doesn't rely on visual imagery alone by adopting the name of a known mythological being and a Japanese word to her name also aided her character to make an impact on the players.

Based on the methodology that Carr uses, we have found that structurally Raiden Shogun serves as both NPC and playable character. Textually in the lore, she is a deity and ruler of her domain which is Inazuma. Her design character revolves around Inazuma's imagination.

Last, due to her usability as a character, her interesting yet stunning design, and the popularity of Japanese culture in popular culture, Raiden Shogun gained such popularity both in the game and in the community's social media.

In the long term, this study has not yet answered the question of the effect of Genshin Impact on the sustainability of visual culture. Although it is safe to say that it may aid in the spreading of a cultural visual pattern through video games media. To find the breadth of influence made by this game, a further study is needed not only to focus on one character but the whole roster of the characters and Genshin Impact's aesthetic as a whole.

REFERENCES

Agung, L., Kartasudjana, T., & Permana, W. (2021). Estetika Nusantara dalam Karakter Gim Lokapala. *Gorga Jurnal Seni Rupa*.

Anon., 2022. Genshin Lab. [Online] Available at: https://www.genshinlab.com/genshin-impact-revenue-chart/

Carr, D., 2017. *Methodology, Representation, and Games*. Games and Culture, 14(7–8), pp. 707–723.

Lune, H., & Berg, B. (2017). *Qualitative Research Methods for the Social Sciences*, 9th Edition. England: Pearson.

Museum, J., 2016. *Johnson Museum of Art*. [Online] Available at: https://museum.cornell.edu/sites/default/files/Johnson%20Museum%20Visual%20Analysis%20101.pdf

Wright, W., 2020. *How to Design a Video Game Character*. [Online] Available at: https://www.masterclass.com/articles/how-to-design-a-video-game-character#quiz-0

The phenomenon of digital wedding invitations: Its potential and cultural shift in Surakarta

S.A. Sari, I. Azhari & M.I.P. Koesoemadinata
Telkom University, Bandung, Indonesia

ORCID ID: 0000-0002-2308-8709

ABSTRACT: The Covid-19 pandemic has given rise to various cultural phenomena, one of which is the digitization of wedding invitations. The rise of this phenomenon is detected in shifts in behavior patterns and cultural values, as happened in Surakarta. As the center of Javanese culture, of course, there needs to be research that examines this phenomenon to the extent of its impact on the cultural change regarding changes in existing cultural values. The analysis mix method was used using a combination of questionnaires and interviews of respondents aged 18–25 in Surakarta as representatives of the premarital group and Enzowedding as an experienced wedding invitation service provider in Surakarta with 102 thousand followers on Instagram were selected as representatives directly affected by the digitization of wedding invitations. The data will be analyzed with the Social Change Theory approach to obtain new findings regarding the cultural shift of the Surakarta people from the use of digital invitations. The findings obtained are digital wedding invitations that have the potential to be developed by the creative industry side in the hope of maintaining local cultural ethics and manners.

Keywords: Covid-19, cultural shift, digital invitations

1 INTRODUCTION

Before the Covid-19, people had chosen something efficient. For example, in the series of marriage customs. In Java, marriage starts from *Siraman* which is a bathing ritual intended for the bride-to-be to be spiritually clean and holy-hearted. Then, in the evening *Midodareni* was taken. In the morning, an *ijab qabul* ceremony happens (Negoro 2001). However, many passed through a series of such wedding ceremonies. Only the *ijab qabul* and the reception were short.

After Covid-19, it has also changed the style of society, which is increasingly more efficient. For example, how to invite guests to a Javanese traditional wedding, and how to invite guests for a wedding also have their own rules. In Javanese custom, inviting someone older to attend a wedding is called *punjungan*. *Punjungan* is an interaction that uses a food symbol carried out by the Javanese people when approaching celebration events such as marriage and circumcision (Fitriana 2021). *Punjungan* has the value of symbolic interactionism, which is an interaction used by humans using a symbol that has meaning (Fitrianan 2021). The meaning of *punjungan* is to show respect and courtesy toward their elders and other relatives. Supported by fitriana's opinion (2021), the meaning of the *punjungan* tradition shows a sense of affection, respect, friendship, social solidarity, and social integration in it, because it establishes harmony and togetherness of family, neighbors, and the surrounding community with the owner of the event. How to invite can also be symbolized by the presence of a printed invitation. Printed invitations have attractive and elegant characteristics and have a variety of forms and shapes. Thus, there is speculation that not everyone understands digital technology, leading to the use of printed invitations.

If you look at today's times, these traditions are rarely carried out. This is influenced by the presence of Covid-19. How to invite guests with a bed or printed invitation is also distracted by digital invitations. Inviting in this way is felt to be a breakthrough that is more effective and efficient. Given its advantages that are considered effective and efficient, many weddings use digital invitations. Digital invitations are free of charge for printing, and there is no need to prepare food as a symbol of the platform. The concept of printed form invitations, the more people are invited, the more costs are incurred to print, and vice versa. This is not the case with digital invitations. The costs incurred are only focused on purchasing video design services and purchasing domains that can be accessed within a certain time. That is, any number of invited relatives does not affect the costs incurred. But usually, the domain will be lost according to the period that the user bought.

In terms of delivery, it is also considered fast and energy-saving. Because of the ease of technology to communicate and the increasing popularity of social media, the news is quickly spreading. The impact is that people are increasingly technologically literate from this shift. The features offered are quite diverse, namely, the link to the location of the wedding event, sending gifts, and how to join the zoom room provided. That convenience provides more information through digital invitations to users.

However, the impact caused is also increasingly diverse, which is related to sustainable cities and communities that promote a cultural shift in Surakarta City in a sustainable manner, namely *adab* and manners. Giving happy news to the family is more afield if visiting in person. Ask for prayers of blessing and strengthen the cords of friendship. The existence of this point in "manners" is still closely held in society. Invitations in digital form are only sent using social media, which is considered inappropriate for some Surakarta people who uphold the value of manners. The city of Surakarta is also considered one of the centers of cultural values in Java. In line with Primasasti (2022), where the term Surakarta City, known as the center of Javanese culture, appeared among the community because the palace civilization has made the pattern of life of the people of Solo City thick with Javanese cultural culture.

This has an impact on the cultural values of the Surakarta people. The condition of Covid-19 is getting better. This shift in the form of digital invitations also needs to be questioned. Whether stick to the form of digital invitations or return to the form of print. This can be answered by using a mixed method by distributing questionnaires that have been distributed and interviewed directly. The target respondents are aged 18–25 years because generally this age pattern is considered appropriate as the premarital age. Therefore, the author wants to know the picture of the chosen form of invitation. This can show the existence of the potential produced. An interview with Enzowedding a wedding invitation service provider in Surakarta with a following of 102 thousand on Instagram was chosen as a representative directly affected by the digitization of wedding invitations.

According to Auguste Comte, this research examines design objects from a cultural perspective and relates to the theory of social change. This concept departs from the theory of evolution, meaning a movement of change that runs in such a way that it is not felt by the supporting society. First, society develops in the same direction, that is, from a primitive to a more advanced society. Second, the evolutionary process experienced by society results in changes that have an impact on changes in values and various assumptions adopted by society. The three subjective views of value are blended with the ultimate goal of social change, namely that the modern society to which it aspires has good and more perfect labels, such as progress, humanity, and civilization. Fourth, the social change that occurs from a simple society to a modern society takes place slowly, without destroying the foundations that build society, and that's why this transition takes a long time (Martindale in Sihabudin 2011 17–18). Social change has three different dimensions: structural, cultural, and interaction. The structural dimension refers more to changes in the shape of the social structure. The cultural dimension refers to the culture in the community environment. These changes include cultural innovation, diffusion, and integration changes. Finally, the interactional dimension refers to the changes in social relations that occur in society.

From the theory presented by Martindale. Same with the digital invitations that appear. First, with the use of digital for wedding invitation designs today. People who originally glorified printed invitations now use digitally designed invitations. This shows that society is heading in a more advanced direction than before. Second, printed invitations have a tangible value, which is evident in the delivery process. Communication, interaction, and friendship are established between the host of the event and the invitee, whereas digital invitations do not offer this, thus eliminating the value. Third, the social values contained in friendship, communication, and interaction that were previously exchanged in person are now through social media. It is increasingly efficient but also forms new emerging values that shift existing socio-cultural values. Fourth, is the existence of digital invitations that suddenly appeared after Covid-19 and not everyone immediately implemented digital invitations. This proves that this phenomenon requires indirect time.

This analysis of digital invitation design is generally related to the visual form of the invitation design. It is similar to the case with Erlistia's (2018) research, entitled Web Design and Instagram Wedding Invitation "Anytime Wedding Invitation", discusses the breakthrough made through an exclusive image that will be built, namely offering a vector-based vintage-style design with the concept of designing online media using web media and Instagram. Halik (2022), with the title Training on Making Digital Invitations in the Covid-19 Era, discusses the provision of knowledge and innovation skills starting from making digital invitations, promoting the results of invitation designs that have been made with an interesting theme. The two studies have not mentioned the impact posed by the existence of digital invitations.

Therefore, the formulation of the problem is the potential development of digital wedding invitations and how the impact on the cultural shift has been caused, especially in Surakarta City. The author can answer this by using the mixed method with the theory of social change. So, the author chose the title *the Phenomenon of Digital Wedding Invitations: The Potential and Shift of Its Culture in Surakarta.*

2 RESEARCH METHODS

This study used the mixed method. Sugiyono (2011: 18) mixed methods, which are research methods by combining two research methods at once—qualitative and quantitative—in research activity. Thus, the author conducts the dissemination of questionnaires and subsequently conducts direct interviews.

The primary data is in the form of questionnaire results that the author got directly, namely from the Surakarta community at the age of 18–25 years because it is considered the premarital age. The number of respondents was 30 people with a simple random sampling method. The author interviewed a well-known invitation printing service provider in Surakarta, namely Enzowedding. Having a total of 102 thousand followers on Instagram makes this service provider experienced and competent in their field.

Based on the nature of the problem, this research uses developmental research. It was answered by interviews with invitation provider services that were directly affected by the number of invitations booked before 2019, 2020, 2021, and 2022. The writer will answer about the potential development of digital invitation design and its impact on the cultural values of the people of Surakarta.

3 RESULT AND DISCUSSION

a. *The potential of digital wedding invitations*

Figure 1. Wedding invitation design selection plan.

Figure 2. Respondents' opinions on the effectiveness of the invitation design.

As shown in Figure 1, the result shows that 43.3% chose to keep using printed invitations. Meanwhile, those who choose to use digital invitations are as much as 40%, and the remaining people combined printed invitations and digital invitations as their choice. Some choose *punjungan* to invite them. When viewed from this percentage, the print invitation design is superior and is still an option, used as a medium in inviting someone to a wedding event.

Of the 30 respondents, 73.3% felt more efficient using digital invitation designs (Figure 2). Meanwhile, 23.3% felt efficient with the design of printed invitations. The rest choose *punjungan* as an efficient medium to invite someone to a wedding.

In Ngafifi's research (2014), technological advances and human life patterns from a sociocultural perspective resulted in discussions, For instance, increasingly rapid technological advances have introduced a digital society and an instant generation, as well as prioritizing effectiveness and efficiency in behavior and actions. In line with the questionnaire results obtained, the characteristics of digital invitations are easy spreading, as well as reducing production costs, making digital invitations considered effective and efficient to use. Moreover, the use of digital media forms effective and efficient patterns and has resulted in human behavior that is increasingly digitally literate.

As a result of the interview with Enzowedding, in 2019, the quantity of printed invitation orders ranged from 500–1000 pcs per pair. However, it has changed greatly during Covid-19. Many couples choose to postpone the wedding. In 2021, the situation has begun to improve, and wedding events have begun to be held, but the quantity of invitation orders has decreased, which is around 200–300 pcs/couple. It is still valid in 2022. Enzowedding provides digital invitations in the form of videos. For ordering, printed invitations include video invitation designs. That offer resulted in a decrease in the number of bookings on printed invitations. Many couples suppressed the ordering of printed invitations and replaced them with digital invitations, thus reducing production costs.

From the above discussion, there is a potential use for digital invitations. If you look at digital development, the current pace of development is very fast. Although the questionnaire data is superior to the public, who will still use the printed invitation design. These results are also very thin from the percentage perspective.

b. *There is a shift in cultural values in the Surakarta community*

The potential development of digital invitations will have an impact on existing values. Especially the cultural value that is generated. The Surakarta community, which is close to the demographic conditions of the Palace, results in the Surakarta community being considered rich in cultural values. However, the rampant digital development that exists, will change the behavior of the social patterns caused. Following the theory of social change presented by Martindale, there are four stages in the social shift resulting in the digital invitation that suddenly appears after the Covid-19 pandemic. Not everyone immediately applies for digital invitations, which proves that this phenomenon requires indirect time.

In line with research conducted by Ngafifi (2014), technological advances give birth to an instant generation but also prioritize effectiveness and efficiency in their behavior and actions. The people of Surakarta are famous for their inherent manners. According to Suyanto (1990), Javanese people strictly maintain the habit of manners. Either to the older one or his friends, or even the younger ones all at once.

Digital invitation possesses much potential to be developed. However, because it is delivered through digital media which facilitates public accessibility, the use of this invitation eliminates direct communication which is usually delivered individually. This way is not in accordance with Javanese cultural values, especially of the people of Surakarta. Based on the questionnaires that have been disseminated, many are aware of this shift in cultural values. Thus, it is related to sustainable cities and communities which promotes a cultural shift in a sustainable manner.

4 CONCLUSION

The result of this study is that the digital wedding invitations caused after the Covid-19 pandemic are a breakthrough of digital innovations that have the potential to be developed. However, the impact caused by the shift in the cultural values of the Surakarta Community is related to sustainable cities and communities which allows there to be a cultural shift in a sustainable manner, namely in the way of communication to invite someone to a wedding which further has an impact on the assessment of *adab* and manners. Using digital invitations is less polite to reward guests because of the non-exclusive way of communication because it is easy for all parties to access. The hope is that more efficient digital invitations can be developed while still maintaining the ethics and manners of the local culture.

REFERENCES

Ariyani, Nur Indah. 2014. Digitalisasi Pasar Tradisional: Perspektif Teori Perubahan Sosial. *Jurnal Analisa Sosiologi* April 2014, 3(1): 1 – 12.
Erlistia (2018), Perancangan Desain Web Dan Instagram Undangan Pernikahan "Anytime Wedding Invitation". *Jurnal ISI* Vol 21, No 1 (2018).
Halik, Sri Asfirawati. 2022. Pelatihan Membuat Undangan Digital Di Era Covid-19. *Jurnal Universitas Galuh: Pengabdian Masyarakat* Vol 4, No 1 (2022).
Martono, Nanang. 2012. *Sosiologi Perubahan Sosial: Perspektif Klasik, Modern, Postmodern, Dan Postkolonial*. Jakarta: PT. Raja Grafindo Persada.
Negoro. (2001). *Tata Cara Pernikahan Adat Jawa Tengah*. Yogyakarta: Pustaka Belajar.
Ngafifi, Muhammad. 2014. Kemajuan Teknologi Dan Pola Hidup Manusia Dalam Perspektif Sosial Budaya. *Jurnal Pembangunan Pendidikan: Fondasi dan Aplikasi* Volume 2, Nomor 1, 2014.
Primasasti, Agnia. 2022. *Solo, Kota Berbudaya the Spirit of Java*. URL: https://surakarta.go.id/?p=23248.
Sihabudin, Ahmad. 2011. *Komunikasi Antarbudaya, Suatu Perspektif Multi Dimensi*. Jakarta: Bumi Aksara.
Sugiyono. 2011. *Metode Penelitian Kombinasi (Mixed Methods)*. Bandung: Alfabeta.
Suyanto. 1990. *Pandangan Hidup Jawa*. Semarang: Dahana Prize.

ature. This research showed three main results, namely medium, technical approach, and
Computer generated photography: Still image to moving image

A.P. Zen* & C.R. Yuningsih
Telkom University, Bandung, Indonesia

I.M. Miraj
Lampung University, Bandar Lampung, Indonesia

*ORCID ID: 0000-0001-7782-7880

ABSTRACT: Computer generated photography is a process for making photos or images using artificial intelligence neural networks. By using software such as Deep Nostalgia, still photos can be made as if they are moving to produce images that seem as if they are alive. This study tries to analyse how this moving image becomes a new artificial intelligence-generated photography process and its possible application in human life in the future. The method used in this study is a qualitative research method with a phenomenological approach regarding the use of software on still photos that are generated using artificial intelligence. The sample of this research is in the form of moving photography works produced from Deep Nostalgia software. This research showed three main results, namely medium, technical approach, and aesthetic. In terms of medium, this study has found that with the help of artificial intelligence, moving photos can make it possible to create a new medium for creation. Technically, this method of implementing artificial intelligence-generated photography is something new and has never happened before. Aesthetically, this study found that the photos from artificial intelligence-generated photography can produce images that have a different point of view from traditional photography because the brain in making this photo is a machine. This study can also discuss the creative process from Ai's point of view.

Keywords: Artificial Intelligence, Moving Images, Photography, Deep Nostalgia

1 INTRODUCTION

Photography is the process of creating images that evolve. Starting in the era of pure photography where the image process is created directly through a camera mechanism that does not have an analogue or digital image processing to the development of contemporary photography, namely the process of photography with the help of Artificial Intelligence (AI) or software. The era of Pure photography is a genre that encourages the art of photography to obtain unique creative and aesthetic effects photography only by relying on on-camera techniques (Xue, 2016). In other words, there is no digital touch that comes into play in creating the image. This development was then followed by digital technology in the form of software or hardware which gave rise to the era of photography with digital creation by computers. A clear example is AI Based Editing Photography which has been widely applied to digital devices such as smartphones and other forms of devices. Due to these developments, the era of pure photography has slowly begun to disappear and has been replaced by the era of computer-generated photography, including the transformation of still images to moving images.

Deep Nostalgia is software created by a company called MyHeritage which has a patent on a unique technology that can animate photos into videos. This process is carried out with Ai in the Deep Learning process so that it can produce moving photos in the form of short videos that look realistic (MyHeritage Ltd., n.d.). This technique is used by MyHeritage to animate faces in historical images or family photos and produce high-quality, realistic video footage.

Figure 1. Sample photos of Abraham Lincoln using deep Nostalgia in MyHeritage Youtube.

2 RESEARCH METHODS

In this study, we will discuss three things, namely the technique of applying images by Deep Nostalgia software. Then this study will also discuss the medium in the manufacturing process and discuss the resulting aesthetics. Researchers will use descriptive analytical research method with a photographic approach. The data presented in this study is data from photo samples produced by Deep Nostalgia. The sample will be analysed with a photographic approach by examining the visuals produced from the images that have been produced.

3 RESULT AND DISCUSSION

3.1 *Medium of creation*

Digital photography affects the way users see a photo with a digital medium. Photos are often uploaded by users into social media or other web forms which are then enjoyed by users. The development of this photography includes the use of photography by relying on software, forming a new category of photography genre, namely Computational Photography. Computational photography is something new and growing rapidly because there are still many kinds of digital processing that can be applied to other fields. In its application, computational photography is applied in this digital world, either in the virtual world or in the process of creating images on the camera.

Digital photography is building a new medium for creating photos. An example of this developing application is the GAN software for creating images as well as painting on a canvas. The only software that has been formed is software by Nvidia called "Paint me a Picture" or "Paper Dreams" which uses Deep Learning Neural Networks to be able to produce processed data with high accuracy (Zhou, 2019). where the user can take a photo just by verbally describing it and AI Generative Adversarial Network (GAN) will directly process it into a realistic photo. In another process, the user can draw on the canvas as well as paint which is then processed into a realistic photo. This technology can support and assist users in creating images that are easier than using photography. In its current position, photography has started to lose its existence due to the new AI. In its development, the new AI can process photos into live images thanks to software developed by MyHeritage, namely Deep Nostalgia.

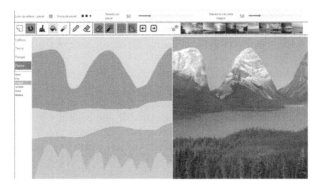

Figure 2. Paint me a picture software by Nvidia.

Deep Nostalgia's goal is to make users feel nostalgic with family members through the medium of photography. The results of family photos are compressed and uploaded by the user which is then processed using software on the web. This website is managed by the MyHeritage company which allows users to create family trees. It is enough to enter a portrait photo of the user and it will be analysed automatically by Deep Nostalgia.

Figure 3. Myheritage website frontpage.

3.2 *Image processing techniques*

In creating moving photos, Ai Deep Nostalgia will analyse the sample images that will be made to move. AI uses complex neural networks and machine deep learning that requires various layers to process input and output which can ultimately recognize a pattern or object (Howard, 2019). Objects that have been analysed from photo samples are combined with a system on AI that already has authentic movement patterns such as gestures and natural human movement patterns. The output is a photo with moving images in video format.

AI requires samples and layers to be able to recognize the images it processes. An image can be represented as a matrix, with each matrix member having pixel colour information. The matrix is sent to the neural network as input data and then processed as output (Pandelea, Budescu, & Cobvatariu, 2015). In this process, AI detects faces in photo samples by looking for facial features such as eyes, nose, ears, and hair. From these results, the face shape is then processed by comparing the layers that have been studied as a reference for the shape of the human face and then processing its movement.

The motion in a moving photo that has been applied by the software is determined by the system based on the direction of the face. The system will detect the movement pattern that will be applied so that the movement becomes natural. The exact facial movement depends on the output issued by the AI, so the accuracy of the AI in detecting faces is the main factor in determining the part of the face. If the AI can't detect the face in the photo, then the video output can't be processed either.

AI with GAN (Generative Adversarial Network) is a method for studying and detecting objects in photos (Brownlee, 2019). With this method, AI can classify objects on the face so that it can determine the appropriate movement of the face in the photo. GANs that are trained to use a large sample of photos and create a framework for photos can increase the speed of processing images (Santos, Castro, Fernandez, Patino, & Carballal, 2020). The process in this Deep Nostalgia software takes about 20 seconds to be able to process still images into moving images.

3.3 *Aesthetic analysis*

Researchers will display two sample photos to be analysed. The sample photos were taken randomly by researchers which will then be processed by Deep Nostalgia. The processed photos will be compared and analysed. To get a different output, sample photos will be taken into two different categories of portrait photos, namely colour portrait and black and white portrait.

Aesthetics in photography is more concerned with visual elements such as lines, textures, and other forms that are harmonious with each other so as to create photos that are more beautiful and pleasing to the eye. According to the Oxford dictionary, aesthetics is a visual engagement with beauty and art, also in the form of an appreciation of things that are interesting or arranged in an artistic way so that they are pleasing to the eye (Oxford University Press, n.d.). Aesthetics can be subjective and other people's views of aesthetics in photos will be different. People's views of beautiful photos can be in the form of texture, lighting conditions, contrast in photos, or other forms of view (Datta, Joshi, & Wang, 2006). With this subjective aesthetic view, the researcher will describe from the side of Ai's natural processing in Deep Nostalgia.

The first photo sample is a portrait photo sample with coloured natural lighting. When viewed from the realistic results, the researchers found that the moving photos had a slightly different facial structure from the original photo. The Ai learning process requires a variety of layer classifications on different photos to produce more accurate output (Dong, Wei, Chen, & Zheng, March). While the photo sample consists of only one photo layer, Ai to study and detect faces in the photo does not have high accuracy and produces photo processing that is slightly different from the original photo.

Figure 4. Processed results of deep Nostalgic photos, black and white photo samples.

The resulting output is video. The video produced in several frames does not appear to have the same facial proportions. The facial gestures and emotions produced, it turns out that Ai can make processed photos that create new expressions. If the photo is a normal expression, in Deep Nostalgia the usual expression can be changed into a different expression. This can happen because of the processing in the software that has studied various layers in the form of facial gestures so that the results obtained have a variety of different expressions.

Face movements in moving photos do not have dynamic movements. Only brief movements such as looking up and to the side. This movement is processed in the same way as facial gestures, namely Ai has studied the movement layer on the face in his software and applied it to the photo sample.

In these two photo samples there is no significant difference in processing still images into moving images. Both have almost the same movement characteristics. Facial gestures that are processed also have the same gesture. It can be concluded that the colour in the photo does not affect the output of AI because Ai only reads faces by applying deep machine learning to photo samples. Ai reads photo visuals by using only one image as the main reference. Even though the results can be said to look natural, however AI cannot apply other visual characteristics because AI only applies what has been learned by AI itself, not generated through thinking processes.

4 CONCLUSION

AI has certain limitations in processing images. The results obtained can have various variables depending on how the AI learns the layers of the neural networks it learns. However, in this case, the Deep Nostalgia software has succeeded in making a new breakthrough in processing images, especially the creation of photography, which is very different from conventional methods. This makes it possible for artists or photographers to create works with new mediums that utilize this GAN technology.

AI cannot create its own work based on thought and imagination processes like the human brain. To be able to produce original work, AI itself requires specific learning, in contrast to the human mindset that can adapt according to other mindsets and inputs so as to produce something new. So AI often does not have a characteristic in the creation of the work it processes.

Along with the development of virtual technology in the form of virtual reality or augmented reality, this technological breakthrough makes it possible to create a new media that can make it easier for users to digest information from a single photo. Moving photos will allow for the creation of new genres of art or photography. On this occasion, what created a new era of photography with the help of AI, namely computer generated photography, became a major part in the development of photography towards the modern era. With this technology, it is possible to end the era of still photography where still images are not widely used and bring up a new era of moving images.

REFERENCES

Azwin, A. (2932). *Blablabla*. 2009, 233.
Brownlee, J. (2019, June 17). *A Gentle Introduction to Generative Adversarial Networks (GANs)*. (Machine Learning Mastery) Retrieved June 22, 2022, from https://machinelearningmastery.com/what-are-generative-adversarial-networks-gans/
Datta, R., Joshi, D., & Wang, J. Z. (2006). Studying Aesthetics in Photographic Images Using a Computational Approach. *European Conference on Computer Vision* (pp. 288–301). SpringerLink.
Dong, Z., Wei, J., Chen, X., & Zheng, P. (March, 23). *Face Detection in Security Monitoring Based on Artificial Intelligence Video Retrieval Technology*. 8, 63421–63433.
Howard, J. (2019). Artificial Intelligence: Implications for the Future of Work. *American Journal of Industrial Medicine*, 62(11), 917–926.

MyHeritage Ltd. (n.d.). *Deep Nostalgia*. (MyHeritage Ltd.) Retrieved June 21, 2022, from https://www.myheritage.com/deep-nostalgia

Oxford University Press. (n.d.). *Oxford Learner's Dictionaries*. (Oxford University) Retrieved June 22, 2022, from https://www.oxfordlearnersdictionaries.com/definition/american_english/aesthetic_1

Pandelea, A. E., Budescu, M., & Cobvatariu, G. (2015). Image processing using artificial neural networks. Constructji. Arhitectură.

Santos, I., Castro, L., Fernandez, N. R., Patino, A. T., & Carballal, A. (2020, May 22). Artificial Neural Networks and Deep Learning in the Visual Arts: a review. *Neural Computing and Applications*, 33, 121–157.

Xue, N. (2016). *Discussion on Pictorial Photography and Pure Photography in the Development of Photographic Art*. 3rd International Conference on Education, Language, Art and Inter-cultural Communication. 40. Atlantis Press.

Zhou, L. (2019). Paper Dreams: an Adaptive Drawing Canvas Supported by Machine Learning. Massachusetts Institute of Technology. *Department of Electrical Engineering and Computer Science*.

The legal status of metaverse law and its implementation in modern era

A.P. Zen
Telkom University, Bandung, Indonesia

I.M. Miraj
Lampung University, Bandar Lampung, Indonesia
ORCID ID: 0000-0001-7782-7880

ABSTRACT: Metaverse is a digital world in the form of virtual reality where humans can interact in real-time using 3D avatars. Metaverse makes it possible to have new experiences online and get different interactions from online meetings in general. The study attempted to look at the development of the metaverse concerning law enforcement, both private and criminal law. Example of a case that attracted attention regarding metaverse in criminal law is about sexual harassment. On the other hand, cases related to metaverse in private law is data protection and intellectual property rights. This is important to discuss given the extent to which law can apply its jurisdiction on cases that occur in the metaverse world. This study is a review of the application of the law in the metaverse with further analysis of its legal status.

Keywords: criminal law, digital law, Intellectual Property Rights, metaverse

1 INTRODUCTION

Globalization has brought about change to modern society over the past few years. As such, in terms of technology and information, legal system is also heavily influenced by globalization. As a matter of fact, the emergence of E-Court even as far as virtual or augmented reality is clear evidence as to how judicial world these days is trying to adapt to the latest technological development in dealing with legal cases, be it private or public law. In that regard, the emergence of virtual/digital system within judicial system has also created another similarly new and advanced technology called "Metaverse." It is to be determined and analysed much further as how judicial system really stands for and view this so called "Metaverse" in dealing with legal cases. Is justice truly worth it by upholding the law through so called "Metaverse"? How far and effective is the concept of "Metaverse" is feasible in legal enforcement? Is the current international legal instruments or even more specifically national legal system in line with the practice of "Metaverse" in enforcing the law for society? These newly found question regarding the emergence of "Metaverse" as a new technological and digital concept which sooner or later must be answered so that the growth of human population globally as well as the demand for enforcing the law can adjust with this new concept effectively.

2 RESEARCH METHODS

This study uses qualitative methods with a normative sociologist approach. Throughout this method, the corresponding research will be further analysed by observing whether there are already laws that govern specifically crimes taking place in the metaverse. If the respective laws do not exist, the question is how can the current law be applied in the event of a crime occurring in the metaverse. In that regard, in terms of sociology, researchers will take data on conditions and behaviour in society, namely cases of sexual harassment that have occurred, and search for effective legal enforcement in the metaverse.

3 THEORETICAL CONCEPT & LEGA STATUS OF METAVERSE

3.1 *Concept of metaverse*

Metaverse is a compound word of transcendence meta and universe and refers to a three-dimensional virtual world where avatars engage in political, economic, social, and cultural activities. It is widely used in the sense of a virtual world based on daily life where both the real and the unreal coexist. (Park & Kim 2022)

Metaverse delivers people an exceptional experience enough to be used in psychotherapy as well. In order to deliver the best real-life experience, the Metaverse like the real world, it is imperative to be able create interaction immaculately and concurrency in a dimension with existence. In order to maintain a sustainable Metaverse, economic activity between users based on these interactions must continue.

Figure 1. A person entering metaverse using devices.

3.2 *Potential legal issues in the metaverse*

The rising frequency of the metaverse will bring about legal issues across a various number of expertise. At the frontline of these issues, it is likely to be intellectual property. It is clearly exhaustive, not least because the growth of the metaverse for the next few years could potentially pose countless challenges not yet considered. This section will discuss possible reframing of existing law and other solutions in the areas of data protection and privacy and intellectual property rights.

Figure 2. A person enjoying metaverse by using devices.

3.2.1 Falsification on data and privacy

In view of data protection and privacy, the metaverse will create new classifications of personal data for further process. It is evidently clear that virtual reality instruments will be massive as companies will be capable of keeping an eye for facial gestures, physiological reactions and biometric data. Thus, the growth of metaverse instruments opens many unanswered questions, such as liability for data processing, liability for lost or stolen data and approval for data processing. (Cheong 2022).

The protection on Data and Privacy is very much needed particularly in the realm of metaverse where the real world and virtual world are merged into one single real-life dimension. Nevertheless, it is also trough the realm of metaverse whereby falsification/forgery on data and privacy is at its most vulnerable stages in today's world of advanced technology. People utilizing virtual reality technology tend currently to be represented by comical avatars which hold little similarity to the people they are portraying. Nevertheless, this is expected to change as technology utilized for forgery/falsification becomes more sophisticated and more widely accessible. The technology is known as "Deepfakes" technology. Deepfakes allow advanced re- enactment of an individual's face, face exchange between two individuals and even re-creation of an individual's voice. If a deepfake is produced by using existing footage of an individual, the owner of the copyright in the footage may be able to deliver an action for copyright breach, although the owner may not always be the individual appearing on scene, which creates difficulty in the concerned matter.

3.2.2 Intellectual property rights

It is viable to say that land law principles can be broaden to the metaverse. A difficulty may arise as to whether real-world legislation would cover violators on private land in the metaverse. Laws must also evolve if a real- world person makes a verdict to take out a mortgage on their virtual property. Instruments and services must undergo build-in technical defences and periodical process, such as the terms of use, to protect their users. The problem, then, is how brands and individuals ought to protect their copyright as the metaverse broadens itself and gets more international recognition. In that regard, one example might be sufficed. As the metaverse delivers new opportunities for misuse of intellectual property, content owners and licensors must consider the proper scope of keeping an eye on such instruments and upholding their rights. Other difficulty in intellectual property rights would have to do with online content. Life in the metaverse is likely to expand opportunities for creators to play upon their content by delivering new channels for consumption by consumers, but it also raises questions for licensees regarding the scope of rights they may have been bestowed upon in content licences which prexist considerations around the metaverse.

There is an example where a licensee has been bestowed upon in terms of exclusivity regarding a particular use case for the licensed content. We may observe licensors increasingly seek to argue that metaverse use falls outside the scope of the exclusivity, since it was not envisaged at the time the licence was granted. Businesses must consider thoughtfully whether they have the precise rights to capitalize on content in the way they contemplate in the metaverse. (Danks 2022) There will be unlimited arguments about whether a specific metaverse patent is being encroached by other innovation, especially given how quick the innovation is developing, since it will be hard to tell a "novel" creation from a simple alteration of a current one. (Chandrakar & Ravishankar 2021)

4 FUTURE CONCEPT OF METAVERSE

Despite the increasing interest of researchers about the metaverse, there are few explanatory and comprehensive studies about a metaverse in the literature. However, it is thought that this situation will increase especially in recent years with the developments in blockchain technology, sensor technology, the advancement of augmented and virtual reality technologies. (Damar 2021) In other words, metaverse can be best described as persistent, connected virtual realities where people work, play, and socialize: anytime, anywhere, and from any device. It is the convergence of the physical, augmented, and virtual reality where users can interact with each other in real-time scenarios. It's a revolutionary form of digital interaction with endless, untapped potential that holds massive opportunities in the marketplace. (Gupta & Sarvepalli An Overview of the Metaverse its Incredible Potential & Emerging Business Opportunities 2021)

Figure 3. Simulation of a metaverse world.

5 METAVERSE IN 5.0 AND FINTECH

Legal issues related to fintech will arise increasingly in the metaverse, especially as more companies offer digital assets and services for sale. Legal questions will surely arise regarding the proper verification of ownership and potential infringement or conversion of authentic and verified purchases. (Chance 2022) The metaverse has infinite potential as a new social communication space. It provides a high degree of freedom for creation and sharing and provides a unique and immersive experience. (Kye et al. 2021 (eISSN: 1975-5937 Open Access)) In reality, the challenges that corporations will face tomorrow (and are starting to see today) will require the creation of a new dimension, beyond the 2D, 3D, hypertext, and other vertical and horizontal exchanges. (Launay & Mas 2008) Metaverse is the three-dimensional world where avatars are active on behalf of the users in the real world. Usually, the virtual world composed of computer graphics is accessed by users with appropriate personal computers and a special application (a viewer). (Suzuki et al. 2020)

6 METAVERSE AND SEXUAL HARASSMENT

One of the biggest questions asked by those outside the metaverse research community is how it was even possible for sexual misconduct to happen there in the first place. The misogyny and harassment women and girls face daily can be reflected in the metaverse. As the metaverse develops, it will come to include all the criminal and predatory activity we experience in real life. Much like the internet, there will be dark metaverses, niche metaverses and a wide range of digital spaces for humans to engage. While personal boundaries may prevent unwanted touching and kissing, they would not necessarily prevent verbal abuse. And even though you could block someone who makes inappropriate sexual comments towards you, by the time you've done that, hasn't the damage already been done? The issue was not about touching, but the intimidation. The truth is, whether we like it or not, the metaverse is growing day by day and will, likely, soon come to define our online lives. The technology is developing rapidly. Experts predict an increase in VR users wearing haptic clothing that would enable them to actually *feel* sensations in metaverse spaces – scent is already said to be in development (Petter 2022).

7 METAVERSE AND SUSTAINABLE DEVELOPMENT GOALS (SDG)

Metaverse is a very significant aspect of ensuring the implementation of Sustainable Development Goals (SDG) in the future. In that regard, by taking into account the implementation of SDG while enhancing the future development of Metaverse, this would of course be paving the way of improving the quality of Eco Green Environment within our environment. This is since previous studies proposed that a mindset that revolves around sustainable development along with an entrepreneurial spirit and design creativity contribute to innovations and transformations. In addition to vigorously advocating for the United Nations' Sustainable Development Goals (UNSDGs), It is now time to explore systems thinking, innovations, and transformations with resilient skills to engage, design and redesign supply chain/logistics modals to conserve resources under a low-touch and creative economy mentioned by UNESCO. [12]

8 CONCLUSIONS

There are several provisions regarding the conclusion of this research paper which are stated as follows:

1. Metaverse is a compound word of transcendence meta and universe and refers to a three-dimensional virtual world where avatars engage in political, economic, social, and cultural activities in which both the real and the unreal also coexist.
2. In view of data protection and privacy, the metaverse will create new classifications of personal data for further process. It is evidently clear that virtual reality instruments will be massive as companies will be capable of keeping an eye for facial gestures, physiological reactions and biometric data. Thus, the growth of metaverse instruments opens many unanswered questions, such as liability for data processing, liability for lost or stolen data and approval for data processing.
3. 'Ownership' in the metaverse is nothing more than a form of licensing or provision of services. The problem, then, is how brands and individuals ought to protect their copyright as the metaverse broadens itself and gets more international recognition.

REFERENCES

Chance, C. (2022). The Metaverse: what are the Legal Implications? *Chance, Clifford*, *1*(1).

Chandrakar, P., & Ravishankar, P. (2021). An AnalysiS ON Law AND Metaverse. *Indian Journal of Integrated Research in Law*, Volume *II*(Issue I), 1–11.

Cheong, B. C. (2022). Avatars in the Metaverse: Potential Legal Issues and Remedies, *School of Law, Singapore University of Social Sciences*, (https://doi.org/10.1365/s43439-022-00056-9).

Damar, M. (2021). Metaverse Shape of Your Life for Future: A Bibliometric Snapshot. *Journal of Metaverse, Dokuz Eylül Universty*, Volume: *1*(Issue:1), 1–8.

Danks, R. (2022, June 20). *The metaverse – A New World for Rights*. (Law Society of Scotland) Retrieved June 21, 2022, from https://www.lawscot.org.uk/members/journal/issues/vol- 67-issue-06/the-metaverse-a-new-world-for-rights/

Gupta, R., & Sarvepalli, M. (2021). *An Overview of The Metaverse its Incredible Potential & Emerging Business Opportunities*. IT Company. Bangalore, India: https://www.happiestminds.com/wp-content/uploads/2022/04/Metaverse.pdf.

Kye, B., Han, N., Kim, E., Park, Y., & Jo, S. (2021 (eISSN: 1975-5937 Open Access)). Educational Applications of Metaverse: Possibilities and Limitations. *JEEHP (Journal For Educational Evaluation for Health Professions)*, *18*(32).

Launay, Y., & Mas, N. (2008, November). "Virtual Worlds Research: Consumer Behavior in Virtual Worlds" Metaverse: A New Dimension? *Journal Of Virtual Worlds Research*, Vol. *1*(No. 2 (ISSN: 1941-8477)).

Park, M. S., & KIM, Y. G. (2022, January 13). A Metaverse: Taxonomy, Components, Applications, and Open Challenges. https://ieeexplore.ieee.org/stamp/stamp.jsp?tp=&arnumber=9667507, *VOLUME 10* (2022).

Petter, O. (2022, March 20). *Why Is No One Taking Sexual Assault In The Metaverse Seriously?* (Vogue) Retrieved June Thursday, 2022, from Why Is No One Taking Sexual Assault In The Metaverse Seriously?: https://www.vogue.co.uk/arts-and-lifestyle/article/sexual-assault-in-the-metaverse

Suzuki, S.-N., Kanematsu, H., Barry, D. M., Ogawa, N., Yajima, K., & Nakahira, K. T. (2020). Virtual Experiements in Metaverse and Their Applications to Collaborative Projects: The Framework and Its Significance. *24th International Conference on Knowledge-Based and intelligent Information & Engineering Systems.* Tokyo, Japan.

Street furniture: Design and material studies on public facilities

T.Z. Muttaqien, R.H.W. Abdulhadi & T.M. Raja
Telkom University, Bandung, Indonesia

ABSTRACT: Sitting facilities in public spaces are currently a requirement of a city. This facility is known as street furniture. The current street furniture design is very varied, both in terms of shape and material, which must answer the needs of outdoor conditions other than aesthetics, function, and user comfort. Improper furniture design can lead to inefficient use of materials. This article focuses on the study of street furniture design and materials in Jalan Ir. H. Djuanda, Bandung. The analysis intended to determine the suitability of existing furniture with standards design based on local rules, trends, and materials focused on sustainable materials. The method of data collection was observation and literature. This study uses a qualitative description to identify and analyze the existing design and material. With the studies of sustainable materials, the designer considers the design of furniture materials possibilities. The conclusion is effective use of materials and good design is a form of sustainability.

Keywords: Product Design, Street Furniture, Sustainable Material

1 INTRODUCTION

Bandung is one of the tourist destinations in West Java, with nickname Paris Van Java (Puji 2021). It is derived from Paris fashion sale at Braga street and changed its function from a vacancy destination to a food hunter in the present day. Its cold climate attracted people to hang out at the park, cafes, and pedestrians along Ir. H. Djuanda street, now known as Dago, one of the famous streets in Bandung

The Dago name was taken from people's habit who had to wait for each other before they went to the market to avoid wild animals (Aktual 2020). Years after, Dago was filled with offices, restaurants, cafés, schools, public spaces, etc. The government issued a public facilities policy, one of which is provision seating. In the beginning, not all the citizens used it correctly, and few used it for skateboarding obstacles causing fines for the perpetrator.

Figure 1. Dago Park bench 1 (detik.com, 6-15-2022).

The material of the bench is cast iron and wood, produced by local industry. When used properly, it can be long-lasting. The bench area is designed for leisure, chatting, enjoying the view, having selfies, etc. The designer had to understand the way to produce the product and who is using it. Collaboration between material knowledge and design principle results in a sustainable product.

1.1 *Public facilities*

Public facilities written in Indonesian rule of law, In Article 1 of Law number 25 of 2009, it is stated that Public Service is an activity or series of activities in the context of fulfilling service needs by following laws and regulations for every citizen and resident of goods, services, and/ or administrative services provided by service providers public (BPK RI Database Peraturan, 2009). It provides public services, such as highways, streets, sidewalks, and pedestrian bridges. The purpose is to make it convenient for the community, we had a responsibility to maintain it together. The design of public facilities had to accommodate the goal, such as convenience and ease of maintenance, function, durability, and aesthetics as well. The amount of seats is 292, spread over several locations including Asia Afrika, Turangga, Braga, Dago, Banceuy, Naripan, and Sudirman street.

1.2 *Sustainable material*

The material is sustained if it is produced according to user needs and can be renewed without depleting and destroying natural resources (Himdi 2021). Sustainable material is also biodegradable, compostable, low-cost manufacturing, and environmentally friendly (Ana et al. 2022). By using renewable materials, sustainability will be maintained. Several factors need to be considered such as: (1) years (2) production (3) energy use. Types of sustainable materials are (1) wood (bred), (2) green concrete, (3) bamboo, (4) palm fiber, (5) hay, (6) hempcrete,, (7) ashcrete, (8) cork.

Figure 2. Hempcrete (HIMDI Binus University, 6-15-2022).

1.3 *Street furniture*

Street furniture Harris and Dines (1998); The elements are placed collectively on a street landscape for comfort, enjoyment, information, circulation control, and protection of the user. It must reflect the local environment's character and blend with it (Mentari, 2022). Street furniture design had to consider this to be functional, aesthetic, and valuable. Types of street furnitures are (1) park benches, (2) trash cans, (3) information boards, (4) park lamps, (5) signage, (6) bollards, (7) bus stop, (8) fountain, (9) restroom, (10) guardrail (Prasetyowati, 2018).

Figure 3. Bollards (grid.id, 6-15-2022).

2 RESEARCH METHODS

The research uses a descriptive qualitative method, focused on sustainable materials of street furniture. Dago street was chosen because of the difference in the material used after the revitalization due to the other place that still uses previous material.

2.1 *Methods of data collecting*

The data collected in this research use non-random sampling with a judgmental approach (purposive) based on observation and literature. The primary data is from surveillance, archives, and literature. The data will be analyzed and used for validation. The data used is for Dago area only.

2.2 *Methods of data analysis*

The descriptive qualitative method is used in this research by processing variables of street furniture, namely years, production, energy used, material sustainability, comfort, circulation control, and protection of the user.

3 RESULT AND DISCUSSION

Based on the data, the Dago's park bench design has changed due to revitalization caused by improper use, first takes the classic style, made from cast iron and wood. The latest is a modern minimalist built from bent iron stirrups. The factors in this research are the

Table 1. Analytical process.

No.	Criterion	Definition	Existing data	Analysis
1.	Years	The age of the natural material used is appropriate, and lifetime of the material.	Material: stirrup iron, medium life	Iron is one of the sustainable materials. The material usage can be long-lasting depending on how to use the coating material, and how to maintain

(*continued*)

Table 1. Continued

No.	Criterion	Definition	Existing data	Analysis
2.	Production	How to produce, waste amount	Weld, grinding, coating. Medium-high waste	Moderate skill, using quite a lot of material, causing moderate waste. Need to reduce the material if it has to be in the sustainable design category
3.	Energy used	How to produce, how to distribute, how to install	Medium electrical energy, medium-low shipping cost, average to medium installation skill	Semi-manufactured product, local industry, ease to distribute, installed with manual and semi-automatic tools.
4.	Material sustainability	Renewable, recyclable, reusable, reduceable, used amount.	Highly recyclable, highly reusable, medium renewable, average use.	Material can be re-melted, or put to another use, using quite a lot of material, there is a chance to re-design. Durability.
5.	Comfort	Ergonomically used	Medium-low comfort	Ergonomically made shape and size but not the material. Not for long-time use.
6.	Circulation control	Space between	High	The size made enough space for pedestrian.
7.	Protection of user	Sharpless corner	Medium	Rounded shape in the corner, but there are still a few dangerous points; the rear leg, causing a stumble while across.

parameters of public facilities, sustainable materials, and street furniture. The analysis result shows that the Dago park bench does not meet the parameters at several points, including the excess of materials, inconvenience for long-time use, and the less safe of the rear leg, but it can be categorized as a sustainable product because the long-time use, energy-efficiency, and the use of recyclable materials.

4 CONCLUSION

Durability, efficient usability of materials, energy saving, recyclability of materials, convenience, circulation, and user safety are some of the criteria for good street furniture. The Dago park bench does not meet it, but it does not detract from the overall score. The new design uses iron to increase strength, change shape, and formation, to obtain a new value for the environment. If the goal is sustainable street furniture, it must be viewed in its entirety, to see how the bench blends with other street furniture elements such as bollards, garden lights, information boards, etc., so they become a unit. After it was combined, it measured whether the street furniture applies the sustainable concept or not. The design of the Dago park bench and the study of materials are only the openings for further researchers to examine other elements such as the environment design or finding new sustainable material for street furniture.

REFERENCES

Ana et al. (2022). Technological feasibility and environmental assessment of polylactic acid-nisin-based active packaging.
Aktual. (2020). Retrieved from Sejarah Nama Jalan Dago Kota Bandung.
BPK RI Database Peraturan. (2009). *Retrieved from Undang-undang (UU) tentang Pelayanan Publik.*
Himdi. (2021). *Retrieved from Apa Sih Sustainable Material Itu?: Binus University.*
Mentari. (2022). Retrieved From Street Furniture Harris Dan Dines.
Prasetyowati. (2018). Retrieved from 5 jenis Street Furniture di Trotoar, Salah Satunya Dipasang Menjelang Asian Games 2018.
Puji. (2021). *Retrieved from Sejarah Kota Bandung dan Asal Muasal Sebutan Paris Van Java.*

Study of second-hand clothing upcycle with mixed media techniques in the context of sustainable fashion case study: Bazooqu Brand

D.Y.P. Longi & M. Rosandini
Telkom University, Bandung, Indonesia

ORCID ID: 0000-0003-0128-5434

ABSTRACT: Fast fashion encourages wear systems with a short service life. Nowadays, many fashion-doers tend to buy fast-fashion products to follow the trend to be considered up to date, so a few of these fast-fashion items end up in landfills and become waste into second-hand clothes. This study used qualitative methods, with a literature study on how fast-fashion trends are developing nowadays, then observing the behavior of teenagers in clothing, and interviewed people who feel the impact of fast-fashion trends. Finally, conducting upcycling experiments mixed media techniques on second-hand clothes. The upcycling method used to reprocess second-hand clothing is reprocessed using the upcycling method made by mixed-media techniques to produce new products that have the same quality, aesthetic value, and economic value, even higher than the previous product. It can extend the life of clothes from fast fashion. In this study, the bazooqu brand will be a case study in research on upcycle and the application of the mixed media technique used.

Keywords: Fashion, Fast Fashion, Second-hand, Sustainable, Upcycle

1 INTRODUCTION

Fast fashion is clothing that is produced at low costs according to the latest fashion trends based on luxury fashion trends and basically encourages a disposability system, and this trend runs so fast that the style of clothing that was yesterday still a trend may end up today in landfills (Joy et al. 2012). In addition, the impact of this phenomenon is directly related to SDGs which discuss sanitation and clean water because the fast fashion industry is the biggest contributor to water pollution after agriculture. In this day and age, fashion actors, especially teenagers, tend to follow fashion trends that are looking attractive and different from others; therefore, teenagers usually buy goods to follow the most popular styles to be recognized as fashionable and up-to-date teenagers (Behesti & Arumsari 2019). Therefore, it can be said that fashion is not eternal. Many second-hand clothes are sold in the market due to changing fashion trends. Second-hand clothing is clothing that has been used before or sent from abroad such as Singapore, Malaysia, and Korea (Karimah & Syahrizal 2014). Quality and brand are also factors that encourage consumers to buy clothes with the status of being used. Second-hand clothes are also of public interest because they want to look different. After all, usually second-hand clothes have brands from well-known brands and have models that are not in the market.

Based on this, people's lifestyles in buying used clothes have an impact on the lack of public knowledge about the use of second-hand clothing which causes hoarding of clothes (Putri & Suhartini 2018). Basically, these imported used clothes can be used and reprocessed into clothes with new designs and have a higher value than the previous value or commonly referred to as upcycling. In a previous study, upcycling is a process of recycling products into different or similar products with increased value, where upcycling is one of the design strategies to extend the product's consumption period (Laitala et al. 2020)

The upcycling method utilizes waste as the main source of material, reduces the impact on the environment caused by the production process, obtains materials locally, and supports local communities (Han et al. 2017) The upcycling process has various techniques, one of which is Patchwork. "Patchwork is an activity of sewing, connecting, and combining pieces of patchwork into a form that has a work of art. Several large retailers such as Uniqlo, Levi Strauss & Co., Zara. Patagonia have already started implementing this method.

If you review second-hand clothing, it has the potential to be reprocessed, many techniques can be used to process it, one of which is mix media. Mix media is a technique that is usually used to make non-functional products, but in this study, the author experimented with using second-hand clothing as the main material and then applied mix media techniques such as patchwork, half-half method, and scraps.

2 RESEARCH METHODS

This research was conducted by using a qualitative approach to obtaining data.Tthe author conducted a literature study on how the fast fashion trend is developing in the present, as well as how to apply upcycling and mixed media techniques to clothing. Data collection was also carried out by conducting literature studies through book and journal sources, as well as observations and interviews with fashion actors who felt the impact of this fast fashion trend.

1. Literature study, the theoretical basis used is the theory of variables related to the topic of the research such as the theory of fast fashion, upcycling, secondhand clothing, mix media so that these theories can be related to each other to support the analysis process later.
2. Observations were made to the Cimol Gedebage market in Bandung, this observation was conducted to find out what kind of waste is generated by this fast fashion trend. And what kind of waste can eventually be reprocessed.
3. Then the author conducted an interview with one of the thrift shop business owners named Zulfa, who carried out the upcycling method using second-hand clothes as the basic material, Zulfa used leftover fabrics from pieces of cloth, parts of second-hand clothes that were different from the main material which is then combined and become a product with a new design.

3 RESULT AND DISCUSSION

Figure 1. Bazooqu logo. Source: personal documentation, 2021.

Bazooqu is one of the thrift shop businesses whose products are made using the upcycle method using the mix media technique. Starting from the owner's anxiety, where she saw the remnants of cloth piled up in a textile production house then she thought of reprocessing it using second-hand clothes as the beginning. The following are some of the products produced by Bazooqu.

Figure 2. Bazooqu product. Source: http://instagram.com/bazooqu.

Bazooqu focuses on using leftover fabrics that are remade into pockets, and mini dresses which will later become accents to add details to second-hand clothes which are the main ingredients in the upcycle method. Then, the design concept of this research is to use second-hand clothing as the basis for applying the upcycle method to provide better value than before to provide economic value (Han et al. 2017). The following table contains technical opportunities that can be used to process second-hand clothing using the upcycling mix media method.

Application Result	Analysis
	**Technique**: Patchwork The results of the application of this technique do not take long because this technique does not require a long process. This technique produces a small amount of patchwork waste from the process of perforating the main media, but the resulting waste can be reused in other techniques and media. This technique can also optimize the use of leftover fabrics/patchwork produced from used clothing waste. This technique can be developed, starting by exploring the techniques used without the main media and patchwork, then the shape, color, pattern, and texture of the patchwork used can also be explored more diversely.
	Technique: Scraps This technique takes a long time in the manufacturing process because it has to put together and re-adjust parts of other second-hand clothing pieces. This technique can optimize the remaining upcycle of second-hand clothing, thereby reducing the waste generated.
 	**Technique:** Half Half This technique does not require a long time in the process, because it only goes through several processes such as cutting and putting back together. This process does not produce second-hand clothing waste at all, because all parts of the main media are used entirely. This technique emphasizes more on the color and texture of the main media to produce harmony and harmony. This technique can be developed by exploring the various types of fibers used.

From the results of the upcycle method analysis using the mix media technique, each technique requires a different time estimate in the process, each technique minimizes the waste produced and does not even produce waste at all and this technique can optimize the use of waste generated by the fast fashion phenomenon.

4 CONCLUSION

The use of the upcycle method with the mix media technique can be one solution to the problem of the fast fashion phenomenon in reducing the waste generated. This technique is able to make the life cycle of a product longer and has a high value for environmental and social welfare. This technique has the potential to be developed starting from the way of processing, from the selection of the main media to the way of selling it. This research aims that these techniques can be implemented from a small scale individually to the largest scale starting from MSMEs and big brands.

REFERENCES

Behesti & Arumsari. (2019). *Pengolahan Pakaian Secondhand Berbahan Denim Untuk Produk Fashion Menggunakan Teknik Surface Textile Design Yang Terinspirasi Dari Jumputan Palembang.*
Han et al. (2017). Standard Vs. Upcycled Fashion Design and Production.
Karimah & Syahrizal. (2014). *Motivasi Masyarakat Membeli Pakaian Bekas di Pasar.*
Laitala et al. (2020). *Environmental Impacts Associated WithThe Production, Use And End-Of-Life Of Woolen Garment.*
Putri & Suhartini. (2018). *Upcycle Busana Casual sebagai Pemanfaatan Pakaian Bekas.*

Reliefs at the Desa lan Puseh Temple in Sudaji Village, North Bali, image of Bali tourism during the Dutch colonial period

I.D.A.D. Putra
Telkom University, Bandung, Indonesia

S. Abdullah
Universiti Sains Malaysia, Penang, Malaysia

ORCID ID: 0000-0003-4978-139X

ABSTRACT: The relief artwork plastered on the walls of the "Desa lan Puseh" Temple in Sudaji Village was made around the 1920s when Bali was entirely under Dutch colonial rule. The reliefs made are different from the traditional Balinese art standards. Relief art no longer uses patterns in *wayang* art. The reliefs are made more realistic both in terms of style and theme. As a work of art, of course, there is a message and a specific purpose and purpose made from the relief work. Through iconographic analysis, including the form of gestures and attributes worn on the figures of each relief scene connected in the context of the socio-cultural history of the Balinese people, this study shows that Dutch intervention in traditional arts for specific purposes has an influential role. Relief art is constructed to form an image of Bali as a "harmonious" and "cultured" tourist destination.

Keywords: relief, image, tourism, Bali, Dutch colonial

1 INTRODUCTION

This paper examines the visualization or imagery and meaning of traditional Balinese relief art displayed on the walls of the Desa lan Puseh Temple in Sudaji, Sawan area, Buleleng, North Bali. The reliefs allegedly made during the Dutch colonial administration were related to their political interests; This is indicated by the iconography of the reliefs that use modern icons or symbols that have never been found in traditional art before.

In terms of themes and imagery, the reliefs displayed at the Desa lan Puseh Temple have left traditional art behind. The embodiment of the depiction of relief figures or figures no longer uses the puppet standard as in the reliefs of East Java temples. Because Bali has inherited Hindu-Javanese culture since the era of Majapahit rule around the 13th–14th centuries AD. (Yudoseputro 2008: 109) Relief also does not use heroic stories from the great Indian epics, namely the Mahabharata and the Ramayana. The report displayed on the temple reliefs showed the Balinese people's socio-cultural conditions when the Dutch colonized them. As for the modern idioms described, the Dutch army is drunk, mechanics repairing jeeps, foreigners (Arabic), and Dutch officials (Covarrubias 2013: 190). Likewise, scenes of festive parties, dances, musicians, and other community activities are presented realistically.

The embodiment of the reliefs, considered modern elements, can penetrate the sacred area or temples (pura); of course, this is related to the power and goals of the Dutch political mission. The Dutch view that the Bali region has exotic cultural and natural potential, which is very suitable as a world cultural tourist destination. Where the initial idea or thought was raised by the Governor of the Dutch East Indies Thomas Stamford Raffles in the 1800s, Bali as a living museum of Majapahit, the remnants of the Hindu-Javanese civilization are still preserved and carried out by the Balinese people until now (Vickers 1996; 21). Therefore, the Dutch rushed to realize this idea and carried out the conquest of the Balinese kings through a long war. They are starting from the north of Bali from 1846–1849 to the south of Bali from 1906–1908.

Since Bali was completely under colonial rule, the Dutch began to pioneer tourism based on Balinese tropical culture and nature. The traces of the development of cultural tourism will be traced from the relief of iconographic aspects, including gestures, attributes, story scenes, and mediums and techniques in the historical context of Balinese cultural development. Reliefs or works of art are expressions of the culture of the people; during the colonial period, of course, traditional art was no longer pure as an expression of the people's values. For this reason, it is essential to reveal and discuss the extent of the Dutch intervention in traditional arts, especially the reliefs depicted at the Desa lan Puseh Temple.

2 RESEARCH METHODS

The study of relief artworks at Desa lan Puseh Temple focuses on the physical appearance (visualization) of the reliefs in the form of gestures or characters in each scene based on iconographic studies that have not been done infrequently so far. Even though this issue is interesting because the idea of the scenes depicted will undoubtedly show a particular meaning based on the situation and social conditions of the Balinese people, the description of gestures and themes will be analyzed through historical frameworks and visual data approaching that period. Visual data were obtained from direct observation and visual documentation owned by several government and private agencies. Then a literature study was conducted on journals, proceedings, archives and lontars at the Buleleng museum. Data were also obtained from interviews with historians to obtain a sequence of colonial events in North Bali.

Generally, the embodiment of relief art carved on the walls of temples or temple gates in Bali uses the "young classic" style, which is a style that continues the Majaphit tradition (Munandar 2011: 175–179). 200) and Clair Holt refers to the "wayang style", as a flat relief that refers to the characters in *wayang* performances that have existed since around the 8th century AD.(Holt 2000: 100). The "young classic" period is a transformation of art back to its true identity, leaving its Indian style behind.

Iconographically, the relief figures depicted at Desa lan Puseh Temple are no longer represented in a *wayang* style; the form shows a more realistic and cartoonish style and daily themes. Therefore, several questions arise, how does this change and shift the meaning of relief based on the depiction of relief? What underlies the change? This study works through iconographic analysis of temple reliefs, supported by other data such as the embodiment of statues, ornament reliefs, interviews, literature studies and existing documentation.

3 RESULT AND DISCUSSION

The visual changes in the Desa lan Puseh Temple reliefs are not the first but are a series of other reliefs involving Dutch intervention. The first relief work that the Dutch constructed was "a man pedalling a bicycle" plastered on the wall of the Madue Karang temple in Kubutambahan Village—built to shape the image of Bali in support of the pioneering tourism that is made. The male figure depicted is a Dutchman named W.O.J. Nieuwenkamp is pedalling a bicycle, which was created in 1904 (Carpenter 1997). The relief was made to show the world in Bali that tourists have come to Bali and form the image of "comfortable" and "cultured", the initial slogan the Dutch echoed.

Orderly or Comfortable as well as having a culture of metal was formulated by Dutch scholars and scientists such as F.A. Liefrinck and V.E. Korn. These two scholars laid the foundation for scientific knowledge about Bali through ethnographic, philological, legal, and cultural studies conducted from the late 19th century to the 1920s. The studies carried out later became the basis for forming the image and core of Balinese culture today, such as the basics of Balinese customary law, understanding of the Balinese people, and the authenticity of their culture. Liefrinck's study of Balinese culture was followed by other scholars and scientists who later came to the island (Robinson 2006: 6).

The relief in the Desa Lan Puseh temple is also a Dutch construction after the Dutch tourism program experienced rapid development around the 1920–1930s. This was also due to the circulation of a photographic book by Gregor Kraus published in 1920, which contained topless Balinese women, which created the image of Bali as an island of "paradise" or a garden of "paradise from the East" so that Bali is flooded by foreign tourists (Vickers 1996: 100). Arabs and Chinese have travelled and traded in North Bali, and even Chinese merchants traded to remote areas of Bali. The colonial government has also prepared the infrastructure for these developments, starting with fixed shipping, travel agencies, publishing bureaus and several other facilities. The conditions and situations described in the reliefs are shown iconographically in the form of a jeep and a mechanic starting the car with several Arab passengers. The embodiment of the British-made Waltham jeep at the Desa Lan Puseh temple illustrates that in North Bali, a travel agency was opened in 1914 using this type of car. This type of car first arrived in North Bali in 1913 (Hinzler 2013: 54), supported by the emergence of a publishing bureau by the "Official Tour Bureau", issuing a brochure illustrating the existence of cultural, natural and tourist charms in Bali. The travel agency arranges tourist visits for three days, starting from North Bali to Buleleng to South Bali and returning to Buleleng for three days (Picard 2006: 31–34).

Figure 1. The relief depicts mechanic, jeeps, Arabs, drinks. Source: Covarrubias 2013.

Reliefs illustrating the habits of foreigners, mainly Europeans, such as drinking and partying with topless women, are plastered on one of the panels on the temple's walls. An image that shows the reality of Balinese life in the colonial era and how free foreign people are while in Bali. The modern lifestyle brought by travellers and colonials has infiltrated the lives of traditional people. The Dutch were aware of the impact of tourism development, which could gradually erode Balinese culture under the influence of modernism. Therefore, the Dutch also created a cultural conservation program called *Balinization* or *Baliseering,* "the Balinese must return to their culture".

The image of "culture" is very clearly shown by the iconography of the dancers' figures with crowns and magnificently dressed, as well as the figures of musicians wearing suits and udeng headbands. In historical records made by Hinzler (2013) the Dutch once constructed traditional Balinese clothing on the uniforms of members of the "sekeha" (group) gamelan Gong Agung North Bali combined with European jackets for performances at the annual festival in Surabaya 1906. The nobles and courtiers also like to wear a jacket or coat with metal buttons called "krongsag", which has become a trend in the North Bali region. When the Balinization program was run, the colonial government also regulated the lifestyle of the Balinese regarding attitudes (Balinese ethics) to the way Balinese people dress and carry out their daily activities (Nordholt 2002).

Figure 2. The reliefs depict gamelan musicians, dancers and accompanists and parties. Source: Dwija 2018.

The embodiment of the reliefs depicting the festivities of the religious ceremony not only imaged "culture" but also described the Dutch colonial program on cultural conservation called *Balinization* or *Baliseering*. The Dutch are aware that the developed cultural tourism will erode the community's traditional culture due to the influence of modernization brought by tourists and the Netherlands itself. The *Balinization* program is not just a preservation of Balinese culture but is to defend the colony, which has successfully absorbed economic benefits from the controlled Bali area. Besides that, it protects Balinese culture from Islamic movements from Java, and Christian *Zending* brought by the colonials.

The success of the Dutch in constructing Balinese art on reliefs to support the image of Bali in the international world is also due to Dutch intelligence in legalizing "chess of colour" into a caste-like the understanding in India so that the *Brahmana* (Brahmins), *Kesatria* (Knights) and *Weisya* (Weisya) groups are given special privileges in a government called the *Triwangsa*. This *Triwangsa* has become the "stedehouder"(stakeholder) extension of the Dutch government, effectively controlling the traditional-based management system. Thus the Dutch were easy to supervise for brahmins or nobles who opposed being exiled outside Bali. At the same time, the last group, the *Sudra* or *Jaba,* are ordinary people (Robinson 2006).

Regarding the choice of the Desa land Puseh temple was the place for the reliefs to be made by the Dutch because the Sawan area and, in particular, Sudaji Village are the residences of sculptors and a place for handicraft souvenirs in the form of small statues. Tourists can also get souvenirs in the form of woven fabrics and silver crafts in the village of Bratan near Buleleng (Picard 2006).

4 CONCLUSION

The description above shows that during the twentieth century, fine arts, mainly traditional Balinese relief art, have changed with the entry of foreign cultures, mainly due to the intervention of the Dutch colonialists. This intervention resulted in an overhaul of the traditional aesthetic mixed with modern rules. From the thematic point of view, there is a secularization of representation; it no longer has a religious function and contains an economic value. Objects take on everyday life, nature, dance and other daily rituals. The strong influence of modern knowledge can be seen in the organization of space in the composition, considering anatomical aspects and slowly eliminating *wayang* figures as standard. The relief artwork made at the temple has expanded its function, Initially functioned as a teaching medium for the local community and as a propaganda image for the general public.

The colonial intervention was not only in relief art but almost all traditional Balinese art, with the excuse of protecting and preserving Balinese culture. The development of modernist ideology brought by colonialism in classical art has opened many opportunities for further research, such as performing arts, *wayang*, masks, drama and other crafts. This can complement the studies that local and international scholars have carried out.

REFERENCES

Carpenter, B. W. (1997). *W.O.J. Nieuwenkamp First European Artist in Bali*. Uitgeverij Uniepers Abcoude.
Covarrubias, M. (2013). *Pulau Bali (Temuan yang Menakjubkan)* (J. Atmaja (ed.)). Udayana Univercity Press. http://penerbit.unud.ac.id
Hinzler, H. I. R. (2013). *North Bali Fist Encounters, Innovations in North Bali, Influences from Abroad & End 19th Beginning 20th-Century*.
Holt, C. (2000). *Melacak Jejak Perkembangan Seni di Indonesia* (T. Rahzen (ed.); Pertama). Masyarakat Seni Pertunjukan Indonesia.
Nordholt. (2002). *Crime, Modernity and Identity in Indonesian History*. (I). Pustaka Pelajar.
Picard, M. (2006). *Pariwisata Budaya dan Budaya Pariwisata* (Pertama). Kepustakaan Populer Gramedia.
Robinson, G. (2006). *Sisi Gelap Pulau Dewata: Sejarah Kekerasan Politik*. Yogyakarta. LKis
Vickers, A. (1996). *Bali Paradise Created* (Ed. Second). Periplus. https://doi.org/0-945971-28-1
Yudoseputro, W. (2008). *Jejak-Jejak Tradisi Bahasa Rupa Indonesia Lama* (Cetakan I). Yayasan Seni Visual Indonesia.

Cultural influences toward society's view to young adult BTS fans (army) in Indonesia and its influence on fans consumption

A. Syafikarani, F. Balqis, A.R. Alissa & I.P. Narwasti
Telkom University, Bandung, Indonesia

ORCID ID: 0000-0002-2260-1849

ABSTRACT: Indonesian K-Pop fans have experienced rapid development growth in recent decades due to K-Pop's fresh concept and good quality music. One of them is dominated by the music group BTS fandom known as ARMY. The act of fangirling, which was deemed inappropriate by the public, resulted in the young adult ARMY hiding its identity as a BTS fan in public. To better understand the factors behind this issue, we conducted a questionnaire that focused on people's perceptions and their impact on ARMY consumption patterns. Although there was only a quarter of societal disapproval, respondents attributed their concerns to the negative responses around them and ignored the positive and neutral responses. This result in their choices while choosing and purchasing not to apparent looking BTS merchandise. Our findings also may be applied to BTS merch business owners when considering items made for young adult ARMY

Keywords: ARMY, BTS, cultural influence, young adult, fans consumption

1 INTRODUCTION

Korean wave is mainly liked by young people who are interested in food, beauty, such as Korean music (K-pop), dramas (K-dramas), movies, and fashion (Tania 2019). This cultural phenomenon is closely related to the complex transnational movement of people, information and capital flow in East Asia (Bok-rae 2015).

Indonesia is one of the countries that is heavily influenced by the Hallyu wave. Many Indonesian teenagers really admire Korean culture. This can be seen from the rapid development of Korean pop or K-pop in Indonesia. The K-pop music industry, especially in Indonesia, is dominated by the amazing idol group, BTS. BTS' popularity began to climb when the song "DNA" entered the Billboard Hot 100 chart, until it was awarded the Billboard Music Awards at the 2017 "Top Social Artist Awards" nomination. After winning the Billboard Music award, BTS has increasingly attracted the attention of the world, including Indonesia.

Based on BTS ARMY census data in 2020, Indonesia occupies the first position with the highest number of fans worldwide. An example of the huge popularity of BTS in Indonesia is during the BTS Meal phenomenon in 2021, until the sales of Chicken McNuggets soared drastically, causing long queues at various McDonald's outlets in Indonesia. In addition, ARMY Indonesia has a big role in providing viewership figures in BTS music videos on their YouTube channel. According to (Statista 2021), Indonesia is the top contributor followed by Thailand, Vietnam, and the United States based on their research in the span of 2018–2019. In 2020, Spotify said Indonesia had the second highest number of listeners after America, with BTS as the most listened to artist.

ARMY (Adorable Representative M.C. for Youth) is estimated to be millions, known as a huge and loyal fan base. They have made headlines in popular media for many different reasons, it has undoubtedly become one of the most powerful and visible fandoms on social media (Park 2021). Henry Jenkins explains in (Nisbett 2018) that fandom is a cultural convergence that dominates a community and its associated media. In addition, fandom has a significant huge impact on social media attitudes and debates. Henry Jenkins also stated that ordinary fans, for the most part, have always been characterized as someone obsessed with everything about a certain celebrity. That includes, their collection, their trivia and often comes across as someone whose attitude is disturbing. Fandom is a socio-cultural phenomenon largely associated with modern capitalist society, electronic media, mass culture, and public performance (Duffett 2015).

According to a survey conducted in 2020, the ARMY BTS Census obtained more than 400,000+ participants from 100+ countries and regions. The survey stated that the majority of respondents came from Indonesia (20%), Mexico (10.6%), and the United States (8.4%). Meanwhile, the age demographic consists of ages under 18 years (50.31%), 18–29 years (42.59%), and 30–39 years (4.24%). The gender demographic consists of women (86.34%) and men (11.3%). For respondents job background categories, were mostly in the fields of education, health, entertainment, hospitality, and the arts. This undoubtedly shows the diversity of ARMY fandom backgrounds and how its complexity has managed to be united by their similarity of interest in BTS.

But on the other hand, the existence of this fandom also forms a certain stigma against them. Quoted from the Kamus Besar Bahasa Indonesia (KBBI), stigma is a negative characteristic inherent in a person's personality because of the influence of his environment. In short, stigmatization is the negative characterization given by the environment to a person. K-Pop fandom in Indonesia is dominated by women. Their passion for K-Pop tends to be viewed negatively. There are 12 feminine traits that are ingrained as gender stereotypes (Schneider 2005). Some examples of feminine traits are talkative, emotionally sensitive, and like to express feelings. According to Schneider's explanation, it can be seen that what K-pop fans do corresponds to the stereotype of feminine nature inherent in women. For example, their shouting reactions lead to an emotional trait that they then express through action. In addition, K-Pop fans also have "negative" stereotypes as excessive, crazy, hysterical, obsessive, addictive, and consumptive diehard fans. This stereotype arises because K-pop fans who tend to be dominated by the fair sex often show intimacy by shouting which is always associated with hysterics, always keeping abreast of the latest developments from their idols who are considered addictive traits, and buying and collecting K-Pop stuff, which tends to lead to consumptive behavior.

One example of public resentment towards the K-Pop community that was widely conveyed through CNNIndonesia.com in their article about K-Pop fans on February 3, 2019, one of which was "The Dangers Behind the K-Pop Opium Phenomenon". In this article, CNNIndonesia.com stigmatize K-Pop fans by emphasizing that the madness of K-Pop fans is a psychological problem that must be stopped. The article even mentions that the impact of liking K-Pop excessively has the same impact as taking drugs. In this article, CNNIndonesia.com also features an interview with a psychologist from Atma Jaya University Jakarta, Vierra Adella, to show that liking K-Pop is a problem that must be understood and solved. Adella stated that some of the acts of fanaticism committed by K-Pop fans were the result of Celebrity Worship Syndrome (CWS) (Silfia 2020).

In addition, there is also a phenomenon of "Sasaeng", Sasaeng comes from the Korean "sa" which means personal or privacy and "saeng" which means life, so it can be interpreted that sasaeng is a fan who is very obsessed and wants to know all the personal life of his idol. Not infrequently from those who will do everything possible to always be close to their idol so that this will have a bad impact on the idol's own life.

The existence of this negative stigma eventually made some people generalize their views on women in the K-Pop community. What they like and all their actions will be underestimated by society. When they express their expressions, stereotypes will affect the perception of people so that their expression is perceived as something superfluous. Including when they buy the things they want, people will regard them as waste because the reason is not in line with people's minds. And when a major media outlet like CNNIndonesia.com release those articles beforehand, it increases the stigma against women's fandom through word selection and the placement of news subjects and objects. These major online media tend to present news that corner female fandom by portraying them as fanatical, crazy, and tend to act without considering their common sense.

Based on the phenomena that have been explained earlier, this study analyzes more deeply about "How does culture influence society's view of young adult BTS (ARMY) fans in Indonesia and its influence on fan consumption". In addition, the keywords that will be generated later on, may be used as a reference or consideration related to fans consumption study and research.

2 RESEARCH METHOD

Qualitative research is a type of research in which the discovery procedure carried out does not use statistical or quantification procedures. In this case qualitative research is a research about a person's life, story, behavior, and also about the functioning of organizations, social movements or reciprocal relationships (Salim 2012). The data collection technique in this study was carried out by distributing questionnaires through an online survey platform known as Google Form. Then the data obtained were analyzed using the theory of stigma and perception.

2.1 *Methods of data collecting*

Data collecting in this study was carried out by distributing questionnaires. Questionnaire is a data collection technique carried out by providing a set of questions to respondents to get answers in the form of information relevant to the research (Sugiyono 2013). The distribution of questionnaires is carried out through an online survey platform known as Google Form. The eligibility and consent of the respondents are stated and confirmed at the beginning of each questionnaire form before proceeding to the next section. Respondents were invited through various social media platforms, such as Instagram, Twitter, and Whatsapp. From the three platforms, we obtained most of our answers from Twitter through student forum accounts with over 892k followers. The questionnaire is divided into two stages, the first stage of the questionnaire was distributed to 143 respondents with an age range of 17–50 years with the criteria of those who knew about BTS and ARMY to find out the Indonesian society view regarding young adult ARMY. In the second phase, the questionnaire was distributed to 113 respondents with an age range of 18–29 years and had the criteria as ARMY to obtain data on the impact of Indonesian society view regarding young adult ARMY to ARMY fandom consumption.

2.2 *Methods of data analysis*

The data that has been obtained is analyzed by connecting each questions, so that conclusions can be obtained as result from the questionnaire. In addition, it is also strengthened by the existence of related theories such as theories regarding stigma and perception.

3 RESULT AND DISCUSSION

A survey-based case study has been conducted to find out the stigma and public perception of the K-Pop community, especially BTS fans. Furthermore, we also conducted a survey containing a series of questions related to this stigmatization to BTS/ARMY fans and how they will respond to it. This research is very appropriate because our findings are to explain some of the social phenomena that happen to BTS fans. Our exploratory survey is designed to better understand two very crucial things, (1) the public's perception of ARMY in the age range of 18–29 years who are still actively involved in fangirling, (2) how ARMY responds to society stigma towards them.

3.1 *Indonesian society view regarding young adult ARMY*

The first questionnaire was conducted to search for preliminary data on the Indonesian society view regarding young adult ARMY. Based on the results of the first questionnaire that has been distributed, very interesting data was obtained from 143 respondents. As many as 75% of respondents gave positive answers and didn't bothered by the fangirling activities carried out by K-Pop fans, especially ARMY in general. The main reason given is that, as long as it doesn't harm themselves and others, fangirling can be used as an outlet for stress by fans aged 18–29. However, this positive response was also accompanied by a negative response, around 25% of respondents associated ARMY with words such as, "toxic", "cringe", and "fanatic". From this data, it can be concluded that the domination of positive remark was due to the development of information and technology which make it more acceptable. This is also supported by the results of the questionnaire which states that 78.8% of respondents know and have an acquaintance of an ARMY, they also ultimately think that ARMY is a big and solid fandom. So, this closeness also affects the reduction of negative stigma against ARMY.

3.2 *Impact of Indonesian society view regarding young adult ARMY to ARMY fandom consumption*

The second questionnaire was carried out as a follow-up to analyze more deeply from the side of the ARMY fandom with the age range of 18–29 years old regarding how the stigma obtained in the results of the previous questionnaire against choosing BTS merchandise and why. From the second survey filled out by 113 respondents, we found that as many as 80% of respondents chose merchandise that did not explicitly show BTS visuals. Most of the responses reveal several reasons, such as being able to be used daily, not being known as BTS fans in public, and still being able to express their fangirling freely. It becomes very interesting to analyze more deeply that the small percentage of respondents who have a negative stigma on K-Pop fans obtained in the first stage of the questionnaire still has a significant effect on ARMY's main factors when choosing and buying BTS-related merchandise. Although it is relatively small, this number of people is enough to make ARMY close themselves off from the public by not using any attributes that directly visualize their idols. Furthermore, data was also found that the majority of ARMY respondents chose not to wear any clothes which could infer their affection for BTS.

Therefore, the results of this design can be considered for business owners who will be involved in the field of K-Pop merchandise, especially if they target the age of 18–29 years old. Based on data obtained from the second questionnaire, the visual characteristics of preferred merchandise are as follows (1) not explicitly displaying BTS visuals (79.6%), (2) it can be used in daily activities (61.9%), such as t-shirts (55.8%), tote bags (46%), and keychains (46%), and (3) illustration and typography based design (58.4%).

4 CONCLUSION

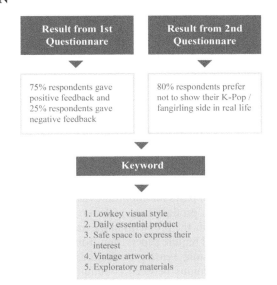

Figure 1. Conclusion flowchart.

Research on the perception of Indonesian society towards ARMY with the relationship between fan consumption patterns, provides a deeper understanding between the two issues. Although the percentage of negative stigma is much smaller in quantity compared to positive stigma, the impact on ARMY is enormous. This is shown by ARMY's preference not to show their fangirling side to the public because of their concerns that are associated with negative perceptions. The final conclusion of this study is that several keywords can be obtained that can be used as a reference, such as safe space, lowkey, vintage, and daily use. Furthermore, this research is also expected to be a consideration for parties who want to produce K-Pop merchandise with a target age of 18–29 years old to present a visually low-key K-Pop products.

REFERENCES

Bok-rae, K. (2015). Past, Present and Future of Hallyu (Korean Wave). *American International Journal of Contemporary Research*, 154–160.
Duffett, M. (2015). *Understanding Fandom: An Introduction to the Study of Media Fan Culture*. USA: Bloomsbury Publishing.
Nisbett, G. S. (2018). *Don't Mess with My Happy Place: Understanding Misogyny in Fandom Communities*. (p. 173). Switzerland: Meditating Misogyny. Springer International Publishing.
Park, S. Y. (2021). Armed in ARMY: BTS and #MatchAMillion for Black Lives Matter. *Proceedings of CHI 2021: ACM CHI Conference on Human Factors in Computing Systems*. Yakohama, Japan.
Salim, S. (2012). *Metodologi Penelitian Kualitatif*. Bandung: Citapustaka Media.
Schneider, D. J. (2005). *The Psychology of Stereotyping*. New York: Guilford Publications.
Silfia, I. (2020). Stigma Media Terhadap Fandom Perempuan dalam Pemberitaan Penggemar K-Pop di CNNIndonesia.com. *Institutional Repository UIN Syarif Hidayatullah*, 87–91.
Statista. (2021, Nov 23). *Distribution of K-Pop Views on YouTube worldwide as of June 2019, by Country*. Retrieved from Statista: https://www.statista.com/statistics/1106704/south-korea-kpop-youtube-views-by-country/
Sugiyono. (2013). *Metode Penelitian Kuantitatif, Kualitatif dan R&D*. Bandung: Alfabeta.CV.
Tania, A. (2019). The Culture of Hallyu Fan Community and its Representations. *Indonesian Journal of Social Sciences*, 33.

— Sintowoko et al (eds)
© 2023 copyright the Author(s), ISBN 978-1-032-44503-8
Open Access: www.taylorfrancis.com, CC BY-NC-ND 4.0 license

The value of Robo-Robo tradition as design inspiration in public spaces in the new normal era

U. Nafi'ah* & M.I.P. Koesoemadinata
Telkom University, Bandung, Indonesia

*ORCID ID: 0000-0002-0195-1743

ABSTRACT: Nowadays, culture plays an important role in the design field. In the social context, adding cultural value to a product can accelerate society's adaptation to the innovative product. This study highlights the problems in the implementation of health protocols in public spaces. The protocols still require supervision from humans because there are often violations such as crowds. In addition, wearing masks and keeping a distance seems to have made social interactions more tenuous between communities. While, on the other hand, there is a culture from Mempawah City named Robo-robo that has the purpose of "protection" as in health protocols but carries several social values with it. This study aims to identify Robo-robo by using "Three Layers of Culture and Design Features" method and trying to initiate a design concept for public space based on the results of these identifications. The result of this research is the design concept of a protection system at City Park based on Robo-robo's kinship value. This study shows that traditional values can be used as design concepts in public spaces because they have the same social context. However, this concept needs further research especially in implementing it as a real prototype.

Keywords: design, public spaces, Robo-Robo, tradition

1 INTRODUCTION

Basically, culture is the way of life in communities (Lin 2009). It lives for years and brings many noble values that are hardly found in nowadays lifestyle. These values should be able to appear in modern design products, as the design is responsible for the formation of a new culture of society. Some research said that culture can inspire design as it grows as a cross-cultural product (Lampel & Lant 2005). The most important part of this idea is to add user experience to the cultural product design process. The aim is to get involved in emotional design considerations (Lin 2018). This way, the new product can resemble values from the culture in a new form. While Lin talks about how a culture can inspire a single product design, this research is trying to use the idea in a wider area that involves social context.

In response to the COVID-19 pandemic phenomenon, the study highlights the problems in the implementation of health protocols in public spaces. The protocols still require supervision from humans because there are often violations such as crowds. Even in some places, there is forced close to prevent crowds. In addition, wearing masks and keeping a distance seems to have made social interactions more tenuous between communities (Napsiah *et al.* 2020). This phenomenon shows that the implementation of health protocol policies is not yet aligned with human needs and natures. Research from Regional Research Council said that local wisdom from culture can be used in social engineering to accelerate the adaptation of new habits in the community (Nurdin 2021). It is said that changing community behavior is not easy and using local wisdom from culture can be a solution.

Using those ideas in responding to the health protocols problems, there is a culture from Mempawah City named Robo-robo that has a similar purpose of "protection" as in health protocols but carries several noble values with it such as kinship (Firmansyah et al. 2021). Therefore, the idea emerged what if Robo-robo could be used in design to implement health protocol in public space in a more acceptable way? According to the question, this research will analyze the culture of Robo-robo and design a product using *"Three Layers and Levels of Cultures and Design Features"* (Lin 2018) as a solution to fix the problems of implementation of health protocols in public space.

2 RESEARCH METHODS

This research was conducted in the following three steps:

1. Collecting data on Robo-robo through literature study qualitatively, then analyzing the structure of Robo-Robo culture based on the concept of *"The Spatial Perspective of Culture"* (Lin 2009).
2. Analyzing the gap area between ideal values from Robo-robo and missing values in the common practice of health protocols.
3. Designing a protection system design in public spaces using *"Three Layers and Levels of Cultural Objects and Design Features"* (Lin 2018).

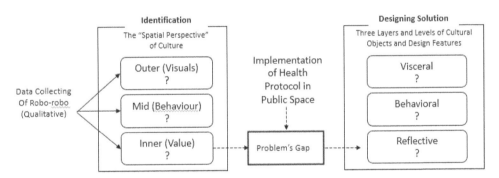

Figure 1. Research's scheme.

3 RESULT AND DISCUSSION

3.1 *Robo-Robo's cultural structure*

In *"The Spatial Perspective of Culture"*, a culture will consist of three-dimensional parts (Lin 2009). The outer part is what we can see, hear, and feel through the five senses. The middle part is what the community does on the basisi of their cultural belief, where the inner part is filled with cultural values. This section will analyze the cultural structure of Robo-robo based on this method.

Robo-robo is a repellent ceremony performed by the people of Mempawah City, West Kalimantan, Indonesia. This tradition is routinely carried out every year on the month of Safar, not only by the Muslim community but also by various ethnic groups in the city (Utami 2017). The community believes by doing the Robo-robo rituals, it can protect the people from perils or calamity (Firmansyah et al. 2021).

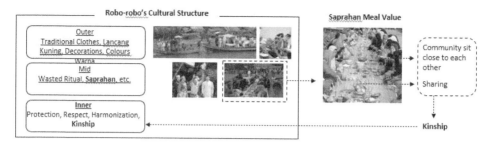

Figure 2. Robo-Robo's cultural structure.

In general, the Robo-Robo ritual procession consists of several activities, such as Repellent Pray, Repellent Bath, Waste Ritual, and Safar meal (saprahan) (Saripaini 2021). One of the most wanted parts of Robo-robo is Safar Meal or known as Saprahan. Saprahan is a meal with all the residents and the people of the palace. The community will sit facing each other, and the drinks and food brought are arranged in the middle and opened intending to flow the blessings of prayer when eaten. As shown in Figure 2, Saprahan is one of the rituals in the middle part of the culture. This ritual brings kinship value to the community, while the other rituals have respect (in repellent pray & bath) and harmonization (in Waste Ritual). In general, Robo-robo is protection value.

3.2 *Value's gap in case study*

As described above, this research highlighted the problem in the implementation of health protocols in public spaces. Research from Central Java shows that public space has fairly high level of the crowd whereas the implementation of health protocols has not been optimal (Sofianto 2022). To prevent the worsening of sick cases, some places are forced close, for example, Digulis City Park in Pontianak, West Kalimantan. Digulis City Park is the biggest City Park in Pontianak, which has many features such as a jogging track, kids playground, city forest, water fountain, and street food. The closing of the park is held due to Large-Scale Social Restrictions in 2020–2021. However, some people still used the park and this action made the government hold the health protocols raid. From this case, it is implied that by closing the public space may not be an effective way to avoid crowds. As we know that public space such as City Park is a need in urban life (Herman 2016). By closing the public space, it creates an empty space that become wasted. Some of public space functions cannot be carried out such us recreation and economic functions. Further, the crowd itself is proof that someone doesn't care about the safety of those around them. While the combination of social distancing and wearing a mask in public places can also make someone seem individualistic.

Compared to Robo-robo rituals and values, the implementation of health protocols phenomenon has similarities. Both have the same value in protecting people, but some behaviors have different values that are opposite each other, especially in kinship-individualism value. Based on this value's gap, the idea is to adapt values (inner level) from Robo-robo by analyzing their rituals (mid-level) and transform it into a new design solution that can be implemented in public space, in this case, Digulis City Park in Pontianak City.

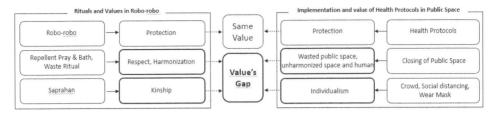

Figure 3. Value's gap between Robo-Robo and health protocol.

3.3 *Design solution: Protection system in City Park*

The focus of this section is to transform values from the inner level of Robo-robo by adapting the rituals (behavior) behind it then creating a scheme as user experience and designing some supporting products. Following the concept of *"Three Layers and Level of Culture and Design Features"* (Lin 2018), designing from the middle level of culture is using behavioral Design while considering the inner values that will be reflected in the interaction design (reflective design). The solution proposed is a protection system for City Park to prevent crowds through human consciousness.

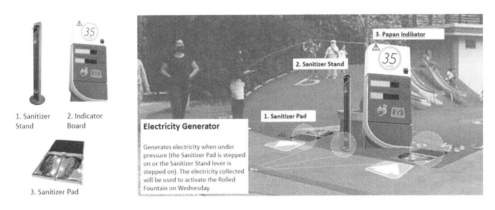

Figure 4. Protection system design in digulis City Park.

1. Protection Concept. The Sanitizer Pad functions to sterilize visitors' footwear, while the sanitizer stand will issue a hand sanitizer to clean visitors' hands. There are two sanitizer pads in each area, namely at the entrance and exit areas. Every visitor sets foot on the sanitizer pad at the entrance, it will automatically fill in the numbers on the indicator board and display the number of visitors in the area, as well as when a visitor steps on the sanitizer pad at the exit, it will automatically reduce the visitor indicator number.
2. Respect and Kinship Concepts. Under the sanitizer pad and sanitizer stand, there is an electric generator that will generate electricity whenever the generator lever is under pressure. Therefore, every time someone steps on the sanitizer pad, a certain amount of electrical energy will be generated, as well as when the sanitizer stand is used. The resulting collection of electrical energy will be stored, and the amount shown on the indicator board. The indicator board will display the number of visitors along with the "security" parameter visually and in audio. When the detected visitor reaches an unsafe number, the indicator board will give a warning in orange (caution) and red (dangerous). When the indicator is in this condition, a loud siren will sound as a warning for people in the area to leave the area. This system is intended to create a sense of care and awareness to share in this context, sharing the opportunity to enjoy the area of public facilities.

Figure 5. Interaction design scheme.

3. Harmonization Concept. The electrical energy collected will be used to activate the Digulis Fountain on Wednesday. This means a "reward" for people who are willing to cooperate following the health protocol when using the Pontianak Digulis Park public facility. This activation is carried out every Wednesday as a tribute to the Robo-robo ritual that inspired the creation of this protection system. It is hoped that people will continue to remember this cultural heritage even though they have not been able to feel it due to the COVID-19 pandemic.

4 CONCLUSION

Kinship, harmonization, and respect values from Robo-robo can be adapted for the design concept in public space in responding to the problems of social interaction in the new normal era. This study shows that traditional values can be used as design concepts in public spaces because they have the same social context. However, this concept needs further research especially in implementing it as a real prototype. More cultural examples and problem cases are needed too in order to get better results.

REFERENCES

Firmansyah, H., Putri, A. E., & Marisah, M. (2021). Implementasi Nilai Budaya Robo-Robo sebagai Penguat Pendidikan Karakter Peserta Didik di Kabupaten Mempawah. *Jurnal Basicedu*, 5(3), 1658–1666. https://doi.org/10.31004/basicedu.v5i3.962

Lampel, J., & Lant, T. (2005). *Towards a Deeper Understanding of Cultural Industries.* https://www.researchgate.net/publication/259760551

Lin, R. (2009). Designing "Friendship" into Modern Products. *Friendships: Types, Cultural, Psychological and Social . . . ISBN: 978-1-61668-008-4 Editor: Joan C. Toller*, 231.

Lin, R. (2018). Acculturation in Human Culture Interaction – A Case Study of Culture Meaning in Cultural Product Design. *Ergonomics International Journal*, 2(1). https://doi.org/10.23880/eoij-16000135

Napsiah, Sri Sanityastuti, & Marfuah. (2020). Perubahan Interaksi Sosial Acara Halal bi Halal pada Masa Pandemi Covid-19 di FISHUM UIN Sunan. *Jurnal Ilmu Aqidah Dan Studi Keagamaan*, 8. https://doi.org/10.21043/fikrah.v8i1.7633

Saripaini, S. (2021). Indigenous Counseling: Karakteristik Spiritual Dalam Tradisi Robo-Robo pada Masyarakat Kecamatan Sungai Kakap, Kalimantan Barat. *Jurnal Studi Agama Dan Masyarakat*, 17(2), 96–106. https://doi.org/10.23971/jsam.v17i2.3052

Utami, N. E. (2017). *Kerajaaan Mempawah Pada Masa Opu Daeng Manambon Tahun 1737–1761 DI Kabupaten Pontianak.*

Vivit Nurdin, B. (n.d.). *Rekayasa Sosial Budaya Dan Kearifan Lokal: Merubah Perilaku Pada Era New Normal Menuju Lampung Berjaya.*

Gender-based design translation: An actor-network theory of MKS Shoes

I. Resmadi* & R.P. Bastari
Telkom University, Bandung, Indonesia

R.A. Siswanto
Universiti Sains Malaysia, Penang, Malaysia

*ORCID ID: 0000-0003-2699-216X

ABSTRACT: MKS Shoes was established in 2013 in Bandung, Indonesia, as a manufacturer of women's footwear. Gender is an important factor in MKS Shoes' growth as a cultural asset in creative industries. This article investigated designing gender-based products as a socio-technical process rather than as a result of cultural and political studies. The lens through which the research for this study was conducted was Actor-Network Theory (ANT). Actor-Network Theory is used in this study to investigate MKS Shoes' translation procedures for human and non-human actors. Gender construction and negotiation as a design process is discussed in this article. Non-human actors play an important role in this study because the design process has historically focused on human/social product actors.

Keywords: Actor-Network Theory, design, fashion, gender

1 INTRODUCTION

The fashion sub-sector is the creative industry prominent in Bandung besides design, information technology (IT) sub-sectors, culinary, arts and crafts market, and performing arts (Gumilar 2015). Since the 1990s, the existence of the youth fashion industry in Bandung has been determined by the capacity of fashion designers to assimilate diverse influences on product innovation (Luvaas 2010). The growth of the fashion industry was also dominated by a highly sticky sort of streetwear associated with the youth culture, such as clothing apparel and distro (distribution outlet), which have a very powerful impact on graphic design works in Bandung's creative industry (Luvaas 2013). Bandung's creative industry will be stimulated by the emergence of items accompanied by footwear as a result of this fashion subsector's future growth. Nonetheless, the growth of shoe footwear products was not as dynamic as graphic clothing.

Product MKS Shoes caters to women, particularly those with careers in the creative industries. The connections between gender and the design process have a significant impact, especially when seen from a societal perspective. Gender concerns in apparel were evident in close-knit feudal societies with class stratification in the background of history (Akdemir 2018). At first glance, the relationship between gender and fashion appears to be quite strong in terms of power relations, social structures, and normative societal concerns. In addition, clothing may serve as a social status indicator, emblem, and marker (Kodžoman 2019). Several researchers have questioned whether gender has a true impact on fashion. Fashion trends that exist in the West will have a significant impact on awareness, as well as context, gender, education, and urban-rural distinctions in the choice of awareness attire. Children and young Sri Lankans have a fashion awareness that is largely impacted by external factors, including social and environmental culture (Rathnayake 2011). The connection between genders more numerous related aspects of consumption and preferences, such as age and gender, have additional implications on a consumer's choice to purchase a product or article of clothing (Ajitha & Sivakumar 2019). When viewed from the perspective of others, the link between gender and fashion reveals that awareness of gender differences is crucial for predicting the behavior and internal demands of consumers (Workman 2010).

Figure 1. MKS Shoes products are targeted at women in the creative industry.

A majority of the aforementioned studies emphasize just the limited connections between gender and fashion in terms of consumption, orientation, gender disparities, attitudes, and cultural stereotypes. There are few analyses of the design process to identify gender-fashion connections. According to Esfahani's Human-Centered Design (HCD)-based study, there is still a lack of user comprehension and awareness, and the design is highly prejudiced toward women. During this interaction between gender and design, a greater number of individuals focused on the issue of power relations. In addition, there is a paucity of gender research in the design process, particularly a failure to view design in a gender context as action-performative rather than only as solid and permanent conceptions. According to Esfahani, the designer plays a significant role as a crucial figure in the visual shape and physique of a product's outline, material, size, and color (Esfahani 2020). The larger perception designer will therefore avoid gender preconceptions. During the design phase, gender perception is essential.

The design process of numerous MKS Shoes also embodies active woman ideals, such as self-trust and empowerment. Gender has a substantial impact on the design process. Through this "gender script" perspective, product design will consider symbols, values, and culture in addition to the material design (Oudshoorn & Pinch 2003). This article will address whether the design process is impacted by gender in a social context (structure social, identity social, or norm social), as well as the relationship between gender and more fashion-oriented sociotechnical characteristics through series negotiation and translation. Thus, the process of character design is dynamic; the character will continue to negotiate and be performative, as opposed to being static and mute (Esfahani 2020). The design process of MKS Shoes will be examined in further detail when the problem's context has been revealed.

2 RESEARCH METHODS

This study employs qualitative research utilizing the Actor-Network Theory (ANT) methodology. We obtained information for this study from a combination of primary and secondary sources. During the observation phase for MKS Shoes, visual data were gathered. Fajar Sadika (Co-Founder of MKS Shoes) was interviewed to get information on the MKS Shoes design process. As part of our literature analysis in previous research, we analyze the relations between gender and fashion/design. As we get a deeper understanding of MKS Shoes via the perspective of ANT, we will find how the translation of human actors and non-human actors interact in MKS Shoes design process. The translation comprises *problematization, interessment, enrolment,* and *mobilization* (Yuliar 2009).

3 RESULT AND DISCUSSION

Use the ANT to study design by analyzing the relationship and formation of humans and non-human actors (Ekomadyo & Yuliar 2015). This article analyzes the translation process in the design process of MKS Shoes which involved *problematization, intervention, enrollment,* and *mobilization.*

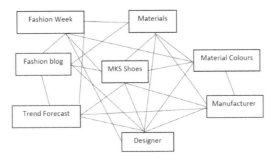

Figure 2. The actor's relation (human and non-human) in MKS Shoes design process.

Problematization

According to MKS Shoes' case study, a woman's need for clothing is based on market needs and values representation for women because youth clothing products (clothing apparel and distros) are more dominated by men. MKS Shoes began the design process to the fact that women in the industry have more creative market opportunities for fashion products that are often dominated by men (apparel clothing and distribution outlets/distro). The designer of MKS Shoes should see the opportunities to design the product based on the trend forecast internationally. But, on the other hand, the manufacturer has limited capability based on materials.

Interessement

Actors have a delegated role in determining whether they are interested in each program or activity at the outset of interest. The actors start involved in the design process of MKS Shoes. The designer Fitria Vidyawati (designer of MKS Shoes) and PT Kreasi Gudang Biru (manufacturer) are intertwined in the phase of actor human interest. Designing MKS Shoes involves non-human actors such as fashion blogs, trend forecasts, materials, and material colors. Blog WGSN.com (trend forecast and fashion blog) and other international fashion blogs, as well as information gleaned from other countries' fashion weeks, are some of the places designers turn for inspiration when creating MKS Shoes. Designing shoes is influenced by global fashion trends during the design process. After that, the designer will discuss with Kreasi Gudang Biru as a company manufacturing facility to interpret the possibilities to create the product. The designer will talk to Kreasi Gudang Biru about whether the process is able or not able to develop the product based on the designer's wants. They will discuss good design, material, and color. Designers and manufacturers must agree on whether they can create a design. In the context of women's fashion, quality garments and aesthetic aspects play a significant role in determining clothing quality (Gitimu et al. 2013).

Enrolment

At this stage, actors' commitment and involvement in the solution network begin to emerge. To make the design process successful, human actors and non-human actors must have the same initiation and objectives. In MKS Shoes, the design process is largely influenced by how the product is manufactured based on negotiation between the designer and manufacturer as human actors and fashion blog, trend forecast, fashion week, material, and materials color as non-human actors. To carry out the design process, there must be aligned with the needs of the designer and the manufacturer. As part of the design process, the designer and manufacturer have already established an intertwined relationship during the enrolment phase. MKS Shoes and PT Kreasi Gudang Biru are still part of the same holding company, making it easier for coordination and negotiations. For designers and manufacturers to work together effectively, they must have access to similar information and knowledge related to product materials. Designing products that are technical and can be manufactured by the company is necessary for a successful design process. As a result, the designer must have the ability to adapt the design to meet the needs of the manufacturing process.

Mobilization

At the mobilization stage, a network has been formed between human and non-human actors. The other actors involved do not need to be visible because they are directly represented by other actors. This MKS Shoes product has already become a finished product at this early stage. Designing MKS Shoes is a series of translations of design knowledge, technical production, material/material, and color. The MKS shoe product goes through a series of socio-technical translations instead of seeing the design process and gender as part of social structure and segmentation.

Table 1. Sociotechnogram of MKS shoes design process.

Human Actor	Non-Human Actor
Designer	*Fashion Blog*
Manufacturing Company	*Trend Forecast*
	Fashion Week
	Materials
	Material Colour

Source: Analyze by author.

4 CONCLUSION

The Actor-Network design process approach views the design process as a 'performative' process rather than gender as a "rigid" and "static" social aspect (identity, norm, and structure). According to the findings of the MKS Shoes case study, the design process of MKS Shoes is influenced by a series of translations and negotiations. Human actors (designers and company manufacturers) and non-human actors (fashion trends, blogs, materials, and colors) influence each other before becoming a finished product. In the context of ANT, translation and negotiation must be intertwined until they become one deal in the design process because if there is no same commitment between one actor and the other, neither the product nor the design will be successful. Design process success MKS shoes are a must as well intertwined the same translation among actors, both human actors and non-human actors.

ACKNOWLEDGEMENT

This research was funded by an external research grant program from Telkom University and MKS Shoes. We would like to express our heartfelt gratitude to all parties involved.

REFERENCES

Ajitha, S., & Sivakumar, V. J. (2019). The Moderating Role of Age and Gender on the Attitude Towards New Luxury Fashion Brands. *Journal of Fashion Marketing and Management*, 23(4), 440–465. https://doi.org/10.1108/JFMM-05-2018-0074

Akdemir, N. (2018). Deconstruction of Gender Stereotypes Through Fashion. *European Journal of Social Science Education and Research*, 5(2), 185. https://doi.org/10.26417/ejser.v5i2.p185-190

Ekomadyo, A., & Yuliar, S. (2015). Social Reassembling as Design Strategies. *Procedia - Social and Behavioral Sciences*, 184(August 2014), 152–160. https://doi.org/10.1016/j.sbspro.2015.05.075

Esfahani, B. K. (2020). Bridging gender and human-centered design: A Design Verification Study. *Procedia CIRP*, 91, 824–831. https://doi.org/10.1016/j.procir.2020.02.241

Gitimu, P. N., Workman, J., & Robinson, J. R. (2013). Garment Quality Evaluation: Influence of Fashion Leadership, Fashion Involvement, and Gender. *International Journal of Fashion Design, Technology and Education*, 6(3), 173–180. https://doi.org/10.1080/17543266.2013.815809

Gumilar, G. (2015). Pemanfaatan Instagram Sebagai Sarana Promosi Oleh Pengelola Industri Kreatif Fashion di Kota Bandung. *Jurnal Ilmu Politik Dan Komunikasi*, V(2), 77–84.

Kodžoman, D. (2019). The Psychology of Clothing: Meaning of Colors, Body Image and Gender Expression in Fashion. *Textile and Leather Review*, 2(2), 90–103. https://doi.org/10.31881/TLR.2019.22

Luvaas, B. (2010). Fashion and the Cultural Practice. *Visual Anthropology*, 26(1), 1–16. https://doi.org/10.1111/j.1548-7458.2010.01043.x.2

Luvaas, B. (2013). Third World No More: Rebranding Indonesian Streetwear. *Fashion Practice*, 5(2), 203–227. https://doi.org/10.2752/175693813x13705243201496

Priyatma, J. E. (2013). Potensi Teori Jejaring-Aktor untuk Memahami Inovasi Teknologi. *Seminar Nasional RiTekTra (Riset Dan Teknologi Terapan)*, 1–7.

Rathnayake, C. V. (2011). An Empirical Investigation of Fashion Consciousness of Young Fashion consumers in Sri Lanka. *Young Consumers*, 12(2), 121–132. https://doi.org/10.1108/17473611111141588

Workman, J. E. (2010). Fashion consumer Groups, Gender, and Need for Touch. *Clothing and Textiles Research Journal*, 28(2), 126–139. https://doi.org/10.1177/0887302X09356323

Yuliar, S. (2009). *Tata Kelola Teknologi: Perspektif Jaringan Aktor* (1st ed.). Penerbit ITB.

Descriptive analysis of graphic layout in interior design catalog

C. Chalik, Andrianto & A.S.M. Atamtajani
Telkom University, Bandung, Indonesia

ORCID ID: 0000-0002-4979-5620

ABSTRACT: Graphic layouts can be determined through the instincts of a graphic designer, and these instincts can be obtained with a lot of practice and experience in designing graphic works. In determining the layout, sometimes we also cannot rely on instincts and subjective desires alone, but also require an assessment of the readers of our work graphic layouts, developed according to tastes, trends, and technological developments. In the world of marketing, catalogs function as a medium for promoting the sale of goods and services that are usually provided to customers. Likewise, in the world of education, especially the design science family, the catalog functions as a media portfolio of works summarized within a certain period of time. The need for a catalog that is informative and communicative in conveying messages visually requires a catalog designer to understand the basics of layout and composition. The research method used in this study is qualitative content analysis. This research uses a qualitative content analysis method with the aim of discussing in depth about the layout content of the interior design catalog contained in the creative market site. This research begins with observing interior design catalogs as initial data in problem formulation, and then developed based on theories related to hierarchy, balance, harmony, proximity, and space. The research shows that the catalog analyzed uses all the basic rules of layout science, which consist of hierarchy, balance, alignment, proximity, and space. The results of this research are expected to be a reference or reference in designing layouts in design catalogs, especially interior design catalogs.

Keywords: Layout, Composition, Portfolio, Graphics, Interior Design

1 INTRODUCTION

Visual publications, in the form of catalogs, magazines, newsletters, and portfolios of works, both in print and digital media are a visual communication system that combines images and texts. There are many demands and also considerations from designers who design works in the format of visual publications because visual publications themselves are not only a single and stand-alone medium like a poster but consist of many pages. Each visual publication has a narrative structure related to the time dimension, such as duration and reading speed, as well as related to the space dimension, such as layout. In designing visual publications such as catalogs, designers are required to manage a large number of texts and images into one system of composition and layout, to create a cohesive whole. Designing visual publications such as catalogs, whether in print or digital format is not a simple or easy job, as easy as operating the basics of text and graphics processing software, but requires a designer's ability to analyze the reading flow and form a visual narrative that is comfortable and also fun to read by the reader. According to Rustan, Suriyanto in his book entitled basic layout and its application (2008) said that layout can be explained as the layout of design elements on a field in certain media to support the concept and message it carries. The layout is one of the processes or stages of work in design. It can be said that the designer is the architect, while the layout is the worker. However, the definition of the layout itself in its development has been very widespread and merged with the definition of design itself, so many people say that me-layout is the same as designing. This research will look at layout analysis in interior design catalogs. The unit of analysis in this research is the interior design catalog pages from the creative market website.

2 RESEARCH METHODS

This research uses a qualitative content analysis method to discuss in depth about the layout content of the interior design catalog contained in the creative market site. Bogdan and Taylor in Moleong (2012) define qualitative research methods as research procedures that produce descriptive data in the form of written or spoken words of observable human behavior. Azwar (2012) argues that qualitative research itself emphasizes its analysis of the deductive and inductive inference process and on the analysis of the dynamics of the relationship between phenomena observed by researchers, using natural logic. Activities included in this research include collecting raw data, recognizing the data that have been collected, grouping and selecting data, reviewing the selection results, and assembling data. The researcher himself understands that qualitative research is a research method that emphasizes more on collecting information, as well as processing information that is explanatory in the form of descriptions and explanations that describe a particular event. Several pages of the interior design catalog contained in the creative market website were analyzed by identifying visual elements, text, and composition.

3 RESULT AND DISCUSSION

Overall, the pages in the interior design catalog contained in this creative market site have used the basic principles of graphic layout and are found on almost all pages in this creative market interior design catalog. The basic principles of layout are also the same as the basic principles of graphic design. The reading order and reading flow in the hierarchy principle refer to the size of the elements. The main reason for placing the largest visual element or the largest text element in a composition is because the largest is the first to attract the eye. This hierarchical principle can be seen on the second and third pages of the creative market interior design catalog, where on the second page there is a photo illustration of the bed which is the largest visual element when compared to the visual elements and text contained on pages 2 and 3 so that the photo illustration of the bed on page 2 is read in the first order, then the headline that says welcome is read in the second order because hierarchically it is the second largest element. Then continued by the sub-headline and body text which are read in the third and fourth order.

Figure 1. Hierarchy by size. (Source: personal documentation 2022).

The second principle is balance. Intrinsically, something that is balanced and can be acted upon is usually balanced and tends to attract the human eye. Rahmat Supriyono in the book

Visual Communication Design Theory and Application (2010) says that balance is an equal division of weight, both visually and optically. The principle of balance can be seen on the tenth and eleventh pages contained in the interior design catalog of the creative market website, where the number of seating images on page ten located on the left and the number of seating images on page eleven located on the right is the same, namely 6 images for each page, the number of images contained on each page, and also the same size frame that limits each image makes the visual composition of the left and right pages balanced.

Figure 2. Left and the right balance. (Source: personal documentation 2022).

The third principle is harmony. The principle of harmony can create harmony in a graphic composition. By applying this principle, all visual elements in a composition will appear to be connected to each other. Harmony in a visual composition can be defined as the formation of elements of balance, order, unity, and also the combination of various visual and text elements that complement each other. The basic principle of alignment is also commonly applied to text and columns, the alignment that is usually used is left-aligned, center-aligned, or left and right-aligned to create a regular alignment. The principle of alignment can be seen applied on the fourth and fifth pages of the creative market interior design catalog, where all visual elements, such as illustration photos, headlines, sub-headlines, and body text, use justified or sometimes called left and right alignment, the application of left and right alignment on the left and right page illustration photo frames also use the same aspect ratio.

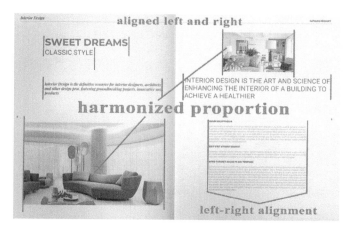

Figure 3. Harmonized proportion. (Source: personal documentation 2022).

The fourth principle is the principle of proximity or sometimes referred to as proximity. Proximity means how the relationship between visual and text elements of a design must provide a visual unity so that the proximity between these elements can show consistency in a composition. If there are several elements in a composition that are related, they should be placed close together. By applying the right proximity principle, it will avoid the clutter of elements in a composition and will certainly make it easier for readers to understand the intended design and what the designer intended. The application of the principle of proximity can be found on pages 20 and 21 of the interior design catalog from the creative market, where the photo illustration on page 20 goes beyond the page, crossing the page boundary to place the rest of the photo illustration on page 21. The distance or margin of the photo illustration on page 20 is only 0.30 cm adrift. The other proximity principle that can be found in the composition of pages 20 and 21 is the margin of the body text frame, which is 0.25 cm against the frame of the second illustration photo and 0.30cm against the first illustration photo. The proximity of 3 graphic elements in the composition of pages 20 and 21 consisting of illustration photo frame 1, illustration photo frame 2, and body text frame produces a connection between all elements.

Figure 4. Proximity. (Source: personal documentation 2022).

The fifth principle, space, like proximity, has a similar function, which is to help separate different elements, unite or bring similar elements closer together, and also organize visual and text elements according to the purpose and function of visual communication intended by the designer. One of the derivatives of the space principle that is commonly used in catalog design is white space. White space is the empty space between or around the layout design elements or on the page. The application of the principle of space in this creative market interior design catalog can be found on pages 8 and 9. This principle of space is needed in composition to give pause to the reader's eyes so that the reader's eyes are not too tired to see a design that is too dense. The more dense and overlapping visual elements and text will cause the reader's eyes to quickly feel tired and cause the perception of visuals that are read as bad visuals. The space between visual elements that are relieved also has an impact on the creation of a neat information structure, both from the elements of photo illustrations, headlines, sub-headlines, and also body text. The principle of space in the layout on page 9 on the right, uses wider columns and longer distances between paragraphs. This can be read as the designer's effort in applying the simple principle of space and has a big impact on making the catalog look more pleasant to read. Besides producing visuals that are comfortable to read, space in a visual composition can create a sense of balance. White empty space also represents the absence of material and emptiness, which is in line with the principle of space that must be empty, and there are no visual elements in it, this is in accordance with what was stated by Andrianto (2021) that color can provide an atmosphere and affect a person's mood, in this case affecting the mood and atmosphere of the reader of this interior design catalog.

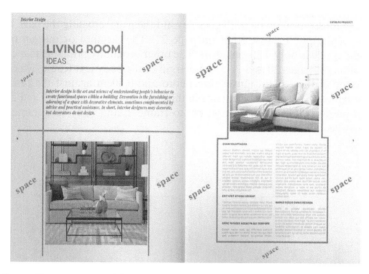

Figure 5. Space. (Source: personal documentation 2022)

4 CONCLUSION

The application of the basic layout principles in the creative market interior design catalog has covered all the basic elements of the graphic design layout principles. This is certainly a good design standard, considering that creative market is a website that provides design template services. Inevitably, the templates sold on the creative market website must be aesthetically pleasing but also in accordance with the rules and ethics of design, one of which is of course the basic principles of layout itself, where there are basic layout principles such as hierarchy, balance, harmony, proximity, and space. Broadly speaking, the layout in the interior design catalog contained in this creative market site applies balance, order, and harmony to the text and visual elements contained on each page. This can be interpreted that the function and purpose of this interior design catalog layout are in line with its basic rules. That is to attract the attention of the reader, giving a comfortable impression to the reader, which can be seen from the layout of visual and text elements. The interior design catalog design from the creative market website can also be used as an example of how to create an attractive and communicative interior design catalog.

REFERENCES

Andrianto, Chalik C. (2021). Perancangan Pembatas Interaksi sebagai Penunjang Kegiatan Bertransaksi di Kasir pada Masa New Normal. *Waca Cipta Ruang: Jurnal Ilmiah Desain Interior*, 7(1), 46–50.
Azwar, Saifuddin. (2012) *Metode Penelitian*. Yogyakarta: Pustaka Pelajar.
Moleong, Lexy J. (2012) *Metodologi Penelitian Kualitatif*. Bandung: PT. Remaja Rosdakarya.

Rakhmat Supriyono, Maria Agustina S. (2010) *Desain komunikasi visual: teori dan aplikasi*. Yogyakarta: Andi.
Rustan, Surianto (2008) *Layout, dasar & penerapannya*. Jakarta: Gramedia Pustaka Utama.
Sanyoto, Sadjiman E. (2005) *Dasar-dasar Tata Rupa & Desain*. Yogyakarta: Arti Bumi Intaran.
Sunaryo, Aryo. (2002) Paparan Perkuliahan Mahasiswa Nirmana 1. Semarang: *Jurusan Seni Rupa Unnes*.

Constructing meaning on Tahilalats comics based on Instagram comments section

P. Aditia & A.N. Fadilla
Telkom University, Bandung, Indonesia

ORCID ID: 0000-0002-5704-4990

ABSTRACT: Tahilalats is a comedy comic strip written and drawn by Nurfadhli Mursyid. This comic was first published on Twitter and Instagram in 2014. Interestingly, some of their meanings lead to multiple interpretations among readers. This research is qualitative research that first groups the Tahilalats comic strips that are considered "multi-interpreted" as well as plot twists, then examines the comments section and looks for how meaning is built from these comments. This research employed the methods of observational, interviews and literature study data collection. The results indicate that the plot twist or multi-interpretation portion in the Tahilalats comic strip. Based on interviews with informants, none of the meanings incorporated into netizen comments are deemed the most correct, yet it is precisely the pluralism of meanings that makes the Tahilalats comic strip not just interesting, but also interactive.

Keywords: comic strip, commentary, plot twist, pluralism of meaning, Tahilalats

1 INTRODUCTION

According to Franz and Meier (1994: 55), a comic is a story that emphasizes motion and action that is portrayed through a sequence of images created specifically with a blend of words. Scott McCloud (2003:9) provides a more comprehensive definition of comics as "a juxtaposed images that serve to convey information or create an aesthetic response from the audiences." All story texts in comics are neatly arranged and interconnected between images (visual symbols) and words (verbal symbols). A comic's images are static images that are organized sequentially to make a story.

In this study, the object to be discussed is the Tahilalats comic strip. Before we get into Tahilalats comic itself, let us briefly define what a comic strip is. A comic strip is a set of images, usually cartoons, arranged by a related panel to illustrate a humorous theme. Comic strips initially appeared in newspapers or magazines (Eisner 2008). Meanwhile, Tahilalats is a comic strip written and drawn by Nurfadhli Mursyid. The comic was first published on Twitter and Instagram in 2014.

As of this writing, Tahilalats has 4.9 million followers on Instagram and has uploaded 3,000 contents. In several comic strips displayed by Tahilalats, there is a characteristic panel with a minimum number of four or by using a carousel technique that contains up to a maximum of ten feeds. Interestingly, not all the Tahilalats' meanings are straightforward. Many of them result in multiple interpretations among readers, as evidenced by the numerous comments that attempt to capture the meaning of each content that is deemed confusing. This online comic is about everyday life and features a "plot twist" that is tough for readers to guess.

2 RESEARCH METHODS

This study was conducted to investigate the meaning generated by Tahilalats comics through various comments in the Instagram comments column. This study will first categorize Tahilalats comic strips that are "multi-interpreted," as well as plot twists, and then analyze the comments area to determine how meaning is built from these comments. Meanwhile, the chosen comments are those that rank first and second in terms of the number of likes received. As a result, this study is qualitative because it deals with meaning rather than providing a certainty that can be calculated positivistically. This study collects data through observation, interviews, and literatures review.

Table 1. Categorization of Tahilalats comic strips and selected comments.

No.	Visual	Title	Selected comments
(1)		Jual	baaaiiimmm Tahu gitu borong tahu... namanya masa depan tidak ada yang tahu.. haifinsa Masa Depan Tidak Ada Yang Tahu, makanyya mahal.. Thanks me later.
(2)		Grup	ste.ady BUT IN THE END, IT DOSEN MATTER reza_arshavin Banyak variasinya thanks to extra panel
(3)		Surat	dickimahardika sebarkan, jangan berhenti di kamu deapypy Jangan berhenti di kamu
(4)		Dunia	pepekomik Siapa yg minta bumi jadi super saiyan! tonysparkfigures Some men just want to see the world burn...

The title of the Tahilalats comic along with the comments taken for this research are found in Table 1.

Meanwhile, the informants for this study were taken from the creator of the Tahilalats comic, Nurfadli Mursyid himself, and numerous netizens chosen using purposive sampling criteria. Criteria for the selected informants are (1) loyal readers of Tahilalat comic and (2) reader of the comment column on Tahilalat comics. For around thirty minutes, all these informants were interviewed online using the Zoom application. Except for the comic creators, the confidentiality of all informants is guaranteed, thus in this instance it will only be written with the code "case ID." The gathered data will next be subjected to analysis.

To strengthen the analysis, this research will also be linked with the idea of plot twist. Plot twist happens near the end of a story, and it may change the reader's perception of the preceding events, or introduce a new conflict that is placed in a different context (Singleton *et al.* 2000). Lehrer (2011) says that revealing a plot twist to readers or viewers in advance is commonly regarded as a spoiler since the effectiveness of a plot twist usually relies on the audience not expecting it.

3 RESULT AND DISCUSSION

According to observations of selected comic strips and comments, Tahilalat's material does not clearly convey the meaning of each content and frequently ends the panel by "hanging." The author sees a gap between panels three and four, which he considers to be a "multi-interpretation." In comic strip (4), for example, the panel concludes with a picture of the planet being burned after an image of a person pleading and saying, "Aku ingin impian semua orang di dunia terwujudkan." (I want everyone's dream to come true). Similarly, the comic strip (3) which shows the image of a person opening a letter in the third panel, without words and immediately showing a worried expression, continues by going to the photocopy keeper and saying: "Mas, photocopy 100 lembar." If parsed together with comic strips (1) and (2), then the analysis can be grouped into the table below, especially with the relation between panel 3 and 4.

Table 2. Relationship between panel 3 and panel 4.

No.	Panel 3	Panel 4
(1)	A time machine inventor showed an excited expression as he said: "Lalu beli banyak emas dengan harga murah, dan kujual kembali di masa depan dengan harga tinggi hahahahaha-haha" (Buy a lot of gold at a low price, and resell it in the future at a high price hahahaha-hahaha")	The inventor of the time machine watched the news on television which declared: "Harga tahu dunia kembali mengalami kenaikan pada pekan ini." (The price of tofu in the world has increased again, this week.)
(2)	A man who is considered to be a lecturer is looking at a student's final project.	The student was declared the best graduate.
(3)	Image of person opening letter and showing anxious expression.	The person's said to the clerk, "Mas, photocopy 100 lembar."
(4)	Image of people pleading and saying: "I want everyone's dreams to come true."	Image of the earth being burned

If it is related to the first and second panels of the entire comic strip, it can be seen that the first to third panels show a more or less linear plot, a storyline that can be understood with a linear framework. But something strange happened when It got to the fourth panel. The plot twist (playing the story) can also be considered of as a juxtaposition or comparison of the two dissimilar objects (objects in this case can be interpreted as the third and fourth panels). According to the author's analysis, this is when the Tahilalats comic becomes difficult to understand. The netizen's comments were then entered as though they were between panels three and four, to explain panel four as the ending panel.

This is implicitly justified by the creator of Tahilalats, Nurfadli Mursyidin, who stated in an interview that Tahilalats was designed in such a way that there is no clear explanation in the plot. As a result, the last panel always depicts an unexpected event, which is non-linear with the first to third panels. The question is, was it all planned from the start? Mursyid stated that everything was planned from the beginning, but he had no idea that netizens' remarks would eventually become a distinctive element to explain the fourth panel. In its development, Mursyid has come to enjoy the netizens' comments and even on some content, he has no intention of inserting a special meaning in the fourth panel and leaves the netizens to build their own interpretation.

Meanwhile, based on interviews with the four informants coded with the term "case ID", the result is as follows:

Table 3. Interviews with informants and significant statements.

Code	Significant statement
Case ID #1	"I never tried to understand the Tahilalats comic because I was more entertained by the understandings of netizens in the comment's column."
Case ID #2	"I often have my own understanding of what is described in Tahilalats, but I keep reading the comments just to be sure."
Case ID #3	"Netizens' comments are sometimes funnier than the comics."
Case ID #4	"I think the Tahilalats comic has succeeded in making netizens provoked to comment and provide explanations. It is this interactive nature that makes Tahilalats comics popular."

The questions posed to the four informants were broadly related to the relationship between the multiple interpretations of the Tahilalats comic and the comments column on Instagram. The four informants' answers, though varied, boil down to one point: the comment column provides a significant meaning construction in interpreting the Tahilalats comic's content. However, none of these meaning constructions can be declared to be the most right because the creators have never confirmed them. What is more, if it's related to the interviews with creators, the netizen's opinions have resulted in a variety of "new meanings."

Based on discussion above, show results as follows:

1. *Plot twist* or the multi-interpretation section in the Tahilalats comic strip derives from a linear structure developed from panels one to three, then the fourth panel / panel conclusion is made in such a way that it deviates from the linear framework. In other words, the author creates a contrast between panels three and four.
2. The creator originally did not have the intention to invite netizens' comments to the interpretation stage, but in its development, the netizen's interpretation can give more value to the meaning of the Tahilalats comic strip.
3. Based on interviews with the informants, none of the meanings built in netizen comments are considered the most correct, but it is precisely the pluralism of meanings that makes the Tahilalats comic strip not only interesting, but also interactive.

4 CONCLUSION

Conclusion based on research above is that Tahilalat builds interaction with its reader through multi-interpretative content. Thus, Tahilalat as a visual narative open for various interpretation written in the comment section.

REFERENCES

Agnes, Tia. 2016. *"Tak Bisa Ikuti Tren Manga, Tahilalats Ciptakan Gaya Sendiri."* Detikhot. Retrieved June 21, 2022 (https://hot.detik.com/art/d-3142356/tak-bisa-ikuti-tren-manga-tahilalats-ciptakan-gaya-sendiri).

Eisner, Will. 2008. *Graphic Storytelling and Visual Narrative*. New York: W. W. Norton.

Franz, K., and B. Meier. 1994. *Membina Minat Baca*. Bandung: PT. Remaja Rosdakarya.

McCloud, Scott. 2003. *Understanding Comics: [The Invisible Art]*. New York: Harper Perennial.

Putri, Ananda Widhia. 2017. *"Nurfadli Mursyid, Kreator Komik Strip @tahilalats."* SWA. Retrieved June 21, 2022 (https://swa.co.id/youngster-inc/self-employed/nurfadli-mursyid-kreator-komik-strip-tahilalats).

Radja, Aditia Maruli. 2016. *"Bincang-Bincang Bersama Komikus Tahilalats Nurfadli Mursyid."* Antara News. Retrieved June 21, 2022 (https://www.antaranews.com/berita/547515/bincang-bincang-bersama-komikus-tahilalats-nurfadli-mursyid).

Singleton, R.S., Conrad, J.A., Healy, J.W. 2000. *Filmmaker's Dictionary*. Le Eagle Pub. Retrieved from https://books.google.co.id/books?id

Digital handmade: A craftsmanship shifts in block printing surface textile design

M.S. Ramadhan* & A. Widiandari
Telkom University, Bandung, Indonesia

*ORCID ID: 0000-0003-0860-4864

ABSTRACT: Block printing is one of the surface textile design techniques which is generally handmade. The process, which is quite time-consuming and requires high hand skills in making printing plates, is often an obstacle for creators to express their ideas. But now, digital technology's development has an opportunity to contribute to producing alternatives to printing plates significantly. Therefore, this study aimed to determine the potential use of block printing plates made with 3d printing technology. This research uses the qualitative method of conducting literature studies, practitioner interviews, direct observation, and explorations by testing several variables. As a result, by comparing some of its visual and technical aspects, the PLA plate made by 3d printing has the potential to be used in block printing techniques. It makes a craftmanship shift to produce contemporary craft textile and fashion products.

Keywords: block printing, digital, handmade, textile

1 INTRODUCTION

It is a necessity that every aspect of human life today will be touched by digital technology. In the realm of creative industries, the role of digital technology can accelerate the process of production and consumption of human works or thoughts. Advances in digital technology also significantly affect the creative industry at the level of institutions, business models, and the creative process itself (Mangematin et al. 2014). Highlighting the creative process in the textile craft and fashion field, we can say that digital technology is still not widely used at the production stage. Allegedly, the tradition of hand skills carried out by textile artisans or designers is still very strong.

Some textile surface design techniques rely on hand skills to produce incredible textile decorations such as embroidery, block printing, etc. For example, in the block printing technique, designers use carved wood and color with ink, then applied to the surface of the fabric by pressing it by hand (Ganguly & Amrita 2013). The wooden blocks used are carved following the shape of the motif by hand using steel chisels or wood cutting machines (Seidu 2019). In addition to wood, conventional block printing generally uses copper or linoleum material as a printing plate. As in the fashion product collection created by Sri Puspitawati in 2019, she uses traditional block printing techniques to provide decorative motifs inspired by Baduy culture on the fabric sheets used. Puspitawati uses linoleum printing plate material in the design process, carving it manually using a chisel and then printing it on sheets of Baduy woven fabric (Puspitawati & Ramadhan 2019).

Figure 1. Block printing motifs on Tinun collection. (Source: Puspitawati & Ramadhan 2019).

However, although the image produced by traditional printing plates is quite successful, the effectiveness and efficiency of using these materials are still lacking because it takes a long time and requires high craftmanship skills in the making process. So that deficiency creates opportunities for creators to look for alternative materials and techniques by taking advantage of technological advances. One of them is using 3d printing technology, which the development in recent years has been very rapid, supported by new software designs, new materials, innovation, and the internet (Lipson & Kurman 2013). So that encourages this technology can print scaled models quickly and in detail (Putra & Sari 2018). Previously in Indonesia, some research has been conducted on using 3d printing technology in the textile craft and fashion field. Still, generally, the study only focuses on the final product, such as clothing and accessories. In this research, further utilization of the potential of 3d printing technology is determined to produce block printing plates that can apply a form of motif to the fabric. This research focused on comparing several visual and technical aspects of printing plates made manually or with 3d printing technology.

2 RESEARCH METHODS

The method used in this research is qualitative, with data collected from several techniques, including observation, interview, and exploration approach. First, the authors interviewed Ammar Farhan Rayudi, an owner of the AF 3D Lab in Samarinda, Indonesia, who is also a member of the 3d Print Indonesia association, to obtain technical information about 3d printing techniques and their potential in the realm of textile and fashion crafts. Based on interviews and observations conducted in his place, Rayudi said that Indonesia has a lot of possibilities to develop 3d printing technology in various fields. However, Rayudi added that 3d printing technology in the fashion sector is still sparse, and there are still many opportunities for development because so far, the products produced are accessories such as necklaces, earrings, bags, and shoes. On the technical aspect, Rayudi said that 3d printing can generally uses several kinds of filament. Still, he suggested using Polylactic Acid (PLA) filaments, which are safer, easier to use, and have minimal risk due to size shrinkage compared to other filaments.

To discover more about the potential use of 3d printing technology in making block printing plates, the authors collected data with an exploration approach to the printing plate material used in this study. This approach intends to obtain factual data from the results of making printing plates to understand the potential, strengths, and weaknesses of the variants of printing plates to produce image on fabric. The authors divide the exploration stages: (1) Initial exploration: making printing plates. At this stage, manually and digitally make printing plates using linoleum rubber, Medium Density Fiberboard (MDF) wood, and PLA materials. Linoleum rubber and MDF wood represent the general material used as printing plates on block printing. While PLA as a potential material to be compared with the two previous. (2) Advanced exploration: printing plates on fabric materials. Print the results of making plates on fabric materials using the block printing technique at this stage. The authors used cotton fabric materials as a fixed variable to get a clear and comparative print result from the three kind of printing plates.

3 RESULT AND DISCUSSION

3.1 *Initial exploration results*

The initial exploration stage objective is to produce a printing plate that can apply a motif image on a sheet of fabric using the block printing technique. In this initial exploration, the authors used three materials: linoleum rubber, MDF wood, and PLA. Linoleum rubber and MDF wood are processed by hand carving, while PLA materials use a 3d printer. The result at this stage is a printing plate module which is then compared based on several variables, including (a) design compatibility, (b) manufacturing time, and (c) level of craftsmanship required.

Table 1. Block printing plates and its technical comparation.

Image design	Printing plate	Method	(a)	(b)	(c)
	Linoleum rubber	manual hand carving	high	medium	medium
	MDF wood	manual hand carving	high	high	high
	PLA	digital 3d printing	high	low	low

Note: (a) Design compatibility, (b) Time consumption, (c) Craftsmanship level.

Based on the performance parameters of the printing plate made of three different materials and techniques, all of them can produce a pattern matching the design. However, the MDF material requires time and a high level of craftsmanship compared to linoleum which needs medium time, and PLA, which only requires less time than other materials. So, it can be concluded that in making block printing plates, the 3d printing technique with PLA material has a reason for creating an optimal printing result but with low production time consumption and does not require high craftsmanship skills. Furthermore, in using PLA filaments for printing plates, there is no significant size reduction found from the printed plate, as stated by Rayudi in the previous interview session.

3.2 Advanced exploration results

The following exploration stage is printing the plates on the fabric. This exploration aims to determine how successfully the printing plate transfers the ink used to produce an image on the fabric according to the motif design. The authors used offset ink and cotton fabrics as a fixed/control variable in each experiment at this stage. The ink is applied to the printing plate using a rubber roller and then manually printed on the fabric by hand. The results at this stage are patterned fabric sheets with block printing techniques which the printed images will then compare based on several variables, including (a) image shape, (b) image texture, and (c) print area cleanliness.

Table 2. Print results of block printing plates and its visual comparation.

Printing plate	Print result on fabric	(a)	(b)	(c)
Linoleum		high	low	medium
MDF wood		high	low	low
PLA		high	medium	high

Note: (a) Image shape, (b) Image texture, (c) Print area cleanliness.

Based on the achievement parameters of the printed plate, the result is that the three materials can produce an image matching the pattern on the printing plate. Furthermore, in the printed image using PLA, there is a medium amount of texture, in contrast to the image from linoleum and MDF printing plates which tend to be less. However, in terms of cleanliness of the print area, the PLA printing plate has the highest yield compared to the other two materials. So, it concluded that in this further exploration, PLA printing plates made with 3d printing techniques could produce images comparable to the other two materials and the resulting texture. Moreover, 3d printing plate printing results have a high level of cleanliness in the print area.

4 CONCLUSION

This study comparing visual and technical aspects of printing plates digitally made by 3d printing technology and manually made by hand from linoleum rubber and MDF wood. PLA-type filament is used to produce 3d printing plates; as stated by Rayudi in the interview session, the PLA-type tends to be safer, easier to use, and has minimal risk due to size shrinkage. From the exploration approach of this study, the PLA 3d printing plate has optimal pattern results but a reasonably low production time and does not require high craftsmanship skills to make it. Furthermore, this material can produce a printed image comparable to linoleum rubber and MDF wood, coupled with the resulting texture and print area, which tends to be cleaner. So based on several advantages of these technical and visual aspects, we can conclude that 3d printing technology has a lot of potential in making block printing plates to produce contemporary textile and fashion products. Involving the latest technology is a form of revolution in textile craft and fashion. It encourages us to conduct further research by applying this motif decoration technique, which can shift craftsmanship value.

REFERENCES

Ganguly, D. & Amrita. 2013. A Brief Studies on Block Printing Process in India. *Man-Made Textiles in India*, 41(6), pp. 197–203.

Lipson, H. & Kurman, M. 2013. *Fabricated: The New World of 3D Printing*. Indianapolis: John Wiley & Sons, Inc.

Mangematin, V., Sapsed, J. & Schüßler, E. 2014. Disassembly and reassembly: An introduction to the Special Issue on digital technology and creative industries. *Technological Forecasting and Social Change*, 83, pp. 1–9. doi: https://doi.org/10.1016/j.techfore.2014.01.002.

Puspitawati, S. & Ramadhan, M. S. 2019. Pengaplikasian Teknik Block Printing dengan Inspirasi Motif dari Kebudayaan Suku Baduy. *ATRAT: Jurnal Seni Rupa*, 7(3), pp. 205–214. Available at: https://jurnal.isbi.ac.id/index.php/atrat/article/download/925/695.

Putra, K. S. & Sari, U. R. 2018. Pemanfaatan Teknologi 3D Printing Dalam Proses Desain Produk Gaya Hidup. In David et al. (eds), *Seminar Nasional Sistem Informasi dan Teknologi Informasi (SENSITEK) 2018*. Pontianak, pp. 917–922.

Seidu, R. K. 2019. The art produced by substitute surfaces in hand block printing. *Research Journal of Textile and Apparel*. Emerald Group Publishing Ltd., 23(2), pp. 111–123. doi: 10.1108/RJTA-08-2018-0047.

Eyeglasses as a functional fashion accessory for generation Z in urban Indonesia: Recommendation for sustainable development

E. Mayor & I. Santosa
Institut Teknologi Bandung, Bandung, Indonesia

ORCID ID: 0000-0002-4818-9776

ABSTRACT: Eyeglasses are continually being improved in terms of functionality and aesthetics. Based on eyeglasses development, cultural changes have occurred. Behavior and social norms towards eyeglasses have shifted. The current fashion industry development, increase in eyeglasses wearers, accelerated digitalization, and consumptive behavior that resulted in an increased demand for eyeglasses indicate that further research is needed on this issue. Consumerist culture is also wasteful and unsustainable; therefore, a cultural shift from consumerism to sustainability is needed. The eyeglasses industry must develop solutions to create a sustainable urban culture. Generation Z is consumptive consumers who are becoming major consumers. They utilize digital devices daily, and myopia is becoming more common among them. The study is conducted to identify the factors that motivate Generation Z to buy eyeglasses as a functional fashion accessory and evaluate the contributions of eyeglasses toward sustainable development. The research is quantitative, with the data collection process using a survey strategy and a questionnaire instrument. The research results will be in the form of a study that recommends the best approach for the eyeglasses industry to attract Generation Z and the most suitable sustainable development approach to be applied to the urban Indonesian eyeglasses market.

Keywords: eyeglasses, generation z, sustainable development, urban culture

1 INTRODUCTION

Technological advances have positively impacted human life, and various technologies were created to replace or enhance various organ functions. For example, refractive errors in the eye can be overcome by using eyeglasses (Novitasari 2019). Eyeglasses were initially used as a visual aid, but it is now common for people with perfect vision to wear them due to fashion trends, giving the wearer of glasses a more fashionable look (Steele 2005). Eyeglasses are one of the few tools for disability that have become a fashion statement (Novitasari 2019; Pullin 2009).

It is predicted that by 2050, around half of the global population, or 4.8 billion people, will need corrective lenses (Agmasari 2016). This number had doubled compared to 2010, when the global population who wore eyeglasses was only 28.3% (Agmasari 2016). Grand View Research (2021) estimates that in 2021, the worldwide eyewear market size valued at USD 157.9 billion and has a predicted compound annual growth rate of 8.4% from 2022 to 2030. In addition, Euromonitor International (2021) reported that the increasing incidence of myopia among young Indonesians contributes to the increasing demand for eyeglasses, and the tendency for Indonesians to experience myopia at a younger age is showing no signs of slowing down. The COVID-19 pandemic has accelerated digitalization, which in turn affects the eyeglasses market. The trend of remote work and virtual meetings are likely to continue (Lund *et al.* 2021). As a result, more people will need glasses to prevent eye fatigue and correct vision since they will spend more time using their electronic devices, which affects their vision and increases the risk of developing computer vision syndrome (CVS) (Grand View Research 2021). The growth in the eyewear market size is also due to the shift in the perception of people who wear glasses from being frequently associated with a smart child to being seen as a cool, stylish, and sociable in the student environment (Oktaviani & Gautama 2020). Oktaviani & Gautama (2020) state that appearance is crucial for Generation Z, especially in their social group. Eyeglasses have become communication tools that people wear to convey their identity (Morgan 2019). According to Steele (2005), eyeglasses have fully evolved from being a practical necessity into a fashion accessory, becoming a medium for design, individual expression, and personal appearance enhancement. Another factor causing the growth of the eyeglasses market size is the change in lifestyle and the rise of consumerism. Generation Z is consumptive consumers who utilize digital devices daily (Simangunsong 2018). People who wear eyeglasses anticipate buying new eyeglasses as a chance to try something new and reinvent themselves (Pullin 2009). However, Assadourian (2010) argues that a cultural shift from consumerism to sustainability is needed to maintain stability as the consumptive consumption patterns are unsustainable. In addition, many people in the technology sector are developing new solutions to address global environmental concerns, making sustainability in the eyewear market a reality (Moskal 2021).

The current fashion industry development, increase in eyeglasses wearers, accelerated digitalization, and unsustainable consumptive behavior that resulted in an increased demand for eyeglasses indicate that further research is needed on eyeglasses as a functional fashion accessory. Therefore, the research objectives are to identify the factors that motivate Generation Z to buy eyeglasses as functional fashion accessories, evaluate the contributions of eyeglasses toward sustainable development, and determine the best approach for the eyeglasses industry to attract Generation Z in a sustainable way. Furthermore, the research findings can become a reference for eyeglasses manufacturers and brands in the fashion industry in terms of their design and approach to attract Generation Z in a sustainable way.

2 RESEARCH METHODS

This research is quantitative, an approach that examines the relationship between variables to test objective theories (Creswell 2014). These variables can then be measured using instruments, resulting in numbered data that can be analyzed using statistical procedures; and this approach is deductive, objective, and can generalize and replicate findings (Creswell 2014). This study uses a survey method. The survey approach seeks to describe quantitatively the attitudes, opinions, and trends of Generation Z in urban Indonesia (Fowler 2009).

This research consisted of two systematic stages. The first stage is a review of materials related to research variables. Literature reviews and references are taken from books, previous research, and journal writings. The aim is to collect data as a theory supporting the research. The second stage is collecting data using a survey strategy and using a questionnaire instrument. The aim is to collect data about Generation Z's behavior towards eyeglasses as a functional fashion accessory and the contributions of eyeglasses toward sustainable development. The research was conducted online through Google Forms and used purposive sampling. The respondent criteria include wearers of prescription glasses, Indonesian citizens living in Jakarta as Jakarta is Indonesia's largest city with 10 million inhabitants (O'Neill 2021), and ages 15–26 years as Generation Z was born between 1995–2012, and young adults are 15–30 years old (Stillman & Stillman 2017; Yarlagadda *et al.* 2015). This study collected 63 respondents (32 female and 31 male).

Table 1. Data about respondents.

Indicator	Result
Gender	Female (50.8%), male (49.2%)
Age	20–22 (66.7%), 15–19 (17.5%), 23–26 (15.9%)
Frequency of buying eyeglasses	Every year or less (34.9%), once every 3 to more years (34.9%), once every 2 years (30.2%)

3 RESULT AND DISCUSSION

3.1 *Eyeglasses as a functional fashion accessory*

Based on the history of eyeglasses written by Steele (2005), it can be summarized that the factors that caused eyeglasses to become a functional fashion accessory include customer needs (the need for prescription glasses), eyeglasses design (fashionable design and a variety of glasses styles), media (influence from magazines and celebrities), and the fashion industry (eyeglasses brands promote eyeglasses as a fashion accessory, fashion designers design eyewear lines, and fashion trends). The eyeglasses frame design, quality, and price are three important factors influencing online purchases (Ratanasirintrawoot 2019). Oktaviani and Gautama (2020) found that the primary reasons students wear glasses as fashion accessories are the design of glasses with diverse shapes, the relatively low pricing for glasses, and the variety of colors offered. Other factors include location (accessibility to eyeglasses store) and customer service (Chu *et al.* 2017; Grand View Research 2021; Pullin 2009). In today's era, social media has become an effective tool to attract new consumers, especially those who want to follow current fashion trends (Euromonitor 2021).

Motivation refers to the reason behind someone's actions (Cherry 2022a). It is the process that starts, directs, and sustains goal-oriented behaviors (Cherry 2022a). According to Maslow's hierarchy of needs theory, motivation results from an individual's attempt to meet five basic needs, with physiological needs being at the first level, followed by security and safety, social, esteem, and self-actualization (Cherry 2022b). Eyeglasses fulfill a security and safety need, health and wellness, where eyeglasses wearers need them to see clearly. Humans being social creatures, need to feel loved and accepted by people, and eyeglasses as a functional fashion accessory give a sense of belonging in an individual's social group. Eyeglasses

Table 2. Factors that motivate generation Z to buy eyeglasses as a functional fashion accessory.

Factors	Average point	Factors	Average point	Factors	Average point
Customer needs	6,2	Price	5,3	Fashion trends	4,1
Quality	6,1	Customer service	4,3	Location	4,1
Eyeglasses design	5,7	Brand	4,2	Media	3,3

also satisfy an esteem need, self-esteem, as eyeglasses design promotes self-confidence. Through the survey carried out by the author, factors that motivate Generation Z to buy eyeglasses as a functional fashion accessory were identified as follows:

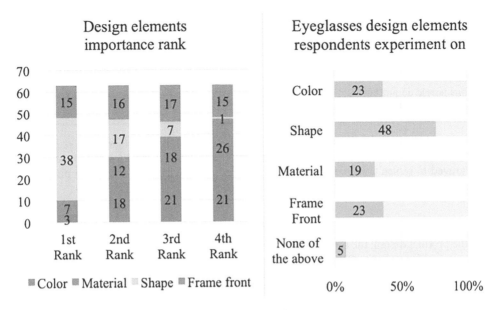

Figure 1. Diagram of data related to eyeglasses design elements.

Factors were tested on a 7-point Likert scale (1: Unimportant, 2: Low importance, 3: Slightly important, 4: Neutral, 5: Moderately important, 6: Very important, 7: Extremely important), and it was found that customer need (the need for prescription glasses) (6.2) is the most important factor, followed by quality (6.1), eyeglasses design (5.7), and price (5.3).

Eyeglasses design is one of the most important factors that motivate Generation Z to buy eyeglasses, and there are several design elements. The survey found that shape is the most

Table 3. Eyeglasses frame design preference.

Indicator	Result
Material	Metal (63.5%), plastic (55.6%), high-performance composites (36.5%)
Color	Neutral colors (87.3%), metallic (36.5%), pastel (9.5%)
Shape	D-frame (58.7%), round (27%), wayfarer (25.4%), square (22.2%), geometric (11.1%)
Frame front	Full rim (66.7%), half rim (50.8%), rimless (19%)

important design element for Generation Z when purchasing eyeglasses and is the element they like to experiment on the most. On the other hand, frame material is the least important design element when purchasing eyeglasses and is the element they like to experiment on the least. Data of respondents related to eyeglasses frame design preference with high percentages are as follows:

3.2 *Contribution of eyeglasses to sustainable development*

Through the survey, the contributions of eyeglasses to sustainable development were analyzed and evaluated. The eyeglasses industry in Indonesia is not sustainable yet, as 71.4% of the respondents believe that the industry is unsustainable. However, there are several sustainable developments that Generation Z has seen in the Indonesian eyeglasses industry. Many of the sustainable development in the Indonesian eyeglasses industry focuses on the frame material, where most of the respondent's state that the industry is offering more sustainable material options, including recycled materials and wood. There is also sustainable development in the transportation and distribution sector as local brands increase in

Table 4. Purchasing behavior towards sustainable products.

Indicator	Result
Reasons people buy sustainable products	To minimize production waste (52.4%), make the environment better (50.8%), lower carbon footprint (33.3%).
Reasons people do not buy sustainable products	More expensive (60,3%), perceived as less effective products (30,2%), seen as a sacrifice (12.7%)

popularity. The carbon footprint is reduced because these brands are manufactured in local factories. There is also sustainable development in the social impact sector, where people are encouraged to return their old frames to be recycled and get cashback. Data on purchasing behavior towards sustainable products are as follows:

3.3 *Best approach for the eyeglasses industry to attract generation z in a sustainable way*

For the product design, the best approach for the eyeglasses industry to attract Generation Z in a sustainable way is to offer a d-frame eyeglasses shape that is full rim with a neutral color. For the material, the author recommends bio-acetate frames as it is a sustainable option and is the most favored sustainable option (17.5%) compared to plant-based composites (6.3%) and recycled material (6.3%). It is lightweight yet strong and durable (quality) and has vast color options (design) (Bartlett 2021; Penczek 2021). Sustainable products tend to be viewed as expensive to buy. However, the survey found that quality and design are more important factors motivating Generation Z to buy compared to price. Therefore, brands should promote the performance aspects of the material to showcase the quality. The eyeglasses campaign must demonstrate that wearing the eyeglasses design can associate them with being fashionable, diligent, self-confident and friendly, as these are the 4 top traits Generation Z would like to be associated with when wearing eyeglasses as a functional fashion accessory. The campaign should focus on the individual self as Generation Z themselves are considered to be the most influential person when buying eyeglasses. One of the main reasons Generation Z buys a sustainable product is to minimize production waste which can be done through considered inventory, a sustainable practice (Moskal 2021). This practice focus on creating eyeglasses in small batch or even offering made-to-order to prevent surplus inventories that usually end up being disposed.

4 CONCLUSION

Several factors motivate Generation Z to buy eyeglasses as a functional fashion accessory, mainly customer need (the need for prescription glasses), quality, eyeglasses design, and price. The contributions of eyeglasses to sustainable development in Indonesia are still limited, and the industry is still considered unsustainable, despite having several sustainable developments that Generation Z has seen in the Indonesian eyeglasses industry. The best approach for the eyeglasses industry to attract Generation Z in a sustainable way is to offer d-frame eyeglasses shape made of bio-acetate that is full rim and neutral in color. Brands can promote the performance aspects of the product and must demonstrate that wearing the eyewear design can associate them with being fashionable, diligent, self-confident and friendly. It is recommended to carry out production in Indonesia to lower carbon footprint and make product in small batches to minimize product waste.

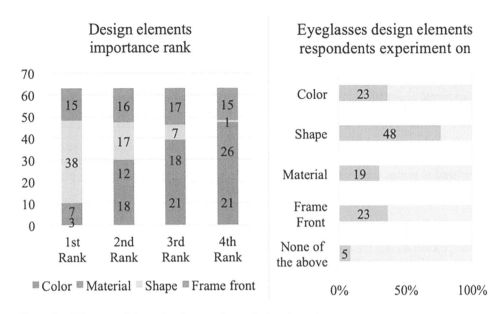

Figure 1. Diagram of data related to eyeglasses design elements.

Factors were tested on a 7-point Likert scale (1: Unimportant, 2: Low importance, 3: Slightly important, 4: Neutral, 5: Moderately important, 6: Very important, 7: Extremely important), and it was found that customer need (the need for prescription glasses) (6.2) is the most important factor, followed by quality (6.1), eyeglasses design (5.7), and price (5.3).

Eyeglasses design is one of the most important factors that motivate Generation Z to buy eyeglasses, and there are several design elements. The survey found that shape is the most

Table 3. Eyeglasses frame design preference.

Indicator	Result
Material	Metal (63.5%), plastic (55.6%), high-performance composites (36.5%)
Color	Neutral colors (87.3%), metallic (36.5%), pastel (9.5%)
Shape	D-frame (58.7%), round (27%), wayfarer (25.4%), square (22.2%), geometric (11.1%)
Frame front	Full rim (66.7%), half rim (50.8%), rimless (19%)

important design element for Generation Z when purchasing eyeglasses and is the element they like to experiment on the most. On the other hand, frame material is the least important design element when purchasing eyeglasses and is the element they like to experiment on the least. Data of respondents related to eyeglasses frame design preference with high percentages are as follows:

3.2 *Contribution of eyeglasses to sustainable development*

Through the survey, the contributions of eyeglasses to sustainable development were analyzed and evaluated. The eyeglasses industry in Indonesia is not sustainable yet, as 71.4% of the respondents believe that the industry is unsustainable. However, there are several sustainable developments that Generation Z has seen in the Indonesian eyeglasses industry. Many of the sustainable development in the Indonesian eyeglasses industry focuses on the frame material, where most of the respondent's state that the industry is offering more sustainable material options, including recycled materials and wood. There is also sustainable development in the transportation and distribution sector as local brands increase in

Table 4. Purchasing behavior towards sustainable products.

Indicator	Result
Reasons people buy sustainable products	To minimize production waste (52.4%), make the environment better (50.8%), lower carbon footprint (33.3%).
Reasons people do not buy sustainable products	More expensive (60,3%), perceived as less effective products (30,2%), seen as a sacrifice (12.7%)

popularity. The carbon footprint is reduced because these brands are manufactured in local factories. There is also sustainable development in the social impact sector, where people are encouraged to return their old frames to be recycled and get cashback. Data on purchasing behavior towards sustainable products are as follows:

3.3 *Best approach for the eyeglasses industry to attract generation z in a sustainable way*

For the product design, the best approach for the eyeglasses industry to attract Generation Z in a sustainable way is to offer a d-frame eyeglasses shape that is full rim with a neutral color. For the material, the author recommends bio-acetate frames as it is a sustainable option and is the most favored sustainable option (17.5%) compared to plant-based composites (6.3%) and recycled material (6.3%). It is lightweight yet strong and durable (quality) and has vast color options (design) (Bartlett 2021; Penczek 2021). Sustainable products tend to be viewed as expensive to buy. However, the survey found that quality and design are more important factors motivating Generation Z to buy compared to price. Therefore, brands should promote the performance aspects of the material to showcase the quality. The eyeglasses campaign must demonstrate that wearing the eyeglasses design can associate them with being fashionable, diligent, self-confident and friendly, as these are the 4 top traits Generation Z would like to be associated with when wearing eyeglasses as a functional fashion accessory. The campaign should focus on the individual self as Generation Z themselves are considered to be the most influential person when buying eyeglasses. One of the main reasons Generation Z buys a sustainable product is to minimize production waste which can be done through considered inventory, a sustainable practice (Moskal 2021). This practice focus on creating eyeglasses in small batch or even offering made-to-order to prevent surplus inventories that usually end up being disposed.

4 CONCLUSION

Several factors motivate Generation Z to buy eyeglasses as a functional fashion accessory, mainly customer need (the need for prescription glasses), quality, eyeglasses design, and price. The contributions of eyeglasses to sustainable development in Indonesia are still limited, and the industry is still considered unsustainable, despite having several sustainable developments that Generation Z has seen in the Indonesian eyeglasses industry. The best approach for the eyeglasses industry to attract Generation Z in a sustainable way is to offer d-frame eyeglasses shape made of bio-acetate that is full rim and neutral in color. Brands can promote the performance aspects of the product and must demonstrate that wearing the eyewear design can associate them with being fashionable, diligent, self-confident and friendly. It is recommended to carry out production in Indonesia to lower carbon footprint and make product in small batches to minimize product waste.

REFERENCES

Agmasari, S., 2016. Tahun 2050, *Setengah Populasi Manusia Butuh Kacamata*. Kompass.

Assadourian, E., 2010. Transforming Cultures: From Consumerism to Sustainability. *J. Macromarketing* 30, 186–191. https://doi.org/10.1177/0276146710361932

Bartlett, J., 2021. Cellulose Acetate for Spectacle Making [WWW Document]. Banton Framew. URL https://www.bantonframeworks.co.uk/blogs/guides/cellulose-acetate (accessed 10.15.21).

Cherry, K., 2022a. What is Motivation? [WWW Document]. Verywell Mind. URL https://www.verywellmind.com/what-is-motivation-2795378 (accessed 8.8.22).

Cherry, K., 2022b. Maslow's Hierarchy of Needs [WWW Document]. Verywell Mind. URL https://www.verywellmind.com/what-is-maslows-hierarchy-of-needs-4136760 (accessed 3.29.22).

Chu, C.-H., Wang, I.-J., Wang, J.-B., Luh, Y.-P., 2017. 3D Parametric Human Face Modeling for Personalized Product Design: Eyeglasses Frame Design Case. *Adv. Eng. Inform.* 32, 202–223. https://doi.org/10.1016/j.aei.2017.03.001

Creswell, J.W., 2014. *Research Design: Qualitative, Quantitative, and Mixed Methods Approaches*, 4th ed. SAGE.

Eyewear in Indonesia, 2021. *Euromonitor*.

Eyewear Market Size, Share & Trends Report, 2021–2028 [WWW Document], 2021. Gd. View Res. URL https://www.grandviewresearch.com/industry-analysis/eyewear-industry (accessed 6.10.22).

Fowler, F., 2009. *Survey Research Methods*, 4th ed. SAGE Publications, Inc., 2455 Teller Road, Thousand Oaks California 91320 United States. https://doi.org/10.4135/9781452230184

Lund, S., Madgavkar, A., Manyika, J., Smit, S., Ellingrud, K., Robinson, O., 2021. *The future of work after COVID-19*. McKinsey Global Institute.

Morgan, E., 2019. *Choosing Eyeglasses That Fit Your Personality and Lifestyle - AllAboutVision.Com [WWW Document]*. Vis. URL https://www.allaboutvision.com/eyeglasses/accessorize.htm (accessed 4.18.21).

Moskal, S.-L., 2021. MIDO 2021: *Digital Edition*. WGSN.

Novitasari, Y., 2019. Pengaruh Kenyamanan Mata, Keamanan Mata, Harga dan Gaya Hidup Terhadap Pemilihan Alat Bantu Penglihatan Kacamata dan Softlens. *J. Perkota.* 11, 162–176. https://doi.org/10.25170/perkotaan.v11i2.779

Oktaviani, I.N., Gautama, M.I., 2020. Tren Kacamata Bergaya: Studi Fenomenologis Pada Mahasiswa Fakultas Ilmu Sosial Universitas Negeri Padang. *J. Perspekt.* 3, 570–576. https://doi.org/10.24036/perspektif.v3i4.310

O'Neill, A., 2021. Urbanization in Indonesia 2019 [WWW Document]. Statista. URL https://www.statista.com/statistics/455835/urbanization-in-indonesia/ (accessed 5.7.21).

Penczek, M., 2021. How Durable Are Acetate Frames? URL https://progressive-glasses.com/how-durable-are-acetate-frames/ (accessed 10.15.21).

Pullin, G., 2009. *Design Meets Disability*. MIT Press.

Ratanasirintrawoot, T., 2019. *A Study of Motives Influencing Thais to Purchase Eyeglasses Frames Online*. Thammasat University.

Simangunsong, E., 2018. Generation-Z Buying Behaviour in Indonesia: Opportunities for Retail Businesses. *MIX J. Ilm. Manaj.* 8, 243–253. https://doi.org/10.22441/mix.2018.v8i2.004

Spectacles in Indonesia [WWW Document], 2021. Euromonitor Int. URL https://www.euromonitor.com/spectacles-in-indonesia/report (accessed 3.24.22).

Steele, V., 2005. *Encyclopedia of Clothing and Fashion*.

Stillman, D., Stillman, J., 2017. Gen Z @ Work: *How the Next Generation Is Transforming the Workplace*. Harper Business.

Yarlagadda, A., Murthy, J.V.R., Prasad, K., 2015. A Novel Method For Human Age Group Classification Based on Correlation Fractal Dimension of facial edges. *J. King Saud Univ. - Comput. Inf. Sci.* 27, 468–476. https://doi.org/10.1016/j.jksuci.2014.10.005

Interior visual identity: Customer interest in interior elements in a retail store

T.M. Raja*, K.P. Amelia, D.D. Prameswari & Y. Sabrani
Telkom University, Bandung, Indonesia

*ORCID ID: 0000-0001-5763-1602

ABSTRACT: In interior design, the identity aspect is an important thing but has not been specifically discussed, this study discusses the visual identity in the interior, where this aspect is one of the interior design aspects of forming a space to continue the identity of a company. In this study, we will discuss the relationship between the interior design of retail and the visual identity of its interior elements on consumer interest. The brand taken as a case study is L'occitane brand. This study aims to determine the perception of visitors from retail regarding the comfort of interior elements from L'occitane. This study uses a descriptive qualitative approach by using analysis methods based on elements of visual brand identity with interior elements. Methods of data collection using a questionnaire and observation. Respondents were selected using a random sampling method. This research can be used to determine visitor interest in the visual retail identity of L'occitane.

Keywords: Interior design, Visual identity, Retail, Customer response

1 INTRODUCTION

The attractiveness of a brand is often obtained from its retail interior design which provides an attractive store atmosphere and describes the brand. One of the components leading to a customer's identification with a company is the attractiveness of that company's identity (Marin L. & Ruiz S., 2007). Through the visitor's impression of a retail brand, visitors' interest in the brand can arise and increase their desire to buy the brand's products (Kotler 2005). The atmosphere of the store itself can arise through the design of retail interior elements, adding brand identity to retail can describe a brand more clearly and build an atmosphere that is in harmony with the brand itself. Through the functional and emotional benefits contained in the value of brand identity, it can strengthen the relationship between brands and customers (Kotler & Keller 2006). Visual brand identity is the identity of a brand that can be seen by consumers and becomes the face of the brand. This identity is usually a logo, color, store design, and others. Therefore, visual identity is the difference between a brand and other brands. The uniqueness of a brand is one of the things that must be considered by a company. Based on the degree of consumers' need for uniqueness relative to brand perceptions, several implications for market segmentation can be drawn (Knight & Kim 2007). The application of a brand identity is important for interior design because it aims to convey or describe the brand's intent to consumers. The application can be done on interior design elements, furniture selection, or display. Wheeler (2017) states that brand identity is an effort that needs to be implemented to build an image to convey an idea or message that the brand wants to communicate.

There is a relationship between the characteristics of the branding process and interior elements as stated by Imani and Shishebori (2014) in their research. Furthermore, in this study, they stated that the elements in interior design that can have an impact on branding are: (1) space, (2) color, (3) material, (4) shape, (5) lighting, and (6) furniture. The relationship between the visual identity of a brand in retail interior elements on visitor interest is related to impulse buying which is based on a person's emotional sense of a product from the retail. Kotler (2005) states that there are factors that can affect impulse buying in a shopping store. These factors are a retail atmosphere that is planned and appropriate by the target market so that it can attract consumers.

The retail brand chosen by the author is L'occitane brand, with consideration that this brand is one of the brands that have an interior concept that applies a visual identity application. The identity concept of L'occitane is sunshine which is taken from the background of the origin of the brand and its trademark that the products are made from flower extracts. This concept describes retail that is spacious, colorful, natural, and delicate with the characteristics of the retail brand. This research will discuss and analyze the visual identity of retail interior design with a case study of L'occitane and its relation to customer perceptions of the application of visual identity to retail interior elements.

2 METHODS

This research used a qualitative method with an emphasis on. visual identity in retail interior elements. The brand chosen as the case study is L'occitane brand. The selection of a case study uses a non-random sampling method with a purposive approach (Kumar 2005). The method of data collection is divided into two: the questionnaire method for primary data collection and literature data from related journals. The analysis method for this study is a descriptive qualitative analysis method by processing indicators which are interior elements components as part of visual identity with the variables which is user perception based on attractiveness, comfort, and uniqueness. These three variables are taken from several statements related to the brand's visual identity.

3 RESULTS AND DISCUSSION

The atmosphere to be achieved at L'occitane retail depicts the atmosphere of the original area filled with flower fields, especially lavender, and the warm weather.

Figure 1. L'occitane retail store, Toronto, Canada. (luciforma.com).

Figure 2. L'occitane retail store, London, UK. (enkimagazine.com).

In achieving this atmosphere, these six aspects are applied through the design of the retail interior elements, which are as follows:

1. Space, spacious is a description of space in retail that is divided into two areas, the tester area and the display area with a centralized layout.
2. Retail Loccitane has a colorful palette of colors, including yellow which is a characteristic that describes the concept of the area of origin, Manosque Provence, France. These colors include white and light brown which are the basic colors and a combination of striking colors that match the concept of warmth and colorfulness.
3. Lighting is an important element to build a store atmosphere. Lighting can give a certain impression, communicate identity, and build customer focus on a certain object. Lighting at L'occitane retail consists of cool to neutral, white-colored lights that produce a space that looks clean, fresh, and bright by the L'occitane concept.
4. Material, the material used is the wood motif which is the dominant material. The goal is to elevate the natural and warm atmosphere in the retail interior.
5. Shape, the use of curves and circles, as well as furniture with blunt angles in retail, makes the interior more delicate so that it suits the target market, the majority are women.
6. Furniture, in L'occitane retail, uses a lot of furniture with blunt angles that are in harmony with a delicate impression, as well as the use of soft colors that are synonymous with a feminine impression.

Analysis of visual identity interest in interior elements

The discussion on interest analysis is done by analyzing the data obtained from the questionnaire. The analysis is divided per component of the interior elements and seen the value of the number of respondents who gave their opinions. The analysis uses a bar graph with instruments that show positive (+), negative (-), and neutral (0) responses from respondents to interior elements.

1. Space

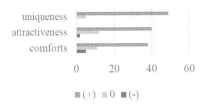

Graph 1. Space element of interior analysis.

Based on the results of data analysis, it was found that public opinion gave a positive response to aspects of comfort, attraction, and uniqueness which was quite large when compared to neutral and negative responses. This shows that the element of the form in the interior of L'occitane retail is quite successful in providing an effect that makes respondents interested in the visual identity that is applied to the interior design of L'occitane retail.

2. Color

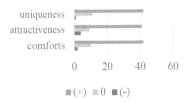

Graph 2. Color element of interior analysis.

Based on the results of data analysis regarding the application of color applied to the interior of L'occitane retail, the majority of respondents agree that in terms of comfort, attraction, and uniqueness, it has been successfully applied.

3. Lighting

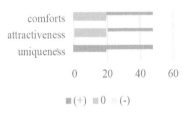

Graph 3. Lighting element of interior analysis.

Based on data analysis related to lighting, it appears that none of the respondents stated a negative response. The positive response seemed very significant from the number of respondents who gave their opinion. Some respondents stated neutrally but not significantly.

4. Material

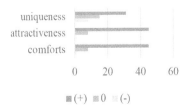

Graph 4. Material element of interior analysis.

Based on the results of the analysis of respondents' data regarding material elements in the interior of L'occitane, it was found that a positive response for the uniqueness indicator was not large enough when compared to other indicators which reached 83% for a positive response. This shows that the material elements are considered not unique enough.

5. Shape

Graph 5. Shape element of interior analysis.

Based on the data from the questionnaire, it was found that the shape elements applied to L'occitane retail have a very positive effect on the comfort and attractiveness indicator. As for the uniqueness indicator, the results show that the shape is not unique enough.

6. Furniture

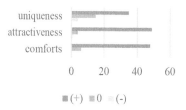

Graph 6. Furniture element of interior analysis.

Based on the results of data analysis for furniture elements, the results obtained were slightly different from other data. The data show that 27% of respondents have a neutral response. The negative response is not much, and the positive response is quite large. This shows the possibility that respondents are used to seeing similar furniture used in L'occitane retail at other retailers.

4 CONCLUSIONS

Based on the results of the analysis, it can be concluded that the interior design of L'occitane retail has been quite successful in implementing its visual identity in the interior elements. Three indicators used as measuring instruments, space, color, and lighting show significant results. This shows that the three indicators have succeeded in attracting the interest of respondents. The other three indicators, which are material, shape, and furniture showed less significant results in a positive response. This show that the application of the three indicators is not unique enough so quite a lot of respondents give neutral or negative responses. This research can still be continued to get maximum results because this research is still at the limit of respondents' perceptions. Although the respondents have been selected with certain criteria, of course, the direct experience factor will have a greater influence if it is used as data.

REFERENCES

Imani, N., Shishebori, V. (2014). Branding with the help of interior design. *Indian Journal of Scientific Research*.

Knight, D. K., & Kim, E. Y. (2007). Japanese consumers' need for uniqueness: Effects on brand perceptions and purchase intention. *Journal of Fashion Marketing and Management: An International Journal*.

Kotler, P. d. (2005). *Manajemen Pemasaran*. Jakarta: PT. Indeks Kelompok Gramedia.

Kotler, P., & Keller, K. L. (2006). Defining marketing for the 21st century. *Marketing Management*, 3–33.

Kumar, R. (2018). *Research Methodology: A Step-by-step Guide for Beginners*.

Marin, L., & Ruiz, S. (2007). "I Need You Too!" Corporate Identity Attractiveness for Consumers and The role of Social Responsibility. *Journal of Business Ethics, 71*(3), 245–260.

Wheeler, A. (2017). *Designing Brand Identity: An Essential Guide for the Whole Branding Team*. John Wiley & Sons.

Analysis of library design in the digital era

F. Mahdiyah* & R. Machfiroh
Telkom University, Bandung, Indonesia

*ORCID ID: 0000-0003-0099-2321

ABSTRACT: Library is an institution divided into regional management levels, that collects and manages knowledge in printed or recorded media. Based on mayoral decree of Bogor City Government, the library has an aim to improve the service system and raise interest in reading among Bogor residents. But only 87% of providing facilities, and 30% in increasing visitors were achieved. One of the factors is inadequate facilities where people in this digital era need facilities to accommodate collaborative work with the use of digital media in their daily lives. A computerized system does tend to minimize human errors that arise due to manual work. The importance of digitization must also be accompanied by an interior with the "Homey" theme that creates a relaxing and comfortable atmosphere, also open space to support work activities. This research was conducted to examine the development of the library through the interior in the current digital era.

Keywords: Design Interior, Digital, Homey Style, Library

1 INTRODUCTION

According to Setyo, the application of this digitization system will also help librarians and users through the available automation functions. Library management with a digital system will be more effective and efficient and can minimize technical or human errors. As a result of advances in digitalization technology, conventional libraries have begun to be abandoned, and people prefer to access collections through digital media (e-books), which are ready and economical to use (Pendit 2003). Another thing that affects the backwardness of conventional libraries is the shift in community work culture, from individual to collaborative, especially for young people in the digital era where startup businesses with collaboration systems are overgrowing. Not only business actors, high school students are considered to have digital skills in the medium category, but "motivation" is in the high category (Machfiroh 2019). So, facilities are needed that can accommodate brainstorming activities. Co-working space has recently become a work trend because it is considered flexible. According to (Aldi 2019), the development of co-working spaces in Indonesia has also occurred quite quickly; within six years, more than 50 co-working spaces have developed until 2016. This is also what causes people to prefer to go to open areas, which supports discussion activities rather than conventional libraries.

Bogor City Library is a city-level library under the authority of the City Government. Based on the Decree of the Mayor of Bogor regarding the Determination of the Bogor City Government Performance Plan, it can be concluded that the government has a performance target to improve the library system and increase the reading interest of Bogor City residents. In the application related to the availability of book collections, the performance achievement has exceeded the target of 118%. Meanwhile, in the category of providing facilities, only 87% of the performance target was achieved, and in the category of increasing visitors, only 30% was achieved. According to the results of a questionnaire distributed to 70 residents of the city of Bogor, especially those aged 15 to 30 years, one of the factors that influence the low achievement of the visitor target is the lack of provision of adequate and comfortable space facilities for collaborative work.

1.1 Literature review

The concept of a library as an individual work area is consistent with the opinion of Atika and Setiamurti. They reveal that most people no longer come to the library to read but to carry out discussion activities and work. The digitization that applies to many things, including book collections, makes most visitors feel no need to visit the library just to read books. So that the form of the library applies a space atmosphere based on the character and behaviour of the millennial generation (which tends toward co-working spaces and discussion rooms). The use of homey, comfortable, relaxed, and open design themes (to support discussion activities) is something more desirable.

One of the efforts to increase the percentage of visitors is to provide facilities that support the convenience of reading and the comfort of working through interiors with a comfortable atmosphere and more computerized technology. This study analyzes the relationship between interior design that carries the concept of relaxing and homey to support visitor comfort in the digitalization era.

2 RESEARCH METHODS

The research method used is a descriptive quantitative method where data is collected through structured interviews with the librarians of the Bogor Library and distributing questionnaires to a sample of respondents who live in the city of Bogor. This study will apply the Design Thinking approach at the empathize stage and relate to the Interior science field, which carries the theme "Homey".

2.1 Methods of data collecting

Data collection will be done through literature study, observation, and distribution of questionnaires to 70 respondents. The research sample was taken based on the average visitors to the Bogor City Library, who were junior high school students and adults aged 15–30 years and living in the city of Bogor.

2.2 Methods of analysis data

The analysis in this study will apply an empathetic approach through the Design Thinking theory. This is done to collect data related to the problems in the Bogor City Library, which are more user-centered through the process of quantitative methods. Data collection will be carried out to understand the needs of facilities, as well as the habits of visitors when they come to the library, including behavior habits, especially in an era where technology has developed rapidly and has become a daily necessity.

3 RESULT AND DISCUSSION

3.1 Visitor behaviour

Based on the results of questionnaires conducted to 70 Bogor residents with an age range of 1530 years, only 41 knew about the existence of the Bogor City Library, and only 80% of the 41 respondents had visited it in person.

The habit of people who come to the library to borrow and read books also shifted along with the development of the role of technology. In this era of digitization, books in digital form will be readily obtained without having to borrow from the library. The emergence of start-up businesses and collaborative work culture also makes people prefer to work outside the conservative office. This theory then continued with the data from the questionnaire revealed that 57% of the respondents preferred to work in a co-working area. In addition, most visitors also come on weekdays (Monday-Friday).

3.2 Facility needs

People in the digital era, especially students and college students, need the library area to be used as a place to stop for work rather than reading. So that the facilities that have to be equipped are technological support facilities, such as Wifi and electricity terminals (Felecia et al. 2018).

This theory supports the questionnaire results, which revealed that on average, visitors came with electronic devices such as laptops and cell phones with a percentage of 59%, gadgets and stationery with 36%, and only stationery with 5%.

The average visitor stays in the library for approximately 1–3 hours, calculated to the answers of 57% of respondents. While in the second rank, visitors more often stay in the library for longer than three hours. This indicates the need for ergonomic facilities to support work activities. In addition, 64% of visitors also tend to come with friends, either for individual work or working in group discussions. So facilities in the form of non-cubical open spaces are also the right solution to support activities in the library.

3.3 Interior concept

In terms of interior layout, the provision of an open sitting area will be suitable to be applied to the library and not only creates a broad impression but also a friendly atmosphere because visitors can gather in one area. According to Quinn, the psychological perception of body contact, the application of furniture, and the continuity of the supporting elements of space can create a comfortable impression. This theory supports the questionnaire results where 44% of visitors feel more comfortable with an ergonomic and harmonious interior, beating the completeness of supporting condiments such as electric plugs and Wifi by 33%, and a complete reading collection by 23%.

According to Titihan, the factors that form the atmosphere between are divided into two, namely:

1. Physical elements include space-limiting elements such as vertical and horizontal elements and space-forming elements such as walls, floors, and ceilings.
2. Non-Physical Elements, including humans as users. So that this element includes social, cultural, and psychological aspects

So that the comfort of space users can be created through a combination of the right elements through lines, colors, textures, shapes, and light spaces. (Wicaksono et al. 2014).

The use of neutral colors in the overall design of the space often creates a monotonous impression that makes visitors feel bored quickly. So we need bright colors that can be used as focal points in several areas to keep the room's atmosphere alive.

(Swasty 2017) divides color categories into two, namely hot colors and cold colors. Hot/warm colors are red to yellow and give rise to active, stimulating, positive, and aggressive traits. While cool colors are the green to purple color family. This color is calm, and is often associated with water, sky, and mountains that create a cool impression.

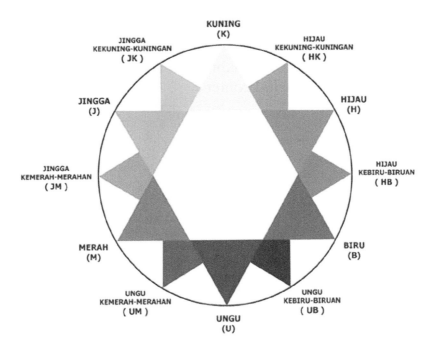

Figure 1. Brewster's color wheel.

4 CONCLUSION

Based on the results of research that has been done, this computerized system does not necessarily erase conventional history as a whole. However, the concept of library digitization needs to be done considering that some people are already living in digital technology. The shift in individual work habits to collaborative work is also one of the reasons for the need to provide facilities that support their behavior when they come to the library, namely the habit of working and practicing. The interior with the form of an open and non-cubical area will create a spacious and intimate impression. While the use of neutral colors will create a memorable and comfortable atmosphere or "Home". However, using neutral colors in the entire space should be avoided because it will create a dull, stiff impression, so cold colors are needed to make some parts the focal point. the application of "Homey" with an open space that supports collaborative work habits is considered appropriate for libraries, especially in the current digital era.

REFERENCES

Aldi, Y. (2019). *Perancangan Coworking Space Dengan Pendekatan Third Place Pada Bangunan Lama di Kota Bandung*. Institut Teknologi Bandung.
Felecia, S. H., Lintu T., & Dian W. (2018). *Library Interior Design for Digital Native Generation*. Knowledge E Publishing.
Machfiroh, R. (2019). *Pendidikan Kewarganegaraan bagi Remaja di Era Digital*. Universitas Pendidikan Indonesia.
Pendit, N. S. (2003). *Ilmu Pariwisata Sebuah Pengantar Perdana*. Pradnya Paramita.
Swasty, W. (2017). *Serba Serbi Warna: Penerapan pada Desain*. PT Remaja Rosdakarya.
Wicaksono, Andie A., & Endah Tisnawati. (2014). Griya Kreasi.Kampanye #PantangPlastik terhadap Sikap Ramah Lingkungan (Survei pada Pengikut Instagram @GreenpeaceID). *Jurnal Komunika: Jurnal Komunikasi, Media Dan Informatika*, 9(1), 40.

Digital unification: The sustainability of Indonesian NFT communities through digital content

R.P. Bastari* & I. Resmadi
Telkom University, Bandung, Indonesia

ORCID ID: 0000-0003-1994-1460

ABSTRACT: Indonesian NFT artists can't stand on their own, and in order to stay sustained in the world of NFT they need to hold on to their supporting community. Currently, there are many Indonesian NFT community which is still creating digital content till today. This study aims to discuss the sustainability of the Indonesian NFT community through their digital content which contained the promotion of many works of Indonesian artists. The case study in this research are 3 Indonesian NFT communities namely, Indo Art Now NFT, Metarupa, and Monday Art Club. The method used in this study is a qualitative approach with the validation from observation and visual data which will be obtained through various internet platforms. The next stage is analysis through comparative study with the theory of field of creativity. The result of this study shows that the community plays an important role in maintaining the artists' network through their digital content and activities.

Keywords: community, creative industry, digital content, digital culture, NFT

1 INTRODUCTION

NFT is a one-of-a-kind digital asset with unique metadata and code in the blockchain network (Sugiharto et al. 2022). The study in the field of NFT is still in the stage of development, as of 2022, there are not many studies that focus on discussing NFT, whether in terms of systems, aesthetics, or social. However, several previous studies showed different conclusions. The first earlier study resulted in the discussion of design models, opportunities and challenges, and some perspective on the systemic and technical components (Wang et al. 2021). Although it's still at the early stage and the study that categorized objects associated with NFTs, this study analyzed 6.1 million trades of 4.7 million NFTs between June 23, 2017 and April 27, 2021 (Nadini et al. 2021) From that number, we can conclude that NFTs growing exponentially in the past 4 years. However, there are still several concerns pertaining legal nature of the field of cryptocurrency and NFTs (Bolotaeva et al. 2019). Although there are still several concerns pertaining to NFTs, since the success of several NFT projects such as CryptoKitties, many new NFT projects started to emerge, younger NFT projects influenced by well-established ones (Ante 2021). The rise of NFTs gave a major impact on artists who began adopting digital platforms. One study concluded that despite the fluctuating price of NFTs, the average sales remain the same, and many artists invited other artists to start adopting the blockchain-based digital platform, and from that point many artists developed similar engagement and sales patterns (Vasan et al. 2022). From this study, we can conclude that there is "gold rush" to the NFT markets.

---

*Corresponding Author: rendypanditabastari@telkomuniversity.ac.id

Despite several risks of this kind of investment, the high volatility drives people to begin investing in NFTs and this action drastically increases the demand of NFTs. One study concluded that the price of NFT depends on token scarcity and aesthetic preferences (Kong & Lin 2021). Several of these studies certainly give a really big contribution to the field of NFT study. However, this study aims to give a different perspective to discern opportunities and development in the field of NFT socially based on several case studies. This is because the large access to the internet, which started in the early 2000s, resulted in the forming of digital communities which gave possibilities to people all around the world to gather. These communities also formed in the NFT world, which support NFT artists. Several of them operate socially on social media such as Twitter and Instagram and conduct interaction with the members on Discord, a social media application where communities conduct direct and real-time interaction. The base for this study is several Indonesian NFT communities namely, Indoartnow NFT, The Monday Art Club, and Metarupa. These communities are well-established and work with great sustainability. Each of these communities has its own digital content which is disseminated through several platforms. Online communities became important in this digital disruption era, where they interact, collaborate, and learn from each other through the digital platform and open up the possibilities for further development in the digital world. By studying their digital content on several prominent platforms that they're currently using and discerning their position socially which is also supported by analysis using relevant theory, we can see how they maintain the sustainability of their community. This study can serve as a basis and reference for studies in the world of NFT and open up opportunities for further studies.

2 RESEARCH METHODS

2.1 *Methods of data collecting*

The Indonesian NFT communities which will be the subject of this study are Indo Art Now NFT, Metarupa, and Monday Art Club. The method used in this study is a qualitative method, with the primary data obtained from observation through internet platforms namely: Twitter, Instagram, and Discord. The first stage is the data search through online observation, and the data obtained at this stage are digital content data which are used to promote artists and other digital activities such as a webinar, exhibitions, and other activities. The second stage is literature study which used the field of creativity theory from Yasraf Amir Piliang, the theory derived from the field of cultural production coined by Pierre Bourdieu. The third stage is the comparative study which will be analyzed using the field of creativity theory. The last stage is concluding the study based on the conducted comparative descriptive study. At the time of writing this, we are active members of each of these communities on Discord, and we also follow these communities on social media, giving us access to a wide range of information from them. This method gave us access through the discovery of experience and open the possibilities of giving descriptive study, as the descriptive method can provide an improved understanding of certain social events (Jackson et al. 2018).

2.2 *Literature study*

Interaction, field, and actors paly key roles in the field of creativity. Every society contains interactions of its factions in the field. The platform of actors to produce ideas, disseminate it, make it to realization, and appreciate ideas is constructed into a structured space called "field". Field is also a fraction of the bigger world called the arena. Every arena contains 4 different fields: (a) field of production, where ideas came to realization; (b) field of expression where ideas are being produced or in thoughts; (c) field of dissemination, where ideas are being distributed or being publicized; and (d) field of appreciation, where ideas are being consumed and appreciated. Every arena also contains its own established rules and hierarchical system (Piliang 2018). In every field, some actors played a role in making the field continue working. The role of each actor has undergone legitimation in several ways: (a) legitimation through their own community which is also called specific legitimation; (b) legitimation through ruled class or institutions which is also called *bourgeoisie* legitimation; (c) legitimation through multiple communities, which is the effect of the popularity of certain actors, this legitimation is also called popular legitimation. To gain this legitimation, actors need their capital; on the other hand, actors can also gain capital through these legitimations. There are several actors; capitals that supported them to work in their field, namely: (1) economic capital, also known as material capital, (2) symbolic capital, non-material capital which gives certain actors a privilege such as social status, class, and authority, (3) cultural capital, which contains a way of thinking and ideology, and this includes every aspect that can produce another idea such as education, language, and art. These capitals are the reason actors have their own habitus. Habitus is the logic of their thinking and their practical sense (Bourdieu 2013). Through these elucidations of field of creativity, we can analyze how certain arenas can sustain while others are not as every part in the field of creativity is synergistic; which means that if there is one field that lacks actors or supports, it will make discrepancies in other fields. As the base of the analysis of Indonesian NFT communities' sustainability, this study will focused on how the actors in every field acquire roles. All the field within the field of creativity will be analyzed to see how they act and engage in their field. The way society works is not individual rather we see it as a whole.

3 RESULT AND DISCUSSION

3.1 *Observation result*

Indoartnow NFT

Indoartnow NFT is not as prominent as Metarupa and The Monday Art Club. Our observation starts with their Twitter page which contains predominantly artists' NFT works promotion. This activity support Indonesian NFT artists, in terms of exposure to the public. Since their Twitter timeline is filled with the promotion of artists' works, they give direct support by announcing work submission forms for artist who want to publicize their works to 7.815 followers on Twitter. Continuing our observation of the Instagram page, we didn't find any contrivances. The content of their Instagram posts consists of the promotion content of Indonesian NFT artists. Their Instagram is not as prominent as their Twitter. However, there is one Instagram post announcing their open discussion on NFT copyright, but compared to the other two communities in this study, they rarely hold discussion events. Our next observation is on their Discord page which comprises mostly artists, other members are several collectors and community managers. One of the most prominent discussion category is "shilling-submission" and "free shill" which allows artists/creator to submit their NFTs to be promoted as Indoartnow NFT digital content. Another discussion in their Discord platform category is a general discussion about art discourse and market discussion. And lastly, Indoartnow NFT shares information pertaining to NFT, from technical information about minting artwork to updated news. Indoartnow NFT is different from Metarupa and The Monday Art Club, and their activities mostly consist of promoting and publicizing NFT artists' works. However, this kind of activity is one of their endeavor to give direct support to artists.

Metarupa

Our first stage of observation is through their Twitter page. Our findings on their Twitter page is that it consists of many promotions of Indonesian artists. Metarupa actively promotes artists' NFT works in Twitter exposing them to 4,590 followers. Although their timeline contains predominantly the works of artists, there are several announcements on their other events such as webinars and open discussions and exhibitions. Metarupa is really active in conducting various online activities such as open discussion webinars on NFTs, online collaborations, and artist talks. These findings are manifested through their Instagram content. Other findings of their online content are the promotion of artists' works. Several of their Instagram posts consist of the NFT works of Indonesian artists, although this type of content is not prominent on their Instagram page. However, their Instagram stories (temporary posts that will be deleted after 24 hours) mostly contain promotion of artists' NFT artworks. One of their biggest activities is a national-scale NFT exhibition which is held in 2021 in collaboration with other actors such as BCA, the Indonesian ministry of the creative economy, and Tezos blockchain. As far as our observations went, Metarupa is mostly actively utilizing Instagram and Twitter for its activities. As for the Discord platform, Metarupa is not as active as Indoartnow NFT and The Monday Art Club. Their Discord members consist of community managers, gamers, and artists. Their most recent and prominent activity in Discord is #hendonesia which is a supportive act toward Indonesian NFT artists to "mint" their works on the Tezos blockchain. Other than that, there wasn't any salient activity.

The Monday Art Club
Monday Art Club mostly uses Twitter, Instagram, and Discord. Monday Art Club appears to be the most active community among the three communities that are the main subjects in this study. In our observation in May 2022 on their Twitter page, they held an exhibition opening party for Indonesian artists called Yusuf Ismail, also known as Fluxcup, which was being held in Decentraland, a new browser-based digital space/metaverse which gave the community an unprecedented experience. Their Twitter contents are different from Metarupa and Indoartnow NFT, and they rarely promote other artists' artwork on their timeline, instead, they share much information pertaining to the NFT world such as NFT market information, upcoming NFT exhibitions, and many other news. Our observation through their Instagram account shows several conclusions. First, most of their Instagram content consists of information about NFTs from tutorials showing how to create artwork, information related to exhibition, and other news. There are several posts on artists' career progress in the digital world, and this kind of post is purposed to promote community activities in the NFT world such as exhibitions other than that, The Monday Art Club also promotes several achievements of individual artists within their communities. The Monday Art Club also utilizes other community platforms such as Discord. Our observation through discord leads us to several findings. The member of The Monday Art Club Discord community consists of artists, general public, managers, and collectors. The online forum has similar content to their Twitter; however, The Monday Art Club Discord page contains several rooms for discussion. There is a room for free discussion on the NFT world between members, job vacancies, Twitter space schedule, and general announcements.

3.2 *Analysis*

The three communities act in the field of dissemination, expression, and appreciation making them a place and channel for Indonesian digital artists and digital art appreciators. Although in this case, the artists act in their respective fields such as the field of production, and in some cases, in the field of expression. They can maintain sustainability in building the NFT community because of the constant support digitally through their content, whether in the form of promotion/publication of Indonesian artist' NFT artworks, discussion forum pertaining to relevant topics, or webinar activities which contain relevant contents. In these three communities, every artist has their respective cultural and economic capital as well as the NFT community managers, with this they have their own habitus. Based on our observation, each of the communities has adequate media literacy to maintain the sustainability of their community.

Indoartnow NFT is very open for artist to submit their NFTs to be promoted. In their Discord platform, this is one of their main advantages, which is absent in the other two communities. However, Indoartnow NFT is not active in holding webinars and exhibitions, but they provided an ample discussion forum on relevant topics in their Discord. Through several of their digital contents, we can conclude that they hold important role in the field of appreciation and dissemination. Although they are also acting in the field of expression when they are active in holding discussions, they are not as big and volatile as the other two communities as they rarely hold webinars.

Metarupa is quite active in promoting and publicizing NFT works of Indonesian artists, especially on Twitter. In addition, they are also active in promoting NFT works on Instagram. Their publication activity makes them an actor in the field of dissemination and appreciation. In addition to that, their Discord is not as active as Indoartnow NFT and The Monday Art Club, but they secure the role in the field of expression as they are quite active in holding webinars which are known as a place for discussion and information sharing. The Monday Art Club is not active in doing promotion and publication of the works of NFT artists on their digital content; however, they are active in providing a place for sharing information and discussion pertaining NFT field; other than that, what makes them different from the other two communities is the new experience they offered through activity in the metaverse. The utilization of new media, especially in Indonesia, makes them an important actor in the field of expression and appreciation. They are not quite active in the field of dissemination as they rarely publicize other Indonesian artists' NFT artworks.

4 CONCLUSION

In the main arena of Indonesian NFT, these communities play important roles in supporting sustainability. Through their online activities such as promotions, discussions, and exhibitions, they implicitly give support to other actors within the Indonesian NFT communities. The analysis concludes that these three communities mostly played a role in the field of dissemination and expression, which can be seen in their digital contents and activities such as discussions (such as webinars events) and promotion (also known as the term "Shilling"). However, in several cases, they also played a role in the field of appreciation which can be seen through exhibition activities as held by Metarupa and The Monday Art Club. Through these exhibitions, actors in Indonesian NFT communities can appreciate ideas through art. However, the field of production is excluded from this study because this field pertains to technical matters which, as far as we are concern, currently consist of the artists themselves, blockchain engineers, and internet providers. Through this study, we conclude that for the sustainability of digital communities, at least certain actors must act in the field of expression and dissemination. Through these two fields, they give a huge impact by giving support to other actors. This study still lacks several data such as interviews with artists, community managers, and collectors, more complete data could improve the result of the study. Further studies suggested the social fields of NFT pertaining to the collectors among the communities which still needs further research. In addition to that, official galleries that entered the NFT world also need further research.

REFERENCES

Ante, L. (2021). Non-fungible Token (NFT) Markets on the Ethereum Blockchain: Temporal Development, Cointegration and Interrelations. *SSRN Electronic Journal*. https://doi.org/10.2139/SSRN.3904683

Bolotaeva, O. S., Stepanova, A. A., & Alekseeva, S. S. (2019). The Legal Nature of Cryptocurrency. *IOP Conference Series: Earth and Environmental Science*, 272(3), 032166. https://doi.org/10.1088/1755-1315/272/3/032166

Bourdieu, P. (2013). *Arena Produksi Kultural: Sebuah Kajian Sosiologi Budaya* (Yudi Santosa (ed.)). Kreasi Wacana.

Jackson, C., Vaughan, D. R., & Brown, L. (2018). Discovering lived experiences through descriptive phenomenology. *International Journal of Contemporary Hospitality Management*, *30*(11), 3309–3325. https://doi.org/10.1108/IJCHM-10-2017-0707/FULL/PDF

Kong, D.-R., & Lin, T.-C. (2021). Alternative Investments in the Fintech Era: The Risk and Return of Non-fungible Token (NFT). *SSRN Electronic Journal*. https://doi.org/10.2139/SSRN.3914085

Nadini, M., Alessandretti, L., Di Giacinto, F., Martino, M., Aiello, L. M., & Baronchelli, A. (2021). Mapping the NFT Revolution: Market Trends, Trade Networks and Visual Features. *ArXiv, arXiv:2106*, 1–34. http://arxiv.org/abs/2106.00647

Piliang, Y. A. (2018). *Medan Kreativitas: Memahami Dunia Gagasan* (Taufiqurrahman (ed.); 1st ed.). Cantrik Pustaka.

Sugiharto, A., Musa, M. Y., & Falahuddin, M. J. (2022). *Regulasi, NFT & Metaverse: Blockchain &* (1st ed.). Perkumpulan Kajian Hukum Terdesentralisasi Indonesian Legal Study for Crypto Asset and Blockchain.

Vasan, K., Janosov, M., & Barabási, A.-L. (2022). Quantifying NFT-driven networks in crypto art. *Scientific Reports |*, *12*, 2769. https://doi.org/10.1038/s41598-022-05146-6

Wang, Q., Li, R., Wang, Q., & Chen, S. (2021). Non-Fungible Token (NFT): Overview, Evaluation, Opportunities and Challenges. *ArXiv, arXiv:2105*. https://arxiv.org/abs/2105.07447v1

Website news media Ayobandung.com user experience

A.D. Pinasti & D.W. Soewardikoen
Telkom University, Bandung, Indonesia

ORCID ID: 0000-0002-5097-6138

ABSTRACT: One of the online news portal websites is Ayobandung.com. This online news media was originally under the auspices of PT Ayo Media Network along with 13 other online news portals, then turned into part of the PT Promedia Teknologi Indonesia ecosystem. To increase the convenience of readers and attract the attention of business relations and sustainability, changes are needed to its user interface. This research uses a design thinking approach, focusing on the empathy stage. In this research, collecting data will be done by using participatory observation of the User Experience Questionnaire (UEQ) method as a research instrument. The results of this research will be proceeded to the define stage.

Keywords: user experience, news portal, design thinking

1 INTRODUCTION

News has become mandatory consumption for the community because through the news they can get various information, both about the latest events and developments. One of online news portal websites is Ayobandung.com. This online news media was originally under the auspices of PT Ayo Media Network along with 13 other online news portals, then turned into part of the PT Promedia Teknologi Indonesia ecosystem. Ayobandung.com is a local news website from Bandung City. Currently, Ayobandung.com continues to contribute to provide news about the city of Bandung and its surroundings, not only that, there are also national and international news. Along with the times, Ayobandung.com wants to continue to make some changes, one of the changes is on its user interface on its website with the aim of creating convenience for a wide audience of readers which will later affect the business processes carried out such as attracting the attention of business relations so that they can cooperate. The business run by Ayobandung.com can be seen on the appearance of its website, namely by a lots of advertisement/programmatic. Previous research on user experience was conducted by Sonia Elisurya, et al entitled "Evaluating User Experience Using Usability Testing and User Experience Questionnaire (UEQ) (Studies on E-Commerce Fashion).

In this study, there are 3 website objects studied. Then the research entitled User Experience Measurement with Usability Approach, (case study of tourism websites in Southeast Asia) by Lilis Dwi Farida (2016) from the STMIK Amikom information system. This study uses a usability approach. Other research on user interfaces is the User Interface and User Experience Usability Testing for Hypertext-Based Thesis E-Reader Applications by Ardiansyah and Muhammad Imam Ghazali (2016) from Universitas Ahmad Dahlan. This study uses a usability user experience approach. From the three studies above, the user experience research of the Ayobandung.com news media website was carried out using a design thinking approach. The user interface is a visual display of a product that serves to bridge the system with the user or users. The appearance of the user interface can be in the form of colour, shapes and writing that are designed as attractive as possible. But in simple terms, the user interface can be interpreted as the appearance of a product that is seen by the user or user.

Meanwhile, user experience is how the user experiences interacting/using a digital product. The design thinking approach in this case is the empathize stage, which is a deep understanding of the people who will use a design. To gain this understanding, it is important to look from the point of view of a potential user. Thus, designers can understand the needs, wants, thoughts, feelings, and motivations of users in using that website. The empathize process begins with a beginner's mindset, meaning that the designer becomes as if they don't know anything about themself. This mindset will clean up the initial assumptions that will 'contaminate' the designer's judgment. Thus the designer can see the needs of potential users clearly. The process of empathizing is like a person looking at another person through a window, not looking at another person through a mirror. Because when you look at other people through the mirror, what you see is yourself. Empathy without a beginner's mindset only results in understanding others according to one's own assumptions (Aji 2018).

2 RESEARCH METHODS

The focus of this thinking approach is on empathy, or using the user's mind set and experience to uncover needs and insights. Data was collected using observation and questionnaires were collected using Google Form with 225 respondents who were readers of Ayobandung.com. Then the objective research stage was carried out to find out how user satisfaction was using the User Experience Questionnaire (UEQ) method.

User Experience Questionnaire (UEQ) is a method to measure the level of user experience from the design side of the Ayobandung.com website. This method is divided into 6 evaluation scales with 26 question items.

Research framework

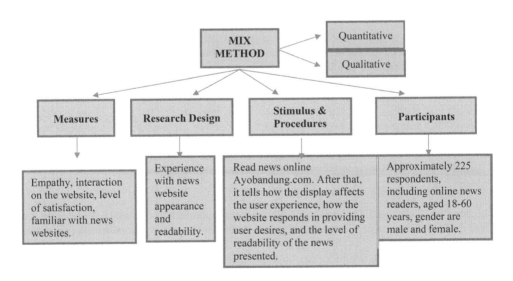

3 RESULT AND DISCUSSION

3.1 Empathy stage

In this stage, the researcher first observes about the product and what the company expects from its product. Ayobandung.com has made a design change to the appearance of its website in 2021. Now the new look is far from showing its clean side by increasing the white space on its website display. But in this new look, it tends to display ads on each side. The ad is called Google Adsense, which is a cost per click (CPC) based advertising program. Through advertising, the Ayobandung.com website will earn additional income. The ads that appear are ads determined by Google, but the content on the ads is relevant to the content on the website. In addition to Google Adsense, this website also has banner ads. This banner ad is a banner at the request of a client in collaboration with Ayobandung.com. This banner has the same system as the advertising column in print media, but this time it is broadcast digitally. If there are no advertisements, this website does not display many design elements. Design elements are only as differentiators and markers between part 1 and other parts.

Figure 1. The appearance of the ayobandung.com website.

The company expects its products to continue to show their existence along with the times and can attract more relationships to collaborate with them. Being the most superior in the Google search engine, a stable and top rating in West Java to the national realm one day is an aspect that is considered. Therefore, observations are made to the user. Users as readers and users of Ayobandung.com are the key to the success of the product. Companies want to listen to what users think, want, and need about the product. Later it will focus on the main thing, namely the user experience on the level of comfort in reading the product.

3.2 User Experience Questionnaire (UEQ) results

The steps taken after getting the results of the questionnaire are processing the data using the UEQ method. The data is processed using Data Analysis Tools from the official website ueq-online.org.

The following are the results of the analysis of the calculation of the questionnaire data using the UEQ method:

Table 1. Results using 6 aspects of the quality system.

Item	Mean	Variance	Std. Dev.	No.	Left	Right	Scale
1	2,0	0,8	0,9	228	annoying	enjoyable	Attractiveness
2	0,8	3,4	1,8	228	not understandable	understandable	Perspicuity
3	0,6	3,2	1,8	228	creative	dull	Novelty
4	0,3	3,7	1,9	228	easy to learn	difficult to learn	Perspicuity
5	0,7	2,9	1,7	228	valuable	inferior	Stimulation
6	0,9	3,3	1,8	228	boring	exciting	Stimulation
7	1,2	2,7	1,6	228	not interesting	interesting	Stimulation
8	1,2	2,7	1,7	228	unpredictable	predictable	Dependability
9	0,9	3,2	1,8	228	fast	slow	Efficiency
10	0,4	3,7	1,9	228	inventive	conventional	Novelty
11	0,4	3,7	1,9	228	obstructive	supportive	Dependability
12	1,1	2,7	1,6	228	good	bad	Attractiveness
13	0,5	3,7	1,9	228	complicated	easy	Perspicuity
14	1,2	2,8	1,7	228	unlikable	pleasing	Attractiveness
15	0,0	3,6	1,9	228	usual	leading edge	Novelty
16	1,2	2,3	1,5	228	unpleasant	pleasant	Attractiveness
17	0,9	2,8	1,7	228	secure	not secure	Dependability
18	0,7	2,9	1,7	228	motivating	demotivating	Stimulation
19	0,4	3,3	1,8	228	meets expectations	does not meet expectations	Dependability
20	1,2	2,5	1,6	228	inefficient	efficient	Efficiency
21	0,8	3,2	1,8	228	clear	confusing	Perspicuity
22	1,4	2,3	1,5	228	impractical	practical	Efficiency
23	1,4	1,9	1,4	228	organized	cluttered	Efficiency
24	0,9	3,9	2,0	228	attractive	unattractive	Attractiveness
25	0,6	3,6	1,9	228	friendly	unfriendly	Attractiveness
26	-0,2	3,4	1,8	228	conservative	innovative	Novelty

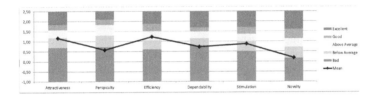

Figure 2. UEQ benchmark diagram.

Table 3. Results of the UEQ benchmark instrument.

Scale	Mean	Comparisson to benchmark	Interpretation
Attractiveness	1,16	**Below Average**	50% of results better, 25% of results worse
Perspicuity	0,58	**Bad**	In the range of the 25% worst results
Efficiency	1,23	**Above Average**	25% of results better, 50% of results worse
Dependability	0,73	**Bad**	In the range of the 25% worst results
Stimulation	0,87	**Below Average**	50% of results better, 25% of results worse
Novelty	0,18	**Below Average**	50% of results better, 25% of results worse

From the results of the evaluation, it can be seen that the results of the scale vary according to the numbers that show the numbers 0 to 1, as well as the diagrams that show the results up and down. The highest scale is on the efficiency scale with the number 1.23, while the lowest scale is on the novelty scale with the number 0.18. The following is the evaluation value standard:

1. **Attractiveness** with a score of 1.16 – Below average but positive research evaluation.
2. **Perspicuity** with a score of 0.58 – Poor with a neutral research evaluation.
3. **Efficiency** with a score of 1.23 – Above average with a positive research evaluation.
4. **Dependability** with a score of 0.73 – Poor with a neutral research evaluation.
5. **Stimulation** with a score of 0.87 – Below average but positive research evaluation.
6. **Novelty** with a score of 0.18 – Below average with a neutral research evaluation.

Table 4. Evaluation value standard.

Average Range	Description
>0.8	Positive Evaluation
−0.8 – 0.8	Neutral Evaluation
<−0.8	Negative Evaluation

3.3 Define stage

Based on the results of data processing and analysis results using the UEQ method, the results show that the news website Ayobandung.com has not met the needs and comfort in reading to the user. Users come from 225 respondents who participated in filling out the questionnaire through the google form. The things that are rated the lowest are novelty and user perspicuity of the website. The novelty in this case can be from the content and also a more contemporary appearance. While perspicuity is difficult for users to understand the content contained in the website.

In addition to collecting data from Ayobandung.com, the research was also conducted by collecting study literature data, namely the previous research of Sonia Elisurya, et al. entitled "Evaluating User Experience Using Usability Testing and User Experience Questionnaire (UEQ) (Study on E-Commerce Fashion). From the results of the UEQ questionnaire, on the test items, XYZ is superior in 3 aspects of system quality, namely perspicuity, efficiency and stimulation. Meanwhile, Vipplaza.co.id excels in the dependability aspect and Berrybenka.com excels in the attractiveness and novelty aspects. From the results of the questionnaire, it can be concluded that respondents tend to be higher in XYZ compared to Vipplaza.com and Berrybenka.com.

The following are the author's recommendations based on 6 aspects tested using the User Experience Questionnaire (UEQ), namely:

1. **Attraction** – The user's overall impression of the product. The evaluation of the research showed a positive response, therefore it continued to increase the attractiveness of users by using a more attractive display or a display that is not rigid so that it can be said that the display is user friendly.
2. **Perspicuity** – understanding of users in using the website. In the evaluation of the research, the lowest score and get it are said to be bad but the value still shows a neutral response, therefore from the display side it is expected to be able to use various icons that can be recognized by the user. This icon functions to make it easier for users to find what they want so that they can speed up use.
3. **Efficiency** – How is the speed of the user in completing his goal of visiting the website. The results obtained show an above average score with a positive assessment, but the developer still maintains the speed and further increases the speed. Avoiding bugs and problems on other websites should also be avoided so that the user experience is more satisfying.
4. **Dependability** – User interaction control. Ratings that are said to be bad but are still said to get a neutral response, therefore it is hoped that there will be changes to the website. The purpose of this accuracy is whether interactions with this website are predictable and

meet expectations. Difficult to predict and not meeting expectations can occur if there are many elements on the website that interfere with users in achieving their goals, for example advertisements on websites that interfere with users when reading articles or wanting to click buttons on the website. Therefore, it is hoped that the website will present an important point to be conveyed to users, if there are other considerations such as presenting advertisements for profit, it is hoped that the position of the ad is patterned (arrange the position of the ad well) in order to create a good user interface.

5. **Stimulation** – Product motivation. The assessment is 0.87 below the average but it can be said that the response is positive on this aspect. The stimulation in question is that users are motivated to return to visit the Ayobandung.com website. If they visit the website again, it can be said that the website has the content that users need. Continuing to improve quality, understandable, and up-to-date content can be an important point in increasing user stimulation for the Ayobandung.com website.

6. **Novelty** – Product innovation and creativity in attracting user interest. In this case the score obtained is 0.18 below the average and the response obtained is neutral. The novelty in this case can be seen from the appearance of the website, the content, and the interactive system. Along with the times, in 2022 it has entered digital-oriented where if the previous year was a year of conventional to digital transition, this year digital technology will experience significant changes. Therefore, it is expected to be able to bring newness to the website, such as a display filled with attractive designs (you can also look for characteristics), up to date content, to interactive systems such as coding for users to interact with it.

4 CONCLUSION

Based on the results of the design thinking approach in the empathy stage, that the company expects an increase in rating and can affect interest in reading the Ayobandung.com website. However, when an analysis was carried out using the UEQ method, the results were still not in line with the company's expectations. The results of the analysis state that there are still shortcomings in the Ayobandung.com website so that the user experience of the website is ranked on average. The level of efficiency is the highest level in this analysis, but the level of novelty and perspicuity is the lowest level. Therefore, it is hoped that this website will update the user interface and the content on the website. The update will later see changes that make the Ayobandung.com website easier to understand, and the clarity aspect will also increase.

REFERENCES

Aji, D. (2018, February Friday). *Darmawan Aji*. Retrieved from https://darmawanaji.com/empathize-tahap-awal-dalam-design-thinking/

Aprilia, P. (2020, April 23). *Getting to know the User Interface: Definition, Uses, and Examples*. Retrieved from Niagahosterblog: https://www.niagahoster.co.id/blog/user-interface/

Haekal, M. M. (2020, May Saturday). Niagahosterblog. Retrieved from https://www.niagahoster.co.id/blog/user-experience-adalah/

Mortensen, D. H. (2020). *Stage 1 in the Design Thinking Process: Emphatise with Your Users*.

Schrepp, M. (2019). *User Experience Questionnaire Handbook: All You Need to Know to Apply the UEQ Successfully in Your Projects*.

Soewardikoen, D. W. (2021). Visual Communication Design Research Methodology. PT Kansus.

Sonia Elisurya, *et al.* (2019). Evaluating User Experience Using Usability Testing and User Experience Questionnaire (UEQ) (Studies on E-Commerce Fashion).

Subverted narrative: On Bandung civic identity within Mufti Priyanka's monography

G. Gumilar
Telkom University, Bandung, Indonesia

*ORCID ID: 0000-0003-1669-2622

ABSTRACT: Bandung today is often depicted as a modern, intellectual metropole where creativity is cherished and flourishes. It has been the pillar where modernity was adopted and implemented during the early year of modern Indonesia. Later, this city is known to be a productive site for contemporary creators to develop, incubate, exercise, experiment, and contemplate their progressing ideas. Within contemporary art practice, their works are known for proposing intellectual merits manifested in sophisticated art idioms, modalities that granted them the status of being 'globally relevant,' albeit individually. Predictably, these commonly circulated qualities and characteristics are carefully curated and constructed, subverting various sincere representations of the city, such as local dynamics and everyday struggles. Giving credit where it is due, this article will explore Mufti Priyanka's illustrations highlighting these subverted narratives of the city. His depictions are somewhat curious, presenting the irony of everyday struggles operating within localism over displaying the 'typical' romantic descriptions. By adopting a specific 90s comic style laden with satirical, deliberate, and raw idioms, Muti's works appear to us as the journal of the truthful representation of the local dynamics of the city.

Keywords: Bandung, identity, illustration, local dynamic, Mufti Priyanka

1 INTRODUCTION

Within our current cultural landscape, Bandung is seen as an already developed, mature city with its particular character and identity. It is still romanticized for its subtle traces and remains of colonialism, a previously *Kota Kembang* (Haryoto 1984) or The City of Flower—for whatever reason—that later curiously transforms into and poses itself as a creative city. During Indonesian colonialism, its beauty granted them the label of *Paris van Java* for how it managed to establish modernist values reflected in their everyday living as apparent within their cultural activity. During those times, various theatre plays, art exhibitions, orchestras, and musical shows were conducted consistently (Siregar 2005). In the early independence, Bandung remained an exemplary site where the integration of modernity within Indonesian society could ideally take place, with a different take than Jakarta, which focused on economic progress and infrastructure development. The slower pace of Bandung and the various existence of higher academic institutions prompted this city to be a site for intellectuals to share, exercise, and contemplate their ideas. Even today, higher education has been one of the primary allures of this city.

Starting from the Indonesian Reformation nearing the end of the 20th century, Bandung positioned itself as one of the centers of the indie (independent) movement. The previously centralized government was replaced by regional autonomy, enabling individual municipalities to develop themselves without intervention (Hujatnika 2011). For the younger generation, this notion of 'autonomy' circulated within bureaucratic circles was interpreted in a rather peculiar manner. They saw it as an encouragement to gather self-reliance and determination in voicing individual concerns and expressions. However, this 'independence' was not evolved within the spirit of activism and anti-conformities. Instead, they developed into a relatively productive jargon for celebrating artistic freedom. Music was perhaps the popular medium they chose; meanwhile, the arts, fashion, and design evolved and catered to specific circles and audiences. This was probably the turning point for the overused 'creative city' that Bandung has proudly promoted to the public: a city envisioning itself as the supportive pillar for our creative economy, a destination where exploration and education are provided, and a site for innovative ideas being contemplated, being exercised and formulated. Admittedly, those claims are not without their problems, representing merely a small part of Bandung, far from their totality. Everyday struggles, issues, and concerns lurking beneath its communities are somewhat dismissed and clouded by the widely admitted 'happiness indices.' In hindsight, problems such as the lack of fair investment that later contributed to the worsening poverty index, irresponsible business practices that aggravated the already problematic environmental issues, and slight conservatism that negated the autonomy of the arts remained persistent. They were left ignored, primarily by the public administration.

Within this article, the first issues of everyday struggles of Bandung citizens, mainly in the southern and eastern parts, would be the center point of discussions. Somehow, the previously mentioned creative, romantic, and intellectual claims only represent Bandung's central and northern parts. Meanwhile, the other sides of the city are often 'operating' within a different 'framework.' Eastern Bandung was once known as *Nagara Beling* or 'dangerous ground' where clashes between groups have sparked and been ignited quite frequently. Meanwhile, the southern part and its outskirts were once industrialization sites where working classes and unskilled laborers resided. It might be said that recent developments have already addressed some of the previously mentioned issues. However, its sociohistorical traces remained apparent in some of their dynamics, perpetuating their socioeconomics and cultural concerns.

I will now introduce Mufti Priyanka, colloquially known as Amenkcoy, a multitalented mid-career artist, illustrator, musician, zine creator, and social media influencer whose works are rendering obsolete those idealized versions of contemporary Bandung by exposing the subverted—often truthful—narrative of the city. As a native 'easterners,' Amenk is sure to have an abundant experience of witnessing the struggles experienced by his community, particularly when we consider his formative experience while living under the New Order of Indonesia. At this perilous time, freedom and autonomy were relatively non-existent. Although Amenkcoy's pieces are often prompted by instantaneous realizations of the social phenomenon in his surroundings—Kiaracondong and its vicinity—his sensibilities are often relevant to the broader Indonesian context. It seems that Amenkcoy practices are almost equivalent to the emic approach of anthropological views, acquired naturally through constantly questioning and challenging the norms circulating within his society. This awareness led Amenkcoy to create works laden with satirical imagery, slang and jargon, harshness and rawness, all being delivered in a casual, everyday manner that directly relates to its audience. He uses rhetorical devices that deny every romantic, creative, and sophisticated feeling commonly found in his peers' works. The presence of various minor narratives he consistently showed in his piece, no matter how trivial, is a sincere manifestation of his relentless effort to spread awareness.

Within the Indonesian contemporary art market, Amenkcoy practice is, and has been, received a relatively 'lukewarm' appreciation. Admittedly, his frequent involvement in group exhibitions during the mid-2000s art booms had put him into the art scene. However, his involvement there has been gradually fading, nearing the first half of 2010. From that moment and after, he chose to cater to a more generic audience, despite his occasional inclusions in select art exhibitions. This move has been, in actuality, liberating him to explore various media and artistic strategies that were once prohibited by current trends and market tastes. Amenk began to explore subversive practices of mass media through zine publication, composing experimental music performances with his band, expressing his visual expression freely through social media platforms, and recently, minting his works as a Non Fungible Tokens and engaging with the crypto communities. Amenkcoy's agency, as a creative professional, has been penetrating deep within the public cultural domain compared to the contemporary art scene, slightly shifting—if not naturally evolving—into an 'honest and truer' artist representative of Bandung.

2 RESEARCH METHODS

Following the primary subject of discussion of peculiar artistic sensibilities, this article will present its research using qualitative methodology. The analysis would be performed solely within Amenkcoy's illustrations—a specific yet primary complex of his oeuvre—carefully selected using purposive sampling. The primary data of this research will be obtained through interviews and observations, while various complementary information will be accessed from open-access archive repositories. In addition, Amenkcoy's portfolio accounts and social media platforms would also be accessed. As for theoretical basis and underpinnings, this article will primarily refer to the content analysis presented in *Visual Methodology* (Rose 2001), complemented by contemporary art critic methodology improved by Terry Barret (Barret 1991).

As for how this article would approach the artwork selections, I will refer to the concept of bricolage, a technique or practice borrowed from the linguistic and literature field introduced by Claude Levi-Strauss (The Savage Mind 1964), in which the author can configure ' ... *fragments ... of previous cultural formations and re-deploying them in (a) new combination.*' (Johnson 2012). Although this approach might embed an inherent contradiction and operate within a different context, this intent is primarily used to conserve the integrity and provide a complete reading of the artist's narratives.

3 RESULT AND DISCUSSION

I will start the analysis by discussing the works (title, year) to provide the reader with a typical condition of struggling creative workers in Bandung, which the artist should

Figure 1. *Tanpa Judul* 2015.

experience at a point in his life. This first illustration depicts a 'makeshift' studio where a 'high-mobility' creator can rapidly adapt to any situation while working on a project. The appearance of a high-end device that symbolizes creativity and productivity, added by the tasteful backpack in the background, seems to signify how maintaining one's image as a creative professional and keeping up with the lifestyle is mandatory. This image is contrasted by the setting, a very modest —if not unfortunate—dormitory room that can barely provide sufficient comfort for the creators. The lack of proper furniture and the appearance of a cheap power strip should be enough to indicate that. We may ignore the figure shown in this picture, as it is assumed to be a random passerby or the housekeeper of the dorm, not the creators I previously mentioned.

Figure 2.   *Romansa Di Era Millenium Ka-3* (2017).

Figure 3.   *Nongkrong Denganku Kuy! (Menjelang Malam Minggu)* (2019).

Figure 4.   *'Kendali Generasi'* (2015).

The depiction of trivial, everyday living would be a recurring theme if we were to explore Amenkcoy's practice. The following works depict habitual events of daily struggles, which humble—if not tacky—love life, cheap alcohol, and cigarettes can be both the source of escapism or further incident. Within these illustrations, we might start to notice the punk subculture that deeply saturated the youngster living in his community and influenced him to some degree. His playful and humorous narrative should be quite evident here, not only in the clever delivery of exploring a paradoxical temporary relief and 'solace' being provided by 'forbidden rituals of drinking' but also in how those rough and heroic anti-establishment ideals of the punk subculture can be eased down by 'love and slavery' dilemmas, particularly in the works of (Figure 2).

Although somewhat amusing, these recurring parodies of everyday dilemmas might evolve into an explicit and deliberate criticism in other iterations of Amenkcoy's works, which often targeted corrupt governance with their fraudulent practices. As shown in the two illustrations below, his disgust and hatred toward these public enemies appear quite apparent, albeit slightly toned down.

These works depict how the concept of 'the man in uniform' might delude someone into thinking they stand at a higher stature in society, even freeing them from public oversight. Instances of police harassment—even for an alleged criminal—as the outcome of heroism syndrome, extended by unhealthy masculinity of displaying power through the practice of adultery exposed by the artist as a sign of his protest. By depicting them with a disgraced portrayal of their unethical deeds, Amenkcoy's work provides insight into how the practice of abuse of power might manifest in trivial, everyday matters. His choice of depicting the

Figure 5. Yesnowave (2014).

Figure 6. Cover of Sleborz Fanzine (2015).

irresponsible officers also indicates his—almost subconscious—sensibilities in capturing daily struggles for how this institution is directly responsible for maintaining civic order.

In various iterations of his works, Amenkcoy's playful narrative can turn into bleak expressions full of angst, agitation, and anger. The previous 'subtle commentary' of exposing trivial instances of power abuse can evolve into a full-blown and explicit criticism. These two works below might reveal some of these expressions: depicting how helpless commoners struggle over violence and power and how the dreams of economic progress provided by capitalism might aggravate social disparity. Various distressful facts and realities left cold the working classes in their sorrow.

Figure 7. *Untitled*, 2013.

Figure 8. *Jawaban Kepada Negeri* (2019).

4 CONCLUSION

These are Amenkcoy portrayals and depictions of the other side of his Bandung, focusing on the struggles and hardships faced by his community. By framing it within his local context as a problematic site of intersection, Amenk can fluidly respond to various societal issues situated within nationalism, tradition, religion, and globalization. As this article previously suggested, the problems of underpaid and overworked labor, power abuse, violence, and persistent poverty lurking within the underside of Bandung are unrevealed, canceling and negating the image of the creative city they desperately want to portray. Amenkcoy also illustrates a classic dichotomy of the oppressor and the oppressed, and his works function as various expressions of the commoners of their struggles against the oppressors.

REFERENCES

Barret, T. (1991). *Criticizing Art: Understanding the Contemporary*. Mayfield Publishing Company.

Haryoto, K. (1984). *Bandung Tempo Doeloe*. Granesia.

Hujatnika, A. (2011). *Negara dan Pasar: Globalisasi dan Dua Dasawarsa Seni Rupa Kontemporer Indonesia*. *MELINTAS, 27*(2), 171–186.

Johnson, C. (2012). Bricoleur and Bricolage: From Metaphor to Universal Concept. *Paragraph, 35*(3), 355–372. http://www.jstor.org/stable/43263846

Rose, G. (2001). *Visual Methodologies: An Introduction to the Interpretation of Visual Materials*. SAGE Publications.

Siregar, A. T. H. (2005). *Lukisan Baru: Setelah Lukisan Non-Representasional di Bandung*. Galeri Kita.

Sustainable modest fashion design based on consumer needs

C. Tifany Fahira* & M. Rosandini
Telkom University, Bandung, Indonesia

*ORCID ID: 0000-0003-3081-3304

ABSTRACT: The world's fashion industry contributes 92 million tons of waste annually, including Indonesia. Modest wear as the largest fashion industry in Indonesia is ranked third out of the top ten Modest Fashion in the world, thus becoming one of the fashion industries that contributes to the high amount of waste. Therefore, it is necessary to apply the concept of *sustainability* in products while still considering consumer needs in an effort to prevent and reduce the negative impact of fashion waste production. This is an opportunity for fashion business people in Indonesia to make sustainable modest wear products based on the needs of their users. In previous research, there has been a lot of development of sustainable modest fashion designs, but it is still rare to make consumers the main consideration in making a design, so this has an impact on the lack of demand for sustainable modest products. Therefore, further research is needed in an effort to explore the needs of consumers to provide suitable products. This is also done to explore consumer needs more deeply so that the products made are not only good by design but also answer the problems of consumer needs, especially sustainable modest fashion products in Indonesia. In this study, the method used was a *design sprint* with a mixed data collection method: (1) descriptive qualitative methods were used to determine consumer tendencies toward sustainable modest fashion products, as well as realize product designs that suited the needs and interests of sustainable modest wear consumers, (2) Quantitative methods were also used to strengthen the validation of data on the suitability of design needs of wider consumers. Consumer responses regarding the final result of this study are the design of modest clothing products according to needs, namely overalls with loose cuts with organic fabric material that is sustainable and has an earth tone color so that it is suitable for daily use.

Keywords: Sustainable, modest fashion, design, consumer needs

1 INTRODUCTION

The largest fashion industry in Indonesia today is modest wear which is ranked third out of ten Top Modest Fashion in the world. This makes the fashion industry cause problems, where the growth in the industry has contributed to 92 million tons of waste annually and also accounts for about 10% of global carbon dioxide (Kulsum et al. 2020). On the other hand, fashion products are primary needs (Rahmawati *et al.* 2021) that must be met.

The phenomenon that occurs in the *fashion* industry can be said to be very worrying. Waste from the *fashion* industry is predicted to increase by as much as 60% between 2015 and 2030 in the world including in Indonesia (Mishra et al. 2021). By seeing this, Indonesia must try to solve this problem. Therefore, referring to the previous explanation, one of the efforts that Indonesia can make in adapting and mitigating climate change is to produce the needs of fashion products with an *eco-fashion* approach. The emergence of *eco-fashion* is backgrounded by the changing preferences of people in developed countries who initially prefer products at affordable prices changing to choosing products by looking at their impact on the environment and society (Yoga & Eskak 2016). In fact, not only in the international market, but local Indonesians have also begun to have a preference for environmentally friendly fashion products. This is in line with the statement of the Chairman of the *Indonesia Fashion Chamber*, Ali Charisma that during the pandemic people will think more carefully and realize that *sustainability* is not only about fashion but also about lifestyle (The Jakarta Post 2006). The majority of people who are willing to spend more *on fashion* are women. It happens because Women pay enough attention to appearance compared to men (Soetjipto & Chandra 2021).

Environmental concern from consumers based on the results of research, has a positive and significant influence on the intention of consumers in Indonesia, to buy environmentally friendly products. This indicates that environmentally friendly products will be widely purchased by consumers if these consumers have environmental concerns as an effort to minimize adverse effects on the environment caused by similar products that are not environmentally friendly. As stated in the results of the research conducted, consumer demand for(McKinsey 2019) *sustainable fashion* is increasing rapidly (Alhally 2020).

1.1 *Sustainable fashion*

Sustainable fashion is a *fashion* that prioritizes the values of the parties involved, especially in the humanitarian and environmental fields. The purpose of *sustainable fashion* is to unite the perspectives of designers, producers, distributors, to consumers to work together in uniting thoughts so that *fashion* can be led in a better direction from production activities to consumption (Henninger et al. 2016; Kulsum 2020).

1.2 *Modest fashion*

Modest wear is a garment made to maintain the decency of the wearer and is made more introverted (Indarti & Peng 2017). The word *modest* itself has a simple meaning. The simple and decent clothes are loose, not dreamy, shoulders closed, and do not resemble the clothes of the opposite sex modest clothing in Indonesia itself is very widely available, although brands that produce modest sustainable clothing are still very few in number. Through the literature study conducted by the author, there is still no study on the design of sustainable modest fashion products based on consumer needs. The research that has been carried out is the creation of modest fashion sustainable designs which are carried out based on the problem of fabric waste so that the results are (FCP 2016). Zero waste designs in an effort to make *sustainable* products (Nisa & Yuningsih 2021).

2 RESEARCH METHODS

The design method used is design sprint. A design sprint is a flexible product or service design framework that aims to increase the likelihood of creating something that customers want (Banfield *et al.* 2015). There are 5 stages in the design sprint, namely map, sketch, decide, prototype, and test (Jake Knapp 2016).

Meanwhile, the data collection method used is qualitative to find out product designs that are in accordance with the needs and interests of *consumers* in a sustainable *fashion*.

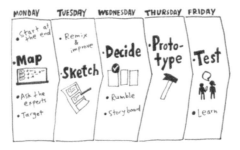

Figure 1. Design sprint process.

2.1 *Qualitative data*

The qualitative data that will be sought comes from literature studies in the form of journals, books, and other scientific articles to obtain data on theories of sustainable fashion, modest wear, and customer-based products. In addition, conducting interviews with several influencers who own Instagram accounts who upload content that cares for the environment, minimalist lifestyle, and content that leads to a sustainable lifestyle as a representative of consumers or the target market of sustainable products to find out consumers' interests and needs for sustainable modest fashion based on its daily activities, in terms of design including materials, pieces, and colors.

The population in this study was women who used modest wear in Indonesia aged 19–35 years and who lived in urban areas in Indonesia. This population was chosen because young consumers under the age of 35 around the world are currently paying more attention to the origin of the product, its composition, carbon footprint, as well as the quality and other consulates of the products they buy by choosing products that apply ethical practices, namely products that apply sustainable principles (CBI Ministry of Foreign Affairs 2019).

The samples used were modest wear users aged 19–35 years who lived in Bandung Raya and Jakarta. Bandung Raya and Jakarta were chosen because based on researchers' observations, they have a sustainable product ecosystem, which can be seen from the availability of shops providing organic, environmentally friendly products, to restaurants and stalls providing vegetarian and vegan food and beverages.

3 RESULT AND DISCUSSION

The result of observations made through social media on 3 Instagram accounts @atiit, @ntsana, and @vendryana is that modest clothes needed for daily use have loose cuts, casual style, are made of cotton, and are basic and neutral colors to be used indoors. From the data on modest clothing needs, it is then processed through the sprint design method to obtain modest sustainable clothing designs that are in accordance with consumer needs.

The 5 stages of the design sprint carried out are as follows: (1) Folder. The purpose of the map stage is to find out the background and maintain the needs of the user. At this stage, the method used is abstraction laddering, which is to expand or narrow the focus to reconsider the problem statement. Through literature studies and observations through Instagram social media to 3 Instagram accounts @atiit, @ntsana, and @vendryana. After the data was processed through abstraction laddering, it was found that there was a user's need for *sustainable modest wear* products for daily activity needs, especially indoor activities; (2) Sketch, which is to sketch with the aim of generating as many ideas as possible. The method used at this stage to make sketches is a solution sketch, sketches are made based on the results of analyzing consumer needs in the form of fashion products. Product sketches are made by considering consumer habits in dressing, it can be seen from the place where ordinary users do activities, pieces of clothing, colors, and materials used. There are 3 sketches of clothing, each sketch represents each user who has been observed. The design shows several pairs of clothes made from organic natural fibers such as Tencel and eco linen, casual style with loose cuts.

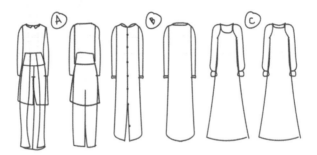

Figure 2.   Design sketch.

Sketch A is a set of clothes in the form of shirt tops and loose trouser bottoms with organic natural fiber material, which can be combined with other outers such as vests made from natural fibers such as wool. Sketch B is a long-sleeved overalls outfit, the length of the shirt to the ankles, with loose cuts that use organic natural fiber material. Sketch C is a set of long-sleeveless overalls and a long-sleeved blouse top with loose cuts that use organic natural fiber material; (3) Decide, that is, choose the best sketch and best suit the needs of consumers. The sketch selection method is carried out through a decision matrix, by comparing user impact and implementation efforts to obtain sketches with high impact / low effort, namely sketch B. Technical difficulty is assessed from several aspects of the production process such as the level of difficulty of making patterns, length of work, to the stages of sewing and the number of supplies used. Meanwhile, the user value is derived from the degree of conformity of the design to Muslim modest wear terms such as less skin-revealing and loose cuts.

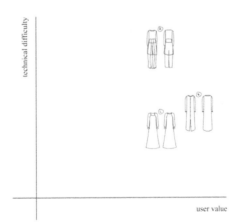

Figure 3.   Decision matrix.

Fourth *Prototyping*. At this stage, the sketch is refined into the final design with the addition of details and coloring as shown below

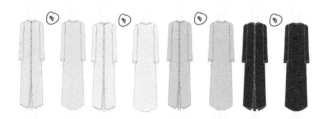

Figure 4. Final design.

The last one is testing, which is aimed at determining the suitability of the design that has been made by the designer to the needs of consumers. Testing was carried out on 5 respondents of sustainable modest wear consumers in the Bandung, Jakarta, and surrounding areas through a questionnaire containing open questions. The results showed that 4 out of 5 respondents agreed that the design made using the sprint design method was in accordance with consumer needs. Respondents judged that the design made had a loose cut that was in accordance with the provisions of the modest itself, seemed minimalist, simple, and elegant, and could be combined with other clothing items such as outerwear. The color used is earth tone so that it is neutral and can make it easier for consumers to choose a veil wana. The material used looks light. However, the design can be made even more visually attractive, by changing the pattern of the piece so that it is not too simple and has accents as a *point of interest*.

4 CONCLUSION

Clothing products needed by sustainable modest wear consumers in Indonesia are clothes with loose, minimalist, simple, and elegant cuts with a little additional accent so that they can be combined with other items. The color used is neutral so it is easy to pair with the color of the veil that the user will use. The material used is suitable, which is made from light natural fibers. The conduct of this research provides more options for appropriate modest wear sustainable products, besides that the modest fashion industry can reduce the contribution of fashion waste that cannot be recycled even though it is the 3rd largest fashion industry in the world. Further research can conduct interviews at the testing stage, resulting in more detailed answers from consumers and getting further explanations about the design responses made and daily clothing needs in terms of cuts, accents, and other details, materials, and colors.

REFERENCES

Alhally, A. (2020). Pengaruh Kepedulian, Pengetahuan, Sikap, Inisiatif Pemerintah, Tekanan Teman Sebaya Dan Spiritualitas Terhadap Niat Beli Produk Ramah Lingkungan Di Banjarmasin. *At-Tadbir: Jurnal Ilmiah Manajemen*, *4*(2), 130. https://doi.org/10.31602/atd.v4i2.3032

CBI Ministry of Foreign Affairs. (2019). *Contents of this page Tip: Tips: 1*(August), 1–21.

FCP, C. (2016). *Modesty*. https://issuu.com/chloelebow/docs/modesty_isuu2

Henninger, C. E., Alevizou, P. J., & Oates, C. J. (2016). What is Sustainable Fashion? *Journal of Fashion Marketing and Management: An International Journal*, *20*(4), 400–416. https://doi.org/10.1108/JFMM-07-2015-0052

Indarti, & Peng, L.-H. (2017). Bridging Local Trend to Global: Analysis of Indonesian Contemporary Modest Fashion. *2017 International Conference on Applied System Innovation (ICASI)*, 1710–1713.

Kulsum, U. (2020). Sustainable Fashion as The Early Awakening of the Clothing Industry Post Corona Pandemic. *International Journal of Social Science and Business*, *4*(3), 422–429. https://ejournal.undiksha.ac.id/index.php/IJSSB/article/view/26438

Kulsum, U., Timmermans, J., Khan, M. S. A., & Thissen, W. (2020). A Conceptual Model-based Approach to Explore Community Livelihood Adaptation Under Uncertainty for Adaptive Delta Management. *International Journal of Sustainable Development & World Ecology*, *27*(7), 583–595.

McKinsey. (2019). *Fashion's New Must-have: Sustainable Sourcing at Scale*.

Mishra, S., Jain, S., & Malhotra, G. (2021). The Anatomy of Circular Economy Transition in the Fashion Industry. *Social Responsibility Journal*, *17*(4), 524–542. https://doi.org/10.1108/SRJ-06-2019-0216

Nisa, N., & Yuningsih, S. (2021). Perancangan Busana Modest Wear Dengan Konsep Zero Waste Menggunakan Teknik Shibori. *EProceedings of Art & Design*, *8*(6).

Soetjipto, A., & Chandra, A. (2021). Fashion, Feminisme dan Hubungan Internasional: Perdebatan Dalam Literatur. *Jurnal Ilmiah Hubungan Internasional*, *17*(1), 17–29.

The Jakarta Post. (2006). *'New Normal' Fashion Trends: The Rise of Sustainable Fashion, Local Pride - Inforial - The Jakarta Post*. https://www.thejakartapost.com/adv/2020/07/04/new-normal-fashion-trends-the-rise-of-sustainable-fashion-local-pride.html

Yoga, W., & Eskak, E. (2016). Ukiran Bali dalam Kreasi Gitar Elektrik. *Dinamika Kerajinan Dan Batik: Majalah Ilmiah*, *32*, 117. https://doi.org/10.22322/dkb.v32i2.1367

Strategy to maintain the existence of the museum in era 4.0 through the virtual Museum of Sultan Syarif Kasim in Bengkalis Regency

M.R. Kurniawan & T. Hendiawan
Telkom University, Bandung, Indonesia

ORCID ID: 0000-0001-8157-6252

ABSTRACT: The digital era of 4.0 has made many changes in people's lifestyles today. The rapid growth of internet use has an impact on all fields, one of which is the tourism industry, the Ministry of Tourism has established the concept of Go Digital or digital tourism as the top Ministry Program. This concept is one of the strategies to maintain the existence of the Sultan Syarif Kasim Museum, the problem found is the low interest and desire of the public to visit this heritage tourist destination directly. The purpose of this study is to analyze the virtual benefits of museums in meeting the needs of today's society. The research method used is basic research with descriptive qualitative processing techniques. Data collection was carried out by conducting direct observations, interviews with 10 respondents, and documentation of problems and field findings. The results of this study show that the use of virtual museums (VR) is an effective modern tourism innovation in an effort to maintain the existence of museums, and the results of the analysis show that virtual reality (VR) museums affect the visiting power of the Museum Sultan Syarif Kasim II Bengkalis.

Keywords: virtual museum, era 4.0, the existence of a museum, the museum of Sultan Syarif Kasim

1 INTRODUCTION

The development of the digital revolution 4.0 era brings very significant changes to people's lifestyles today, people are facilitated to carry out any activities using digital technology in their daily lives, such as cellphones, laptops, and televisions, which are all connected to the internet network so that they can access whatever they want to know through the internet of things (IoT) platform. This is certainly a special concern for the novelty of innovation in the tourism sector so that it can survive and maintain its existence in an already digital environment like today because if it does not develop, then the sector will experience setbacks and even be abandoned by the digital society as it is today. Some of the fields that are considered a sign of the revolution era 4.0 include the internet of things (IoT), big data, cloud computing, cyber-physical systems, autonomous robots, additive manufacturing, augmented reality/virtual reality (VR), and artificial intelligence (AI).

One of the important sectors of the digitalization era 4.0 that will be of special concern in this study is to maintain the existence of museums as one of the sectors that are starting to decrease the number of enthusiasts, the museum of concern in this study is the Sultan Syarif Qasim Museum, which is the only museum in Bengkalis Regency. This museum is located on Jl. Jendral Sudirman, Parit Bangkong, Bengkalis District, Bengkalis Regency. It was established in 1977–1978. Furthermore, this museum was inaugurated and also officially named the Sultan Syarif Qasim Museum by the Governor of Riau in March 1996. This museum contains objects left by the Malay kingdom, which are already in duplicate form.

Previously, this museum was the resting house of the Sultan of Siak. Several relics thick with Malay culture can be seen in the museum, such as royal symbols, golden chairs of the Siak Sri Indrapura Kingdom, royal photos, various royal traditional clothes, royal weapons, jewelry, king currency, art tools, and two ceramics. Known as a historical tourist attraction, this museum does not escape the visit of the community, both local people and people outside the region, to get information (Andini 2016).

Virtual reality (VR) media can be used in visiting museum buildings that are currently in increasingly poor condition due to the low public visit to museums. This research focuses on maintaining the Sultan Syarif Kasim museum building located in the Bengkalis city of Riau Province with a digitalization 4.0 approach from the virtual reality (VR) sector. This is due to the low visiting of local and outside communities to the museum building. According to the manager of the Sultan Syarif Kasim Museum (2022), one of the leading tourism sectors in Bengkalis City is museum, but the visit data encountered experienced a decrease in the number of visitors. It can be seen from the latest 3-year visit data as shown in Table 1.

Table 1. Museum visitor data 2019–2021 source. museum manager (2022).

Kategori Pengunjung	2019 L	P	2020 L	P	2021 L	P
Umum	22	15	14	3	29	8
PNS/Instansi			19	3		1
Mahasiswa	4	4	4	2	2	4
Pelajar (SD/SMP/SMA)	10	8	4	13	8	5
DLL						
Jumlah keseluruhan	63 Orang		62 Orang		57 Orang	

Data in Table 1 shows that people's interest in visiting this museum is still very low. This research aims to provide innovation and create a new ecosystem for smart and sustainable digital educational tourism and increase knowledge and cultural values that can be an advantage for the Sultan Syarif Kasim Bengkalis museum as a tourism destination that will be enjoyed not only by local people but attract tourists outside the country to digitally visit the museum. Based on the background description above, the purpose of this research is to design a concept about digital tourism that will make it more appealing for people to visit actual or virtual museums. This is because the Sultan Syarif Kasim Museum has a variety of issues and weaknesses, necessitating the development of more digitalized innovations.

2 RESEARCH METHODS

This research was conducted using a qualitative descriptive method. According to Sugiyono, (2013) "Qualitative research methods are used in research where problems are clear while based on the research objectives and methods to be used, this type of research is descriptive and verifiable research." According to Sugiyono (2005), "Descriptive research here aims to obtain a description or description of tourists' interest in visiting the Sultan Syarif Kasim museum in the city of Bengkalis." The sample, according to Sugiyono, (2013), is part of the number and characteristics possessed by the population. In this study, 10 samples were randomly distributed to local respondents in Bengkalis City in June 2022.

The object studied is a visual image in the form of a 360° Virtual Reality (VR) sketch, which is designed simply and then conducts experiments on respondents to determine objects that will be more in demand and easier to visit in the era of digitalization 4.0 as it is today.

The data is then collected through observation, in-depth interviews, and documentation of the object under study. The data analysis technique was carried out by conducting in-depth interviews by asking for feedback from respondents who observed directly the visual results of the virtual reality museum about the benefits of digital education media for the people of the city of Bengkalis. The respondents included community leaders, students, and government officials in Bengkalis City.

3 LITERATURE REVIEW

Museum definition

Statutes International Council of Museums (ICOM) is a supra of the museum system that binds museums in one professional container. The bond of the profession is not only based on knowledge, expertise, and similar skills but also because they both have the same ethical and philosophical foundations, which are universal in style. The definition of a museum is broadly taken from that listed in the Statutes of the International Council of Museums (ICOM), which has practically been in force throughout the world. A museum is an institution that is fixed, does not seek profit, serves society and its development, and is open to the public, which collects, cares for, communicates, and exhibits. It is for the purpose of study, education, and pleasure, as well as material evidence of man and his environment. In addition, a museum when viewed as an institution or organization is also a system consisting of various elements or components that each other relate to and interact with, because each component is alive and moving as it functions. The components of the museum system are personnel, buildings, collections, public, and other facilities. However, if the term system is taken, then the other components are related in one work network.

Complementing the definition of the museum as referred to above, ICOM recognizes the following as conforming to the above definition:

1. Conservation institutions and exhibition spaces are regularly maintained by libraries and archival centers.
2. Archaeological and ethnographic relics and natural places, relics and historical places have the style of a museum, because of their activities in terms of acquisition, maintenance, and communication with the community.
3. Institutions that exhibit living creatures, such as gardens, plants and animals, and other creatures and plants.
4. Nature sanctuary.
5. Knowledge centers and planetariums

Virtual Reality (VR)

Virtual reality is a technology that can make a person do a simulation of a real object by making a computer as a tool that will provide a three-dimensional sensation so that it makes the user feel as if they are actually involved (Achyarsyah dkk. 2020). Users will see a three-dimensional world that is actually dynamic images. Virtual reality environments generally provide a visual experience that is displayed on a computer screen. Some simulations include additional information, such as sound through speakers and headphones. With additions such as headphones or speakers, listeners will be able to hear realistic sounds and can add to the real feel.

According to ALA, virtual reality (VR) is a simulation of an image or an entirely computer-generated environment that can be experienced using specialized electronic equipment, which allows its users to be "present" in alternative environments such as in the real world to three-dimensional (3D) virtual objects and information with additional data such as graphics or sound. It takes the form of a 360° video that captures the entire scene where the user can look up, down, and around it and allows the user to interact with both physical and virtual objects. This new "reality" can create unique experiences that expand opportunities and direct user engagement. The following illustration of a unique experience is from a commentary of Jeff Peachey, a book conservator who visited a virtual reality exhibition of the interior of the library resulting from the collaboration of Alberto Manguel/ Robert Lepage "La bibliotheque, la nuit" at the Bibliotheque et Archives Nationales du Quebec, in Montreal, Canada in 2016.

Virtual museums are one of the marketing strategies carried out by museums. The idea of a virtual museum was first introduced by Andre Malraux in 1947. He developed the concept of an imaginary museum (le musee imaginaire), a museum without walls, locations, or boundaries of space, like virtual museums, with its content and information surrounding an object, and perhaps made affordable for all people in the world (Styliani dkk. 2009). Virtual Museums have the advantage of introducing, informing, and promoting museums because they are effective in terms of conveying information and efficient in their costs. On the other hand, with a higher interest in visiting, it is hoped that later it will make a real visit.

4 RESULT AND DISCUSSION

Sultan Syarif Kasim museum is located in Bengkalis Regency, Riau Province, and it was established in about 1977–1978. It was inaugurated by the Governor of Riau in March 1996 under the name Sultan Syarif Kasim Museum. Some of the museum's collections include the golden chairs of the Siak Sri Indrapura kingdom, royal symbols, various royal clothes, royal weapons, empress jewelry, royal photos, art tools, ceramics, and currency. The museum is located on Jalan Jenderal Sudirman, a main road that connects Laksamana Bengkalis Port with the city center.

Gambar 2. Interior museum SSK II (Sumber: self doc., 2022).

Gambar 3. Fasad museum SSK II (Sumber: self doc., 2022).

Virtual Museums or digital museums have not been widely done in Indonesia, in order to maintain the existence of the Sultan Syarif Kasim II museum, then it will be researched and developed it as a virtual museum. Virtual Museum has the convenience of inviting the public and students to visit this virtual museum because its designation as a museum is used as a medium for educational tourism in an area and becomes a reference for visitors to the progress of world civilization through media in the form of digital and physical artifacts displayed in the museum.

The following is an overview of the Virtual Plan Sketch of the Sultan Syarif Kasum II museum, which will be tested on respondents so that it can give an idea that the media will be more frequent and practical to use is in accordance with the progress of the times in this 4.0 era.

Figure 3. Virtual reality sketch of Sultan Syarif Kasim II Museum Source: Processed personal (2022).

Figure 4. Barcode Virtual reality museum Sultan Syarif Kasim II Source: Processed personal (2022).

Based on the results of the research mentioned above, there are several inputs and responses from 10 respondents who became resource persons including museum managers, museum guides, and the general public during direct observation. One of the 10 respondents stated that the existence of this virtual museum is a novelty that is very effective to be applied in the future because according to Ernawati as the head of the UPT Museum Sultan Syarif Kasim said "the difficulty of access to Bengkalis City which has to cross using a Roll on Roll ferry. off (RoRo), so it takes at least one day to visit the museum in Bengkalis City. This is considered by the community to be impractical and complicated".

The existence of this Virtual Museum will slightly alter the visual image displayed in the design. This is because the design displayed in visual Virtual reality (VR) is an ideal and updated museum design in terms of layout and interior appearance. This needs to be supported by applying a visual interior design into the building so that displayed visual and real forms of the building will remain the same.

The sequence of events or the history of artifacts is carried out in the form of a "story line" so that it can make it easier to arrange the circulation of limited rooms and can maximize the movement of visitors who have been directed. This is very different from the existing situation where the museum does not regularly arrange its artifacts so that it will reduce the experience of visiting the museum because it does not get a clear and directed storyline.

From the results of the discussion, several things can be summarized. First, the community around Bengkalis City feels the positive impact of innovation in the digital form of the Sultan Syarif Kasim Museum; second, the community also complains of difficulty in obtaining detailed and concrete information related to the existence of the history of the Sultan Syarif Kasim Museum because currently the history related to this artifact object is explained orally or through a "tour guide" explanation only so that the information conveyed is very likely to be wrong or wrong the delivery.

5 CONCLUSION

From the description of the discussion above, it can be concluded that the use of digital educational media in the form of virtual reality (VR) simulations in the interior space of the Sultan Syarif Kasim II Museum has a positive impact on the sustainability of the main function of the museum as a place for cultural education so that the Sultan Syarif Kasim Museum will still be able to survive and exist. In the era of digitalization 4.0 where everything can be accessed easily by humans only by using fast internet, to balance that speed, a new media in visual communication was created at the Sultan Syarif Kasim II museum located in the city of Bengkalis. Based on the results of observations of 10 respondents who saw directly the visualization of this digital museum, they stated that the museum was more interesting and more flexible to visit so that people outside the city of Bengkalis had easy access to visit this museum virtually.

With this paper, the author offers a solution for the existence of other museums in Indonesia that have similar problems related to the lack of interest in visiting and public curiosity regarding the history of an area that can be learned through the artifacts on display. With the appointment of the case study of the Sultan Syarif Kasim regional museum, it is hoped that it can increase public attention and attract tourists from outside the city and within the city, who are enthusiastic about visiting the museum.

This article is expected to contribute to the study of how important it is to innovate in an effort to maintain the existence of the museum so that people can continue to study and understand the value of tradition and history of an area that has long historical value in the past.

REFERENCES

Achyarsyah, M., Rubini, R. A., Hendrayati, H., & Laelia, N. (2020). *Strategi Peningkatan Kunjungan Museum Di Era Covid-19 Melalui Virtual Museum Nasional Indonesia. Journal IMAGE |* (Vol. 9). Diambil dari https://www.museumnasional.or.id/virtual-

Andini, O. (2016). *Pengelolaan Fasilitas Museum Sultan Syarif Kasim Di Kabupaten Bengkalis. JOM FISIP* (Vol. 3). Pekanbaru.

Styliani, S., Fotis, L., Kostas, K., & Petros, P. (2009). Virtual Museums, A Survey And Some Issues For Consideration. *Journal of Cultural Heritage*, 10(4), 520–528. https://doi.org/10.1016/j.culher.2009.03.003

Sugiyono. (2005). *Memahami Penelitian Kualitatif*. Bandung: CV. Alfabeta.

Sugiyono. (2013). *Metode Penelitian Kuantitatif, Kualitatif dan R&D*. Bandung: CV. Alfabeta.

Analysis of startup business promotion mistakes and solutions through digital media

T.F.L. Adino & M. Wardaya
Universitas Ciputra, Surabaya, Indonesia

ORCID ID: 0000-0002-5182-3969

ABSTRACT: Indonesian start-up businesses that are always required to grow, 90% still fail in the global market competition, because the founders often apply the wrong promotion strategy. In fact, the Indonesian government has facilitated capital and transactions through digital banking services. The purpose of this study is to list these mistakes by collecting data from local and international journals and books, as well as interviewing several start-up business owners. The result was that busyness in production led to a lack of design and copywriting skills. The solution is education in collecting data on the Ideal Customer Profile, determining social media, making copywriting according to the target market, and executing audio and visual designs. This research is important as a reference for start-up business owners to manage digital marketing more measurably and with direction. As well as being a guideline for digital creative agencies to help start-ups execute a brand's promotional campaign.

Keywords: digital media, startup business, promotional mistakes, online marketing

1 INTRODUCTION

Digital transformation is now the most massive business development method (Magnusson *et al.* 2021). Various media are used to deliver promotions, product or service information, and interactive content. Along with this phenomenon, Indonesian startup businesses are also growing significantly. Wiesenberg *et al.* (2020) revealed that startups are different from other businesses, because they must always innovate and grow. Since startups dominate business markets around the world, the government has finally facilitated a supporting ecosystem (Spiegel 2017 in Wiesenberg *et al.* 2020). For example, in Indonesia, there are non-cash banking service facilities that make it easier for startup businesses, which are equipped with e-commerce transaction service features, capital loans, bank assurance, and investment. Quoted from the CNBC Indonesia (2022) page, the National Banking Service has been regulated in the Financial Services Authority Regulation Number 12/POJK.03/2018. The government has also appointed Grab and Gojek, unicorn startups in Indonesia, as providers of People's Business Credit (KUR) for small to medium-sized startups. Even though there are facilities from the government and the sophistication of digital technology, many start-up businesses still fail at the beginning because there are many inhibiting factors in the scope of digital marketing which will be examined in this journal article. Therefore, this study took a real sample in the form of answers to interviews with several start-up business owners regarding the mistakes they made in terms of product or service marketing. Then, the data is synchronized with trusted literature sources to find solutions for each of these mistakes. The collected data becomes a reference in finding solutions to overcome digital promotion strategy mistake.

This journal article explores data related to the following problem formulation. (1) What are the mistakes of startup business promotion strategies that fail to reach the target market at the beginning of the business? (2) What is the solution to overcome the mistakes in the startup business promotion strategy through digital media? So that the objectives of this study can be obtained, namely (1) Collecting data on mistakes in digital promotion strategies from startup businesses that fail to reach the target market at the beginning of the business (2) Finding solutions to overcome mistakes in startup business promotion strategies through digital media.

2 RESEARCH METHODS

This study uses a qualitative approach through literature studies (national journals, international journals, and books) as the theoretical basis for initial data recording. Then, it is matched with data from interviews with several start-up owners who have at least entered the second year of their business being established. Several other books and journals are a source of inspiration for solutions offered from various marketing experts. Secondary data was also obtained through several websites that contained articles related to the topics discussed.

2.1 Methods of data collecting

Primary data were collected using interview techniques. Interviews are a data collection technique through an oral question and answer process that takes place in one direction with startup business owners, meaning that the questions come from the interviewer and the answers are given by the interviewee. Questions are closed questions with the method of asking through social media and live chat.

2.2 Methods data analysis

Analysis of qualitative interview data usually begins with conducting interview transcripts. Analysis of qualitative interview data is often done by coding. Coding against qualitative data makes it easier to interpret. Through coding, it is easier for the interviewer to catch the intent conveyed by the interview subject.

3 RESULT AND DISCUSSION

Startup business promotion strategy mistakes: Literature study

Based on data from the Indonesian Information and Communication Technology Creative Industry Society (MIKTI) quoted from the kominfo.go.id page, the number of startups in Indonesia in 2018 reached 992 startup companies. Now, as of 2022, Indonesia has become fifth out of ten countries in the world that have the most startups in the country, amounting to 2,346. Although the number is very large, 90 percent of startups fail because they do not pay attention to what is the cause of the growth disconnect of their business (Yu 2021:8). According to Cholil (2018:5), corporate identity is a visualization tool that differentiates a business from other businesses with a symbol with a unique philosophy. Through the consistent uniqueness shown in all its promotional media, it will be easier for the public to recognize the business without thinking twice. For example, startups that are engaged in the food and beverage industry and are now growing rapidly, namely Hangry, Yummy Corp, Dailybox, and the like are consistent with color, shape and packaging design, logo color and placement, and design composition on social media. However, due to financial constraints, startup business owners are reluctant to invest their capital to pay for social media design or management services like this example.

Figure 1. Hangry, Yummy Corp, and Dailybox social media design resource: author's documentation.

Furthermore, in addition to the absence of a corporate identity and social media design that does not highlight the aesthetic side and is rarely updated, the second mistake is in the marketing copywriting style that is still too explicit. For example, using the sentence, "Buy our product because it is better than the other product", "Our product is guaranteed to be the best", and so on. According to Cholil (2018:6), marketing is not only interpreted as a product battle but also a builder of audience perception. If we are able to create creative advertisements that touch the subconscious of consumers, they will feel more familiar with our products and have the potential to continue making purchases until customer loyalty arises. According to Gultom *et al.* (2020), customer loyalty means the consistency of a customer to buy the same product at different times. Customer loyalty is inseparable from the satisfaction that has been felt by the customer based on the experience of expectations from marketers' promises compared to reality. The best alternative to create customer loyalty is to create memorable content to develop an emotional connection with consumers. So not only continuously forcing audiences to swallow explicit promotions, startup businesses can also take advantage of storytelling content, product reviews or testimonials, viral entertainment news, and other fresh content.

Startup business promotion strategy mistakes: Interview results

Based on the results of interviews with 10 (ten) start-up business owners who have been running for at least 2 years, a number of promotional strategy mistakes were found. The dominant mistakes made by some startup business owners are too focused on developing the quality of the products or services offered (production stage), busy taking care of clients or consumer orders so they do not have time to do promotions. on the social media accounts of business entities, there is no official website to display products and information thoroughly, and there is still low education about the importance of digital marketing so that the business takes longer to reach the target market.

Marketing solutions through digital media

Quoted from Quesenberry (2019) in Suhartono (2021), brands can bring important social influence through customer experiences on social media, because brands help to realize consumer needs. Furthermore, brands will also provide solutions to consumer problems through their social media content (Miller 2017; Xollo *et al.* 2020 in Suhartono 2021). It is this brand communication that needs to be clarified in the aspect of the message conveyed so that consumers hear it. The way to achieve this is to create messages that are simple, relevant, and repeatable. Messages that tend to be dense, clear, and useful actually attract the attention of the target audience and potential consumers.

Therefore, it is important to understand the target market segment at the beginning because it will determine the design style, language, and platform used. Based on the theory of Neumeier (2015) in Tantra (2021) that the current target market not only wants to seek experience using something but wants to be more involved with certain brands. They tend to look for brands that are relevant to their personality, starting from their speaking style to their point of view (Yarrow 2014 in Tantra (2021). So it is important to resonate with consumer habits regarding certain metaphors, expressions, or characteristics. Yu (2021) notes that based on his experience developing a start-up business, there are several techniques that can be used to record the Ideal Customer Profile (ICP), namely:

1. Using case studies
2. Comparing startups with medium and large businesses
3. Using a vertical industry, namely a market that includes a group of companies and customers who are all interconnected around a particular niche
4. Analyze the market by geography
5. Create a buyer persona
6. Create a product or service user persona

After listing the market segments, then the start-up business owner needs to understand the media that will be used throughout the marketing process. Various digital media, which are dominated by social media, become a means for exchanging information, collaboration, and getting to know each other via writing, audio, and audiovisual (Puntoadi 2011). Some of the recommendations quoted from Ainurrofiqin (2021) are as follows.

1. Facebook Fanpage that can be a purpose of displaying portfolios and other important information
2. A separate business-only Instagram account from a personal account
3. Twitter which can be a means of making post threads for promotion
4. Google Trend, Semrush.com, Keyword.io and other keyword search sites
5. Whatsapp Business or LINE Official Account to communicate further with consumers
6. Marketplace online stores such as Shopee, Tokopedia, Lazada, Blibli, Sociolla, and so on to provide various free shipping services, discount promos, and cashback for consumers
7. Tiktok and Youtube for humanist video content that is more familiar to consumers
8. Google Display Network for tools to target ads effectively
9. Email to create a newsletter about product updates, promos, and other related articles
10. Official LinkedIn website and account for business entities
11. E-wallet account to simplify the transaction process

Some of the media recommendations above can be focused on the needs of the startup business and not run all at once. The priority of using media is as a product display with clear detailed information, a communication tool between sellers and buyers, a provider of interactive content in addition to promotional content, and as a means of business transactions.

The next step is to make copywriting according to the target market. Based on Anindya (2022), there is an AIDA formula (attention, interest, desire, and action) that is often used by content creators. The final step is to execute a design that will have a major impact on the interest of profile visitors. The steps taken in the production of visual designs are divided into four stages, namely the search for ideas, initial visualization sketches, revisions, and finalization for publication (Thifalia & Susanti 2021). Visualization must still prioritize text clarity, layout composition that is not too heavy or empty, and a color palette consisting of a maximum of 3 colors so that it is not too conspicuous in the eye. All designs in one account must also always be consistent to look beautiful in unity.

4 CONCLUSION

Through the discussion above, it can be concluded that the most dominant mistake made by start-up owners is to focus too much on production, product or service development, and product distribution to the market so that digital promotions are not carried out and education related to making attractive designs and copywriting has not been conveyed. The digital marketing solution for these startup businesses is to carry out the Ideal Customer Profile (ICP) analysis stage, compose copywriting content that is suitable for the target market, and start executing visual or audio-visual designs as needed. Another solution that is easier for budding business owners is to consult and leave social media management completely to digital agencies.

This research still has limitations in the duration of data collection, so that the interview data obtained are still relatively small and there have been no direct interviews with trusted startup owners who have succeeded in developing their business on a national and international scale. Through these limitations, further research can consider a longer duration so that the data is more accurate.

REFERENCES

Ainurrofiqin, M. (2021). *99 Strategi Branding di Era 4.0*. Quadrant.
Anindya, W. D. (2022). Strategi Menulis Teks Promosi (Copywriting) di Instagram untuk Meningkatkan Penjualan Onlineshop. *JAST: Jurnal Aplikasi Sains Dan Teknologi*, 5(2), 148–155. https://doi.org/10.33366/jast.v5i2.2720
Cholil, A. M. (2018). *101 Branding Ideas: Strategi Jitu Memenangkan Hati Konsumen - Akmal*. Quadrant.
CNBC Indonesia. (2022, February 18). Apa Sih Layanan Bank Digital? Nih Intip Kecanggihannya! *CNBC Indonesia*. https://www.cnbcindonesia.com/mymoney/20220218091731-72-316386/apa-sih-layanan-bank-digital-nih-intip-kecanggihannya
Gultom, D. K., Arif, M., & Fahmi, M. (2020). Determinasi Kepuasan Pelanggan Terhadap Loyalitas Pelanggan Melalui Kepercayaan. *Maneggio: Jurnal Ilmiah Magister Manajemen*, 3(2), 171–180.
Magnusson, J., Elliot, V., & Hagberg, J. (2021). Digital Transformation: Why Companies Resist What They Need for Sustained Performance. *Journal of Business Strategy*. https://doi.org/10.1108/JBS-02-2021-0018
Puntoadi, D. (2011). *Menciptakan Penjualan Melalui Social Media*. Elex Media Komputindo.
Suhartono, N. P. (2021). Social Media Strategy: The Use of Social Media in Stationery Business to Reach Gen Z. *VCD (Journal of Visual Communication Design)*, 6(2), 109–119. https://doi.org/10.37715/vcd.v6i2.2704
Tantra, G. C. (2021). Building Brand Relationship Through Millennial Generation Behaviors: Marketing Strategy on Instagram. *VCD (Journal of Visual Communication Design)*, 6(2), 97–108. https://doi.org/10.37715/vcd.v6i2.2703
Thifalia, N., & Susanti, S. (2021). Produksi Konten Visual dan Audiovisual Media Sosial Lembaga Sensor Film. *Jurnal Common*, 5(1), 39–55. https://doi.org/10.34010/common.v5i1.4799
Wiesenberg, M., Godulla, A., Tengler, K., Noelle, I.-M., Kloss, J., Klein, N., & Eeckhout, D. (2020). Key Challenges in Strategic Start-Up Communication. *Journal of Communication Management*, 24(1), 49–64. https://doi.org/10.1108/JCOM-10-2019-0129
Yu, H. (2021). *Ascend Your Start-Up: Conquer the 5 Disconnects to Accelerate Growth*. Amazon.

Suroboyo UNESA bus stop design with the iconic design approach

A.M. Nurkamilah, D.D.A. Utami, I.F. Fauzia & I. Sudarisman
Telkom University, Bandung, Indonesia

ABSTRACT: Surabaya is one of Indonesia's cities with a high use of private vehicles. Therefore, the Department of Transportation (Dishub) in collaboration with the Surabaya City Government aims to reduce the number of traffic jams that occur in the city of Surabaya, one of the ways is by providing public transportation facilities that can make it easier for people to travel and carry out their activities, thus creating the Suroboyo Bus transportation. The UNESA Suroboyo Bus Stop is located on the UNESA campus. The problem was reviewed because, at the previous stop, there was a lack of information listed at the Suroboyo bus stop. This stop was not only used as a place to wait for buses, but many users took advantage of this stop for various purposes, one of the functions was to be a meeting point. The method used in this study is qualitative, using primary data by conducting observations, direct analysis in the field, and surveys in the form of questionnaires. The data collected is factual and original to determine the quality of the shelter and the need for redesign. This design is useful to make it easier for users of the Suroboyo UNESA bus stop when they meet because it applies an iconic design and can be used as a source of learning and proposals for interior studies.

Keywords: Bus stop, Iconic, Ujung Galuh Batik, Sawunggaling Batik, Sura and Baya Statues

1 INTRODUCTION

Halte in the Big Indonesian Dictionary is a place where trains, trams, or buses (usually there is a waiting room with a roof, and smaller than the station) (KBBI 2019). According to the Law of the Republic of Indonesia, No. 22 of 2009 concerning Road Traffic and Transportation, a bus stop is a place for public motorized vehicles to pick up and drop off passengers. Another definition of a bus stop is a place for pick up and drop off of bus passengers, usually placed on the service network bus transportation in the city which is equipped with buildings so that users are protected from rain and sunshine. Bus stop users consist of various groups including office/shop employees, students, traders, professionals as well as housewives using public transportation as their mode of transportation. The city of Surabaya is one of Indonesia's cities with very high use of private vehicles. As reported in a study (Sebulan et al. 2018), the number of vehicles in Surabaya is estimated to reach more than 17,000 units per month for both motorcycles and four-wheeled vehicles. Therefore, the Department of Transportation (Dishub) which is interconnected with the Surabaya City Government aims to reduce the stagnation that occurs in the city of Surabaya, one of the ways is by providing public transportation facilities that can make it easier for people to travel and do activities, then the Suroboyo Bus transportation was created. According to Broadbent (1980), the Iconic Approach is one of the approaches in designing that is based on social agreement (Andayani et al. 2019). The Iconic Architectural Principles are: magnificent building scale, attractive shape, easily recognizable, location at a strategic location, has a repeating element (continuous rhythm), and has a consistently strong structure (Pawitro 2012). Iconic architecture as a place marker or marker of time/age. In the development and history of architecture, architectural works are divided or grouped in a certain period of time. In the course of historic architecture, there are great architectural works that can be used as Signs or Markers era of human culture. With these great architectural works, there is also a connection with a 'Sign' or 'Icon' in the form of a building or building that is used as a place marker (Pawitro 2012). The UNESA Suroboyo Bus Stop is located on the UNESA campus which is not far from the hospital, restaurants, cafes, shop houses, and surrounding housing. Thus, users vary greatly in terms of gender, age, and profession. This Suroboyo Bus Stop provides facilities in the form of a bench for passengers waiting for the bus to come, a bus travel path map, and barcode information for the GOBIS (Golek Bus) application to facilitate transportation services, on the application. In the last facility, there is a route information board bus, but it doesn't work properly because the writing on it has faded and isn't rewritten.

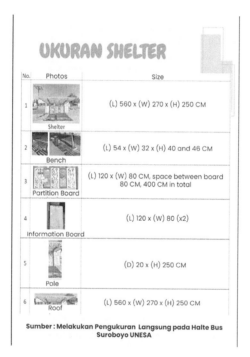

Figure 1. Physical appearance of UNESA Suroboyo bus stop source: Researcher.

This bus stop is also not only used as a place to wait for buses, but many users use this stop for various purposes like as a meeting point. However, the drawback of this stop is that the design is not iconic, so it doesn't have a characteristic and doesn't stand out when someone is looking for a stop. Therefore, a redesign was made at the Suroboyo UNESA bus stop.

2 DESIGN METHOD

The method used in this study is qualitative, using primary data by conducting observations, direct analysis in the field, and surveys in the form of questionnaires. The data collected is factual and original to determine the quality of the shelter and the need for redesign.

3 RESULTS AND DISCUSSION

3.1 *Design concepts*

The UNESA Suroboyo Bus Stop has design problems that have not implemented the iconic design, that's why the iconic redesign is necessary. Transformation is the process of changing form or structure through a series of separate permutations and manipulations and responding to a particular environment or set of conditions without losing identity or concepts (Ching 2008). By using the shape transformation method from batik motifs including Sura and baya statues, Ujung Galuh batik motifs, and Sawunggaling batik motifs, it is possible to redesign the bus stop. From the three alternative design concepts, a derivative concept is made as follows.

3.2 *Pattern concept*

The concept of the pattern in the first alternative, namely from the sura and baya statues, and the pattern is taken from simplified crocodile skin. It consists of a rectangular plane and an acute triangle that envelops the body. The patterns are combined like enveloping or encircling the crocodile's body, so that it forms a hexagon where it has a flat and sharp plane. Some of the sides are removed to open up space and there is a change in the size of the shape. Pattern further from the white shark's curved and repeated gills, from the explanation There are several points, namely:

1. Static, angular, enveloping or encircling, repetition with the transformation of the shape's size
2. Organic, curvy, and has rhythm. So we get the following pattern:

Figure 2. Exploration of the shape of the statue of Sura and Baya.

The concept of the pattern in the second alternative, the pattern contained in the typical Ujung Galuh batik motif Surabaya, namely geometric shapes and curves that are repeated, resulting in a pattern as follows:

Figure 3. Exploration of the shape of the Ujung Galuh batik motif.

In the third alternative design, the sawunggaling motif uses a dynamic and regular pattern. Then there is the focal point, namely, the rooster which in the surrounding area is surrounded by various motifs of flowers, leaves, and tendrils. So that the patterns that can be created on this sawunggaling batik motif are:
1. Repetition of shapes that have a focus
2. Harmonious and dynamic
3. Overlapping or layering

Figure 4. Exploration of the shape of the sawunggaling batik motif.

3.3 *Shape concept*

The concept of form in the first alternative, the form of the bus stop, is a simplification of the form derived from the statues of Sura and Baya. It consists of two basic shapes, namely, the left side with a static, rigid, and sharp initial rectangular shape and triangular corner inspired by crocodile skin, and the right side of the basic curved line shape that is more organic and repetitively inspired by the white shark's gills. With distinctive shapes and colors from each source of inspiration, namely green and white, the bus stop design will be easy to remember.

Figure 5. Alternative design 1 – bus stop perspective view.

According to Broadbent (1980), the Iconic Approach is one of the approaches in designing shapes through empirical and customary practices based on social agreement (Andayani et al. 2019). The principles of Iconic Architecture are: the scale of the building is magnificent, the shape is attractive, can be easily recognized, is located in a strategic location, has elements of continuous rhythm, and has a strong structure (Pawitro 2012). The UNESA Suroboyo Bus Stop has a repeating shape with an interesting shape. The roof of the bus stop has a repeating wave shape that makes the design iconic for the city of Surabaya when viewed by the human eye.

Figure 6. Alternative design 2 – bus stop perspective view.

The iconic bus stop applies a magnificent building form, with attractive shapes, is easily recognizable, is located in a strategic location, has a continuous rhythm element, and has a strong structure. The concept of outdoor space planning, is where the formation of the building takes the form of layered chicken feathers. If it will be applied to the Suroboyo UNESA Bus Stop, the roof of the stop can be made to resemble the shape of a rooster's beak plus the incorporation of feathers with a layering system.

Figure 7. Alternative design 3 - bus stop perspective view.

3.4 *Material concept*

The main material in the first alternative is ACP (i.e., Aluminum Composite Panel) which is a mixture of PE (Polyethylene) material that is coated with aluminum on both sides. This material has fairly good resistance to weather and climate and is easy to shape. Finishing on this material is a PVDF coating paint that adds resistance to sunlight, weather changes, color fading, and cracking. Between the walls and the roof of the shelter, there is tempered glass which is strong enough, even if it is broken, it will not be dangerous because it becomes small and not sharp shards. Tempered glass is coated with window film to reduce the sun's heat so that the area underneath does not overheat. For paving, concrete was chosen because it is strong and not slippery. Good materials for shelters (bus stops) are those that are weather-resistant, durable, and easy to maintain. According to Guidelines for the Location and Design of Bus Stops, usually, panels and roofs use frosted plastic, but plastic is easily damaged. For the side panels, tempered glass is usually used which is better than plastic in terms of visuals, but if it breaks, it is not safe for bus stop users (Townes et al. 1996). For the Suroboyo UNESA Bus Stop itself, the roof of the stop uses spandex metal material coupled with silicon material so that the roof is more flexible and does not collapse easily. Then for the side of the bus stop, use glass surgery, which is transparent clear glass and also has the advantage of absorbing the sun's heat. Threats to users while waiting at the bus stop. Those are applied in the second alternative design. The material concept of the third alternative, in the paving area, uses concrete because the surface is not slippery. Then the Sawunggaling batik motif can also be applied to coat the existing partitions at the bus stop as a decoration element. Then the roof uses steel material which is painted white and combined with plastic material. Clear plastic can make the interior of the shelter visible from a distance and increase the level of security of the shelter if the shelter is closed, it will pose a threat to users while waiting.

3.5 *Color concept*

The colors for the first alternative applied to the bus stop are taken from the green color of crocodile skin, white sharks, and gray from the snaking part of the statue. By using these colors as a symbol of the hope of the city of Surabaya, which is always peaceful and free from all threats, colors are also expected to provide comfort for bus stop users. The second alternative design uses the colors contained are dominated by blue color. This color is known to give a calming impression. Blue symbolizes patience and understanding, so people often feel comfortable around a blue room. Besides blue and white being the color used, white itself gives the impression of being clean, bright, and natural. The color applied to the third alternative design, the bus stop follows the color of the sawunggaling batik motif. The color of the batik motif used is usually bright colors or complementary colors such as red, yellow or gold, and blue. Chocolate blends with nature, and White gives the impression of being clean and healthy. Gray gives the impression of being formal and strong.

3.6 *Lighting concept*

The lighting concept applied to this shelter uses natural lighting in the morning and LED lights whose energy source comes from solar panels that are used at night. This solar panel is installed right next to the bus stop.

3.7 *Furniture concept*

The concept of furniture in alternatives one, two, and three, all furniture is built in or cannot be moved. The material used for the seating facilities is iron material and is finished using a special iron paint and is equipped with a backrest. Then the trash can also uses iron material. The use of this iron material is assessed because it is not easily damaged and can last longer.

4 CONCLUSION

From the analysis that has been carried out both from field studies and through the questionnaire method, data obtained shows that the UNESA Suroboyo Bus Stop is not only used as a place to wait for buses but also as a place to wait for online motorcycles, taxis, and private vehicles. Therefore, it is necessary to redesign the shelter using an iconic design approach so that the stop can be easily recognized by users. The iconic design approach has several characteristics, including a magnificent building scale, an attractive shape, an easily recognizable, strategically located, a continuous rhythm element, and a consistent and strong structure. It can be used as a point of interest in the city of Surabaya. The redesign of this bus stop is based on the local cultural wisdom of Surabaya, namely the use of *Sawunggaling* batik motif, *Ujung Galuh* batik motif, also Sura and baya statue as concepts applied to the UNESA Suroboyo bus stop. By using the shape transformation method, a new design form for the UNESA Suroboyo Bus Stop was made.

REFERENCES

Andayani, G. S., Dermawati, & Puspatarini, R. A. (2019). Implementation of Iconic Architecture Approach to Façade of Pasar Johar Area Building Semarang. *Prosiding Seminar Intelektual Muda #2, Peningkatan Kualitas Hidup Dan Peradaban Dalam Konteks IPTEKSEN, 5 September 2019, Hal:161-167, ISBN 978-623-91368-1-9, FTSP, Universitas Trisakti.*, 161–167.

Ching, F. D. k. (2008). *Architecture: Forms, Space, and Order Third Edition.*

Pawitro, U. (2012). *Perkembangan "Arsitektur Ikonik" Di Berbagai Belahan Dunia.*

Sebulan, Kendaraan Baru di Surabaya Ada 14 Ribu Un Surabaya Makin Macet. (2018, July 15). Surabayapagi. Com. https://surabayapagi.com/read/surabaya-makin-macet#

Townes, M. S., Barnes, L. E. E., Blair, G. L., Millar, W. W., & Monroe, D. O. N. S. (1996). *TCRP Report 19: Guidelines for the Location and Design of Bus Stops.*

Transit Cooperative Research Program. (2006). Guidelines for The Design and Placement of Transit Stops. Washington, DC: Author, 2010

United States of America. *Washington Metropolitan Area Transit Authority.* (2009). Guidelines for the Design and Placement of Transit Stops. Washington, DC: Author, 2010

The influence of visual visibility on the behaviour of visitors in the cinema

Imtihan Hanom*, M.D.D. Ar-Rasyid & A.M. Auliarahman
Telkom University, Bandung, Indonesia

*ORCID ID: 0000-0002-3790-536X

ABSTRACT: This research departs from the phenomenon of rampant immoral acts that occur in the cinema room. Immoral acts are acts that violate Law No. 44 of 2008 and also cause inconvenience to other visitors. This research will examine behavioral regulation related to cinema visitor activity system. According to Haryadi and Setiawan (2010), the activity system is a series of behaviors that are intentionally carried out by a group of people. The research method used in this research is qualitative, by conducting a literature study to find out how to prevent immoral acts in cinemas associated with the visual sensibilities of visitors. The results of this study focus on 2 variables, namely, the difference in level height between floors and seat layout. It is hoped that further research will get results related to human perception obtained from the auditorium room simulation image by combining the variables that have been obtained from this study.

Keywords: Visual Visibility, Behaviour Setting, Cinema Room

1 INTRODUCTION

Cinema is a place to watch movie shows using a wide screen. The word cinema comes from the Greek words bios which means life and skopos which means to see. The development of cinemas in Indonesia is quite rapid, it can be seen from the number of cinema openings in major malls throughout Indonesia. This is a response to the community's need for recreational facilities where watching is currently an alternative for human recreation. A cinema usually has various types of auditoriums to pamper visitors so that they can choose the type that suits their wishes while watching a movie. We can see that CGVBlitz has 6 types of auditoriums, namely, regular class, sweetbox, screen x, 4DX, gold class, and velvet. Likewise for the type of auditorium at Cinema XXI, namely, regular/deluxe, dolby atmos, IMAX, and The Premiere. However, another phenomenon emerged related to the existence of cinema as a means of entertainment. Many visitors take advantage of the dark and private atmosphere of the cinema to commit immoral acts. The occurrence of immoral acts that occur in the cinema area is a separate problem for the sustainability of cinemas today. As quoted from idntimes.com (2017) that the Palembang City Government temporarily closed the Velvet studio from Blitz for fear of inviting immoral acts. This is because the velvet study is equipped with pillows and blankets and is intended for two users. However, the phenomenon of immoral acts has actually been widely discussed, not only in the velvet-type auditorium. Seeing this phenomenon, it can be concluded that visitors tend to ignore the presence of CCTV in the cinema room. This may be because CCTV footage is not considered effective in certain corners of the room. For this reason, this study will analyze how to make visitors feel that they are constantly being watched by others in order to avoid immoral acts. The sense of being constantly noticed by local visitors can also be presented from interior variables, including planning the layout of the space and the height of the auditorium floor level. Of course, by preventing visitors from committing immoral acts, it will indirectly provide a sense of comfort for visitors who want to enjoy watching activities at the cinema.

1.1 *Behavior setting*

A behavior setting is an interaction between an activity and a specific place (Haryadi & Setiawan 2010). According to Barker (in Haryadi & Setiawan 2010), the study of behavior setting is how we can identify behaviors that constantly or periodically appear in a certain place or setting situation. A simpler understanding is with behavior setting, then we can examine human behavior related to the setting system. The behavior that will be analyzed is related to the occurrence of immoral activities in a public space. This behavior arose due to the condition of the auditorium which stimulated visitors to commit immoral acts, such as a dark room, the position of the CCTV which may have been far away or in a blind spot, the tight arrangement of the seats, and the minimal difference in the height of the front and rear seats.

1.2 *Visual visibility*

The visual visibility of visitors is related to the human ability to see an object as a whole. One that can affect human visual abilities is scale and proportion (Kartika 2008). That way, the scale and proportion of the sitting area in the auditorium become the object to be observed. The difference in floor height between levels in the auditorium is something that affects the visual quality of humans. Likewise with the height of the backrest on the audience seat in the room. O, Newman (in Zhinchenko 2013) argues that an important characteristic of a sustainable space is the possibility to informally observe what is going on (who is on the streets, what people are doing, etc.). In accordance with the results obtained by Zhinchenko (2013), if a room can be observed automatically by people around, then they can be eyewitnesses if something happens, so that it is expected to suppress the perpetrators' intentions, especially for immoral acts in an auditorium cinema room.

1.3 *Anthropometry*

In this study, the visual ability of visitors is associated with human anthropometry, namely the ability of human sight distance. The study of the standardization of the auditorium space related to human viewing distance to the cinema screen is not the focus of this study. The main thing that is of concern is the ability of the human eye to be able to pay close attention to the surroundings. As shown in Figure 1, the eye range is at an angle of 30° up from the vertical line and 40° down from the vertical line (Panero 2003). For the range of eyes above that angle, it can be considered that it has exceeded the ability of the human field of vision.

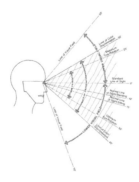

Figure 1. The visual visibility of human point of view Source: Panero 2003.

2 RESEARCH METHODS

In this study, the author uses a qualitative research method with a literature study method. This literature study method begins with collecting data by looking for sources and reconstructing from various sources such as books, journals, and existing research (Fadli 2021). Fadli (2021) added that the advantages of the qualitative method through this literature study because the design is described in a comprehensive manner that is easily understood by researchers and academics. Crewel, 2008 (in Fadli 2021) divided 6 stages carried out in research with qualitative methods. Based on this, the researchers also divide 6 stages as follows:

1. Problem identification: Researchers must study phenomena or issues related to cinema that can develop into a problem.
2. Literature review: At this stage, the researcher must look for articles related to the phenomenon of immoral acts in cinemas. After that, researchers can provide updates or differences from existing research.
3. Determine the research objectives: At this stage, the researcher identifies the intent or purpose of this research so that it is hoped that the novelty that will be produced will be seen.
4. Data collection: Researchers must pay close attention to the research object and determine the data collection method. Data collection was obtained from observations in several auditoriums of CGV and XXI cinemas, as well as conducting interviews with related parties, namely one of the supervisors who worked at the Cinema. Data collection in qualitative research cannot be separated from the main things of concern, namely what, where, when, and how.
5. Data analysis and interpretation: The data that has been collected is then analyzed to produce new ideas or theories.
6. Reporting: The researcher descriptively makes his research report so that it is easily understood by the reader.

Qualitative research places more emphasis on holistic descriptions, which can explain in detail about ongoing activities or situations rather than comparing the effects of certain treatments or explaining a person's behavior (Fadli 2021). The selection of the literature study method for this research is because it is quite relevant to analyze what interior variables affect human actions, especially in a cinema.

3 RESULTS AND DISCUSSION

Before the researchers went directly to the field observations, the researchers limited the auditorium to be studied. The type of regular auditorium is an auditorium consisting of 100–600 standard seats as the object of observation. This is because this type of auditorium is the most widely available auditorium in several malls in Bandung. From the results of an interview with one of the supervisors who work in the cinema, it was found that the size of the floor level height in the auditorium room was made according to the standard of the auditorium room. Then from the observations of several cinema auditoriums, it was found that the height of the front and rear seats was about 20–25 cm. This is in accordance with the standard distance between the chin and the top of the head (JDPK) for viewing comfort, which is 21 cm. However, with the difference in floor height between the chairs, the researcher felt that visitors could still perform immoral acts without being noticed by other visitors. This is because the difference in the floor height of the front and rear seats is not so significant. The height difference between the floor of the chair and the floor of the back seat should be adjusted to the maximum visual ability of humans so that visitors can really feel strange movements to prevent perpetrators from committing immoral acts. In accordance with the description of human visual abilities, humans have the ability of 30° up from the vertical line and 40° down from the vertical line (Panero 2003). If the floor height is adjusted to the maximum human visual ability, then visitors can see each other or monitor the movements of other visitors. Given that the main activity in the cinema is watching, humans tend to be aware of movement or sound even in a state of focus. Moreover, with the ease of visual ability to monitor the surrounding area without being hindered by anything, it can deter perpetrators of immoral acts from carrying out their plans. It can be seen in Figure 2 that when human visual abilities are maximized, it is clear that each visitor can observe each other, serving visitors from back to front as well as from front to back. From this picture, there is no way to hide when sitting in a chair while performing immoral acts.

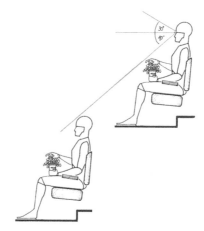

Figure 2. Analytical sketch that describes the size of the height between the audience seats in the cinema auditorium.

Then, for the layout of the seats in the cinema area, it has a form that is lined up in one row and repeated to the next row behind. It also makes the perpetrators of immoral acts feel invisible to visitors from the rows of seats in front or behind them, so that the perpetrators of immoral acts feel safe to carry out their actions in the cinema. To enlarge the visual ability of visitors, it is better to pay attention to the layout of the seats in the auditorium of the cinema. According to Suptandar 2004 (In Pambudi 2007), the arrangement of seats is made alternating between the front and rear seats to widen the viewing area. It can be seen in Figure 3, which is the layout of alternating chairs creating a free line of sight. This, of course, indirectly makes visitors feel cared for more easily, thereby suppressing immoral actors from carrying out their activities.

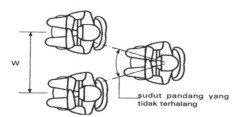

Figure 3. Seat configuration to create a free line of sight. Source: Suptandar 2004 (in Pambudi 2007).

4 CONCLUSION

From the results of the analysis that has been carried out in accordance with qualitative research methods, namely library research, it is found that interior variables can influence human behavior. In this research, the results of the initial analysis are related to what variables affect human behavior, namely the difference in height between the front and back seats in the cinema auditorium and the seat layout in the cinema auditorium. By considering the maximum human viewing ability in an auditorium, it will pressure the perpetrators to commit immoral acts because they feel that they are easily monitored by nearby visitors. Then by taking into account the pattern of alternating seat configurations in the auditorium room, it is also expected to increase the visual ability of visitors which leads to the prevention of immoral acts. By emphasizing the maximum visual ability to all visitors, it is hoped that it can reduce the occurrence of immoral acts. That way there is an increase in the comfort of visitors to be able to enjoy while watching in a cinema auditorium. This research is an early stage to find out what variables affect the visual ability of visitors. Further research will combine the variables that have been obtained to be processed into an auditorium room stimulus in order to determine the perception of visitors.

REFERENCES

Fadli. M.R. 2021. *Memahami Desain Metode Penelitian Kualitatif*. Humanika Kajian Ilmiah Mata Kuliah Umum, ISSN:1412–1271. Vol.21.No.1.

Haryadi, Setiawan. B. 2010. *Arsitektur Lingkungan dan Perilaku*: Pengantar ke Teori Metodologi dan Aplikasi. UGM Press. ISBN: 978-602-386-886-5. Yogyakarta.

Kartika, Felicia F. 2008. Pengaruh Activity Support Terhadap Penurunan Kualitas Visual Pada Kawasan Kampus UNDIP Semarang, Studi Kasus; Koridor Jalan Hayam Wuruk Semarang. *Universitas Diponegoro. Semarang*.

Pambudi, A. 2007. Perencanaan dan Perancangan Interior Bioskop IMAX dan Home Theatre di Surakarta. *Universitas Sebelas Maret. Surakarta*.

Panero. J. 2003. *Human Dimension and Interior Space Handbook*.

Zinchenko Y.P., Perelygina E.B. 2013. A Secure City: Social-Psychological Aspects. *Procedia. Social and Behavioural Science*. Elsevier. Retrived from https://www.idntimes.com/opinion/social/erwanto/opini-penutupan-velvet-suite-blitz-palembang-semesum-itukah-orang-indonesia?page=all

Promoting animation technology in preserving and reintroducing *Wayang Cina Jawa* to younger generations

A. Lionardi, D.K. Aditya & I.G.A. Rangga Lawe
Telkom University, Bandung, Indonesia

ABSTRACT: Wayang Cina Jawa (*Wacinwa*) is a traditional shadow puppet show (*wayang*) in Java, Indonesia, in which the characters are an acculturation result of Javanese and Chinese culture. *Wacinwa* is a cultural artifact from Yogyakarta and the surrounding areas. *Wacinwa's* acculturation can be seen in the design of the puppet, which is dominated by Chinese nuance and is performed using Javanese ways to deliver the narration. Unfortunately, this one-of-a-kind show has been discontinued without a successor since 1967. Because *Wacinwa's* assets system is similar to animation, this research suggests that animation technology could be an alternative way to reintroduce *Wacinwa* to younger generations. This study concentrates on the animation's character design. We conduct an in-depth interview with people who are familiar with Gan Thwan Seng (a puppeteer and writer of many *Wacinwa* scripts from Yogyakarta, Indonesia) and his work. Data were collected from interviews dan literature. This study found that Kho Ping Hoo's script can be adapted in a new form as we designed as 2D character animation in the *Wacinwa* style.

Keywords: Wayang, Wacinwa, Peranakan, cultural, character design, animation

1 INTRODUCTION

Wacinwa is an acronym of *Wayang Cina Jawa*, which is also known as a Chinese-Javanese (*Peranakan*) shadow puppet. This art was born in Yogyakarta. People also call this art as a *Wayang Thithi*. *Wayang thithi* refers to the name that comes from the sound of the music that accompanies this puppet performance. Historically, its art was first introduced by a puppeteer and also a performer, known as *Dalang* Gan Thwan Seng. Shortly, *Dalang* Gan Thwan Seng is the leader of *wayang* who was born in Jatianom, Klaten, Central Java in 1885. His interest in creating this *Wayang* art stemmed from his puppetry experience, friendship with Javanese people, and his grandfather's knowledge of Chinese culture and literature (Mastusti 2016).

The usual Javanese *Wayang Kulit,* shadow puppet show, in which the puppet, is made from animal skin such as cow and goat. The form is the heads and the bodies of the characters that can't be separated. The heads and bodies of the characters in the *Wacinwa* puppet are made separately. The reason is to follow the puppet master to create several characters by changing the heads and bodies of the characters' roles to be played. Thick cardboard is the main material to make *Wacinwa's* puppets. This Chinese-Javanese *Wayang Kulit* performance is held in public (people's homes), especially those who hold birthday parties and weddings. Performances are also carried out at temples on big day celebrations such as Chinese New Year and *Cap Go Meh*. Another unique function is a means of ritual to worship gods and ancestral spirits.

Unfortunately, the historical traces of *Wacinwa* cannot be traced any further since 1967, after Gan Thwan Seng passed away. There are many factors causing the *Wacinwa* to lose its existence. First, it is the issuance of government regulation no. 14/1967 by the Soeharto regime, which banned all forms of cultural activities of the Chinese community in public under the pretext of a comprehensive assimilation policy for the descendants of China with the indigenous people (Suryadinata L. 2016). As a result of the closure of Chinese schools and the prohibition on using the Chinese language, the generation of *Peranakan* (Javanese-Chinese) born during the New Order era (Soeharto regime) lost their ability to speak Chinese while also preserving the previous traditional culture (Mastuti 2014) (Wirasari *et al.* 2021). It includes *Wacinwa* and *Wayang Potehi* (also puppet show) performances. Second, it said there were Gan's students that could be his successor, but they had passed away before Gan Thwan Seng. Meanwhile, Gan's original puppet scripts were burned by his family because they claimed they didn't know how to perform (according to some experts, Gan's family was afraid of the authorities in the 1970s who watched any cultural performance due to the total assimilations policy) (Mastusti 2016).

1.1 *The research goals*

Because of *Wacinwa's* asset systems, which are similar to modern two-dimensional animations. Thus, the current study seeks to describe how modern animation technologies can digitally preserve *Wacinwa* and perform popular stories and narration in order to reintroduce this type of *Wayang* to the next generation.

2 RESEARCH METHODS

The qualitative research method was used to reflect on cultural phenomena because the targeted audience of the present study is the younger generation who are disconnected from the previous culture. Researchers use the visual research method approach to examine data related to objects, shapes, and forms of *Wacinwa's* puppet. Understanding 1) visual culture; 2) design thinking; and 3) reference data used are all closely related to visual research methods (Soewardikoen 2019).

2.1 *Methods of data collecting*

Because of the Covid-19 pandemic, the researcher participated in some online webinars and talk shows hosted by Rumah Cinwa (the community who try to preserve *Wacinwa*). Gathering information about the history, narrations, and how Wacinwa performed are needed. Therefore, gaining more comprehensive data about these artifacts, particularly on puppet form was strongly conducted as well. The uniqueness of *Wacinwa* puppet is found in texture, silhouette, and color from previous studies such as books, journals, and documentation. This study observed existing new media about *Wayang*, such as illustrations and comics for a specific purpose.

2.2 *Data analysis*

Since the researcher's goal is to make a new interpretation of *Wacinwa* in the modern look of the animation, the researcher gathers the whole data about *Wacinwa* performance. They are useful to deconstruct *Wacinwa* old form to become a new concept which suitable for animation production. We use descriptive analysis to analyze the written document and documentation of the *Wacinwa* puppet. The flow of the data analysis can be seen in this graphic:

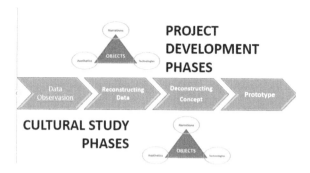

Figure 1. Data analyisis used in this research (Source: D.K. Aditya 2022).

3 RESULTS AND DISCUSSIONS

Gan Thwan Seng, as *Wacinwa's* creator had wrote many of his *wayang* scripts in Javanese language, for example the story of King Thig Jing's Marriage was rewritten as *Rabenipun Raja Thig Jing* (in Javanese). The names of the characters in *Wacinwa's* performance, are pronounced in the *Hokkien* dialect, while terms such as rank, and title, use Javanese terms such as *pangeran* (prince), *tumenggung* (landlord/nobleman), and *raja* (king) (Mastusti 2016) The story that is often to plays in *Wacinwa*, tell a lot about the heroic story of Sie Djin Koei/ Si Jin Kwi (spelled in Hokkian's dialect). Sie Djin Koei is a myth written during the Tang Dynasty. This story itself was quite popular among *silat's* (one kind of martial art in Indonesia) comic readers in the 1950s. It was published through Star Weekly magazine, and the story was written and illustrated by Siauw Tik Kwie, another Javanese-Chinese artist (Aditya 2018) (Dermawan T. 2016).

Figure 2. Carboard made *Wacinwa* puppets with separated head. (Source: https://wacinwa94.blogspot.com/2019/01/wacinwa-sim-bol-akulturasi-budaya-cina.html)

The shape and appearance of the *Wacinwa* are clearly visible from the depiction of the characters and attributes [clothes, weapons and ornaments]. One of the hallmarks of *Wacinwa* puppet is that the heads of the characters can be separated and combined with different bodies to display different characters. In addition, there are also several puppets made of cardboard the use bamboo as a *gapi* (*wayang* holder) (Figure 2).

From the information above, the research studied that *Wacinwa* itself was created as a modified artefact. The creation that Gan Thwan Seng made, showed the emerged and blended two cultures within, the Chinese's narrations and visuals with the Javanese's artefact (*wayang*). What we will discuss here are the acculturations between the two cultures, and how they blended. This acculturation teaches us that we can preserve our cultural artifacts, like *Wacinwa* in the same way that Gan Thwan Seng started. The research has studied, that many other academics in art and culture try to reconstruct and recreate *Wacinwa* like what it used to be. Some of the previous research that the research team found are willing to recreate *Wacinwa*, because of the rarity and exclusiveness. The only museum that keeps *Wacinwa* as their collection is Sonobudoyo Museum located in Yogyakarta (Kiswantoro 2019). The previous research made a copy of the puppet with some modifications for performing an old narration related to Sie Djin Koei's epic story. Many roadshows, exhibitions, and talk shows have been held to reintroduce *Wacinwa*, but the audience is limited to academics and cultural observers who want to learn more about *Wacinwa*.

Figure 3. First Mockup of *Wacinwa* Puppet For Short Animation Project Based on The Research (Source: D.K. Aditya 2022).

This research sees many possibilities to reintroduce *Wacinwa* as a heritage artifact to teach unity in diversity values to younger generations. With the same methods, this research will use popular media that is enjoyed by the young generations. So the research chooses short animation as the medium. The animation will accommodate the concept of *Wacinwa*, because the asset systems created by Gan Thwan Seng are similar to asset systems that are used in animations. For the narrations, this research will develop a narration based on a popular local *Wu Xia* (Romance Martial Arts) story written by Asmaraman Sukowati or Kho Ping Hoo.

For decades Kho Ping Hoo's stories have been reprinted and loved by his fans, all around Indonesia and lately, one of his works translated and printed in China. With the huge numbers of his fan base, this research could collect more data about Koo Ping Hoo's works, and choose which story can be performed with this new incarnation of *Wacinwa*. We choose a story entitled Dendam Membara (Burning Revenge). This story tells the story of a young man who wants to avenge his mother who was killed by her employer. This young man was helped by a young woman who has the same revenge on the same person. This two main characters will be the first character design of this animation.

4 CONCLUSION

Although this research still does some reconstructions to the data and still developing some model and mock up with new incarnation of *Wacinwa* concept that described above

(Figure 3), the researchers believe that we can find any efforts to preserve and reintroducing *Wacinwa* as cultural heritage that teach us about respect to diversity and blend together as one identity. The technology nowadays can be explored as a new media to do the effort. The traditional culture could blend with the technologies, in order to reincarnate itself to prevent the values behind the artefacts from being extinct in future generations. Like what Gan Thwan Seng innovated, to merge Chinese culture with Javanese popular culture in the past; this modern *wacinwa* incarnation should bring the same spirits, not only to preserve the artefact but spreading the values to young generations.

REFERENCES

Aditya, Dimas Krisna. 2018. "Pengaruh Ilustrasi Poster Film Shaw Brothers Pada Ilustrasi Cover Cergam Dan Novel Silat Di Indonesia." *Kalatanda: Jurnal Desain Grafis dan Media Kreatif* 1(2): 183.

Afif, R. T., Prajana, A. M., & Prahara, G. A. (2020). Analysis of Character Design and Culture in the Laskar Cima Animation. In Proceeding *International Conference on Information Technology, Multimedia, Architecture, Design, and E-Business* (Vol. 1, pp. 410–414).

Cohen, S. (2006). *Cartooning: Character Design: Learn the art of cartooning step by step*. Walter Foster Publishing.

Crossley, K. (2014). *Character Design From the Ground Up*. Ilex Press.

Dermawan T., Agus. 2016. *Melipat Air: JurusPendekar Tionghoa: Lee Man Fong, Siauw Tik Kwie, Lin Wasim*. Kepustakaan Populer Gramedia (KPG).

Gumelar, M. S. (2018). *2D Animation: Hybrid Technique*. An1mage.

Kiswantoro, Aneng. 2019. "Perancangan Wacinwa: Sang Manggalayudha." *Wayang Nusantara: Journal of Puppetry* 3(1). https://journal.isi.ac.id/index.php/wayang/article/view/3055/1203.

Mastusti, Dwi Woro Retno. 2016. "Wayang Kulit Cina-Jawa Yogyakarta." In *Tionghoa Dalam Keindonesiaan: Peran Dan Kontribusi Bagi Pembangunan Bangsa*, ed. Leo; Didi Kwartanada Suryadinata. Jakarta: Yayasan Nabil, 109–18. www.nabilfoundation.org.

Mastuti, Dwi Woro Retno. 2014. *Sinar Harapan Wayang Potehi Gudo*. Jakarta: PT. Sinar Harapan Persada.

Soewardikoen, Didit W. 2019. *Metodologi Penelitian Desain Komunikasi Visual*. ed. Bayu Anangga. Jakarta: PT Kanisius.

Soenyoto, P. (2017). *Animasi 2D*. Jakarta: Elex Media Komputindo.

Suryadinata, Leo. 2016. "Kebijakan Terhadap Tionghoa Dan Pembangunan Bangsa." In *Tionghoa Dalam Keindonesiaan: Peran Dan Kontribusi Bagi Pembangunan Bangsa*, Yayasan Nabil, xxvi–xcii.

Wirasari, I. *et al.* 2021. "Featured Animation Design for Cultural Respect and Understanding in Tjap Go Meh's Narration." In *Dynamics of Industrial Revolution 4.0: Digital Technology Transformation and Cultural Evolution*, 101–4. https://library.oapen.org/bitstream/handle/20.500.12657/50303/9781000441017.pdf?sequence=1#page=116.

Small classroom design adaptation as a response to teaching and learning activities in the new normal conditions

A. Farida & R. Fawwaz
Telkom University, Bandung, Indonesia

I. Subagio
Parahyangan Catholic University, Bandung, Indonesia

ABSTRACT: COVID-19 pandemic shifts learning methods and classroom conditions with new normal requirements. We need better classroom design to ensure that learning activities will be held in a healthy, comfortable, and safe environment. Classroom and study room with limited space, as in schools or dormitories need further arrangement. The limited space classroom conditions require a new layout arrangement following social distancing guidelines, furniture design that prevents the spread of the COVID-19 virus, and the fulfillment of other facilities. Through interviews, field studies, and literature studies, the classroom design has been developed to meet the needs of students and teachers' activities within these new normal conditions. The result shows that the new classroom design can be adapted to the requirements of formal learning, group activities, and furniture alternated design. With this new classroom design, a solution is expected to prevent the spread of COVID-19 and increase the comfort and safety of students during teaching and learning activities.

Keywords: Adaptive, Furniture, New Normal, Small Classroom

1 INTRODUCTION

According to the Ministry of Education and Culture, starting January 2022, all education units in the Enforcement of Restrictions on Community Activities level 1, 2, and 3 areas are required to implement limited face-to-face learning. The local government may not prohibit limited face-to-face learning for those who meet the criteria and may not add more stringent standards (Kemendikbud 2022). Face-to-face learning in the new normal period requires adjusting the learning space so that it can run optimally without compromising the safety of students and teachers. However, currently, face-to-face learning readiness in Indonesia is still relatively low. Based on the face-to-face School Readiness Checklist made by the Ministry of Education and Technology, in 2021 only 41.28 percent of schools answered the face-to-face learning readiness of 537,029 schools (Kusumo 2021).

According to research that has been carried out, it is stated that there is an impact on educational conditions during the pandemic. There are many conditions of students who are less motivated so they are less active or passive when online learning is carried out (Ananda Putri et al. 2021). The biggest obstacles felt by students when learning online are a lack of concentration, lack of facilities, and lack of privacy (Farida 2020). In addition, the condition of education in Indonesia during the COVID-19 period with this online mode is feared to experience learning loss. Where the concept can be interpreted with reduced understanding and academic skills (Andriani et al. 2021).

These conditions that have been stated above, make the adjustment of the classroom for the implementation of face-to-face learning must be carried out immediately. Unfortunately, not every classroom will meet these health protocol standards. Many classrooms whether in regular or in boarding school sometimes have limitations as a matter of their size. To proceed with social distancing in such a small classroom area will be quite challenging. A new way to design a classroom should be studied more to find a solution. New classroom design, including layout, materials selection, and interior design, must be developed to support the learning activities during the new normal period.

2 RESEARCH METHODS

The research method used is the qualitative method to describe how classroom arrangements meet the need of the pandemic situation. It starts with observations, surveys, and interviews to find the classroom problems when it's used for face-to-face learning based on the pandemic requirements, then continues with a literature study. After mapping the problem based on the field and literature study, the data were been analyze to get some input for the classroom design. After the problem is known and there are inputs for the design, this study starts to develop classroom design as an alternative solution to solve the problems of the pandemic classroom.

2.1 Methods of data collecting

Two types of data were used in this study, namely, primary and secondary data. Primary data were obtained through direct surveys in one of students dormitories in Bandung, observations of classroom activities and methods, and the results of interviews with 5 teachers who had teaching experience in the new normal period and 10 students who had gone through the process. In comparison, secondary data was obtained from printed and electronic literature studies.

3 RESULT AND DISCUSSION

To simplify the design process, we took a case study in one of the student dormitories in Bandung, which also has classrooms for learning in the New Normal era. This case study was taken because of the size of the classrooms which are limited to 27 m² to 30 m² per unit with a capacity of 15 students for formal classrooms and 20 students for communal classrooms. The design that will be produced consists of 2 types of classrooms, formal classrooms, which are used to convey theory, and communal classrooms, which are used for practicum classes and group learning.

3.1 Formal classroom design

In the design of formal classrooms, there are several aspects to consider, namely social distancing, the capacity is reduced by 50 percent to meet the requirement. In a formal classroom, the classroom's atmosphere will be proper and comfortable but still apply a design following the precaution of the health protocol during the COVID-19 pandemic.

As shown in Figure 1, the layout of the formal classroom that the distance between students' desks is set to match the minimum social distancing distance of 100–150 cm. The distance between tables is adjusted to the conditions and class size, which is about 60–70 cm so that the distance between table users can reach 100–150 cm. This has met the rules of social distancing in the classroom. It is necessary that the classrooms also have natural ventilation in the form of windows.

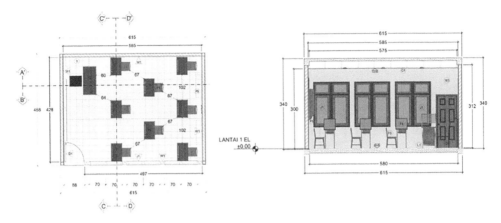

Figure 1. Layout and elevation small classroom desain (formal).
Source: *Personal source.*

3.2 Communal classroom design

In the design of communal classrooms, there are also several aspects to consider similar to the formal classroom, such as social distancing, and the capacity is also reduced by 50 percent to meet the requirement. The classroom atmosphere is made more comfortable to have a discussion but still implement designs following the provisions of health protocols during the COVID-19 pandemic.

As shown in Figure 2, the layout of communal classrooms, furniture is placed on the right and left sides alternately, increasing the distance between students, but students can still interact with one another. Then the table uses a partition with transparent material, so students do not meet face to face directly. Classrooms also have natural ventilation in the form of windows. Some of these things were done to minimize the possibility of exposure to the COVID-19 virus.

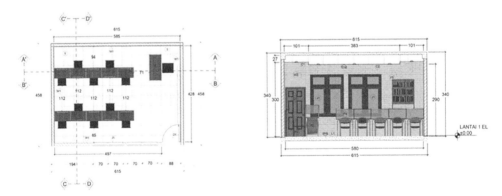

Figure 2. Layout and elevation small classroom desain (communal).
Source: *Personal source.*

3.3 *Application of furniture materials in classroom*

To further maintain social distancing, the table design uses transparent partitions. The material used in the table partition is glass with a thickness of 3 mm, and for the frame, it uses stainless steel pipe material. For the table surface, it uses HPL antivirus material. All the materials used considerably have shorter COVID-19 persistence.

Figure 3. Perspective small classroom desain.
Source: *Personal source.*

4 CONCLUSION

The adjustment of the classrooms for the implementation of face-to-face learning must be carried out immediately. But in schools and dormitories that have classrooms with a limited space, the application of social distancing is quite challenging; therefore, interior design recommendations that can support the learning process during the new normal period will be greatly needed. Several treatments can be done on the classroom design appropriate for the new normal period. The first is to apply the concept of social distancing in the furniture placement in the room. This is done by determining the minimum distance between student desks and the teacher's desk so that each person has a minimum distance of 1m. The second way is to consider the selection of materials to be used in the furniture in the classroom. The choice of material should be a material that has shorter COVID-19 virus persistence. Furniture design using barriers also helps maintain social distancing and reduces exposure to the COVID-19 virus, and make sure that every classroom has natural air ventilation. This research is still preliminary research, and it should be continued with evaluation after the design implementation.

REFERENCES

Ananda Putri, W., Magdalena, I., Khotimah, K., & Putri Syahra, N. (2021). Pengaruh Pembelajaran Jarak Jauh Terhadap Prestasi Belajar Siswa Kelas 3 SDN Sudimara Timur. *Cerdika: Jurnal Ilmiah Indonesia*, 1 (3), 321–327. https://doi.org/10.36418/cerdika.v1i3.47

Andriani, W., Subandowo, M., Karyono, H., & Gunawan, W. (2021). Learning Loss dalam Pembelajaran Daring di Masa Pandemi Corona. *Prosiding Seminar Nasional Teknologi Pembelajaran Universitas Negeri Malang*, 2, 485–501. http://snastep.com/proceeding/index.php/snastep/index

Farida, A., Liritantri, W., & Hanafi, M. S. (2021). Planning Private Spaces for Design Students to Support the Optimization of Online Learning. In *Dynamics of Industrial Revolution 4.0: Digital Technology Transformation and Cultural Evolution* (pp. 53–59). Routledge.

Fiorillo, L., Cervino, G., Matarese, M., D'amico, C., Surace, G., Paduano, V., Fiorillo, M. T., Moschella, A., La Bruna, A., Romano, G. L., Laudicella, R., Baldari, S., & Cicciù, M. (2020). COVID-19 Surface Persistence: A Recent Data Summary and its Importance for Medical and Dental Settings. *International Journal of Environmental Research and Public Health*, 17(9). https://doi.org/10.3390/ijerph17093132

Kampf, G., Todt, D., Pfaender, S., & Steinmann, E. (2020). Persistence of Coronaviruses on Inanimate Surfaces and their Inactivation with Biocidal Agents. *Journal of Hospital Infection*, 104(3), 246–251. https://doi.org/10.1016/j.jhin.2020.01.022

Kemendikbud. (2022). *Semua Sekolah Wajib Melaksanakan PTM Terbatas pada 2022*. https://ditpsd.kemdikbud.go.id/public/artikel/detail/semua-sekolah-wajib-melaksanakan-ptm-terbatas-pada-2022

Lewis, P., Nordenson, G., Lewis, D. J., & Tsurumaki, M. (2020). *Manual of Physical Distancing*. Columbia: Columbia University.

Pane, M. Sn., S. F. (2021). Dampak Pandemi Covid-19 Mengubah Konsep Tata Letak Furnitur Desain Interior Ruang Belajar di Perguruan Tinggi. *JSRW (Jurnal Senirupa Warna)*, 9(2). https://doi.org/10.36806/jsrw.v9i2.120

Rizki Kusumo. (2021). *Sekolah Dibuka, Bagaimana Kesiapan Pembelajaran Tatap Muka?*. https://www.goodnewsfromindonesia.id/2021/08/30/sekolah-dibuka-bagaimana-kesiapan-pembelajaran-tatap-muka

Suman, R., Javaid, M., Haleem, A., Vaishya, R., Bahl, S., & Nandan, D. (2020). Sustainability of Coronavirus on Different Surfaces. *Journal of Clinical and Experimental Hepatology*, 10(4), 386–390. https://doi.org/10.1016/j.jceh.2020.04.020

UNICEF. (2021). *Classroom Precautions During COVID-19*. https://www.unicef.org/coronavirus/teacher-tips-classroom-precautions-covid-19

Adaptive reuse strategy at Maison Teraskita Hotel Bandung

A.D. Purnomo, N. Laksitarini & L.C. Lase
Telkom University, Bandung, Indonesia

ABSTRACT: The city of Bandung preserves many historical buildings. Several historic buildings are in neglected condition. Some were knocked down because they were not in accordance with the current development of the city. These problems are often experienced by many historical buildings in Indonesia. The purpose of this paper is to examine adaptive reuse strategies in hotel interiors in Bandung. The research methods used are qualitative methods with descriptive analysis. Maison Teraskita hotel was originally the Waskita Karya office building. The building is a historic building with a grade B in the city of Bandung, which now switched its functions into a four-star boutique hotel. The adaptive reuse strategy is implemented on the building. The old building and the new building are combined. The old building is used as a lobby and coffee shop, while the new building consists of hotel rooms and meeting rooms. The hotel's interior design follows the Art Deco style. This style connects the old and the new buildings. The adaptive reuse strategy has become a solution for preserving historic buildings. Historic buildings can be preserved and even generate economic, social, and cultural value.

Keywords: adaptive reuse, heritage, hotel, interior

1 INTRODUCTION

The construction of new hotels in the city of Bandung sometimes sacrifices the old buildings. Old buildings are considered incompatible with today's demands, so they are often torn down and replaced with new buildings (Rahadian 2019). While the existence of the old building has been regulated in Law Number 11 of 2010, concerning cultural conservation. In addition, the demand to create a sustainable environment as mentioned in the Sustainable Development Goals (SDGs) has become a must. Considering number 11 which says sustainable cities and communities, it creates cities and settlements that are inclusive, safe (resilient), and sustainable, including building preservation efforts. Building preservation that responds to the development dynamics of the times (Martokusumo 2021; Sutopo *et al.* 2014). For this reason, strategic steps are needed in responding to these challenges. One solution is through an adaptive reuse strategy.

Adaptive reuse is a building preservation strategy with reference to the reuse of old buildings for new functions that offer economic, social, cultural, and environmental innovation (Lewis 2016; Soewarno *et al.* 2017). Adaptive reuse is also interesting and challenging in the field of architecture and interior design. Adaptive reuse is considered an important strategy for the preservation of cultural heritage (Gewirtzman 2017; Plevoets *et al.* 2018).

One of the new hotel buildings in the city of Bandung is Maison Teraskita. The hotel applies an adaptive reuse strategy. Constructing a new building and also preserving the old building. The old building was maintained by the Waskita Karya office. The building has been registered as Class B Cultural Conservation Building. While the new building is located adhere behind the old building.

This paper aims to examine the adaptive reuse strategy in the interior of Maison Teraskita in Bandung. In addition, this study aims to determine the role of interior design as a liaison between new buildings and old buildings. It is important to note that Maison Teraskita consists of two different buildings, namely the new and old buildings. Therefore, interior design plays a role other not only as a practical function but also creates an image of the interior.

This paper is the result of initial research about the application of adaptive reuse at Maison Teraskita. And it's hoped that it can be developed for further research. In addition, the research results can add references related to the application of adaptive reuse in the interior design of historical buildings in Bandung.

2 RESEARCH METHODS

The research method uses a qualitative descriptive method. Primary data through direct observations in the field and interviews with hotel managers. First visit for initial data collection. While the next visit is to complete the previous data. Secondary data through literature studies include: books, scientific journals, proceedings, and videos (vlogs) about reviews of guests who have stayed at or visited the hotel. The obtained data were then analyzed. Analysis related to the history of the building, aspects of changing functions, and forms of interior design styles. And at the last stage, conclusions were drawn.

Maison Teraskita is located at Asia Afrika Street number 55, Bandung. Asia Afrika Street is one of the areas designated as the Core Area of Cultural Conservation (Area I) in Bandung. The hotel's interior design concepts are heritage, luxury, and tropical. In the implementation of the hotel development, it applies an adaptive reuse strategy.

3 RESULT AND DISCUSSION

Before becoming Maison Teraskita, it was originally the NV. Volker Aanemings Maatschappij office. This company was later nationalized to become PT. Waskita Karya in 1961. NV. Volker Aanemings Maatschappij building is located on the Asia Africa street and was built in 1913. Initially, this building was used by the Bandung branch Bank Unit from 1917–1924. Then, in 1926, it was used by a German company, namely Siemens & Halske. The company is engaged in telegraphy, telephone, radio, x-rays, measuring instruments, and others. During the 'Bandung Sea of Fire' incident in 1946, the building also caught fire. From 1961–2017, the building underwent several renovations. The rear was added with a new building in 1965.

During construction, the hotel applies adaptive reuse. The original building in front is preserved and became an important element related to its history. While the additional building (on the back) was torn down because it was not the original building. This is still in accordance with the provisions that apply to class B cultural heritage buildings. The facade has been restored to the condition of the building in 1920–1940. The architecture is symmetrical with a shield-shaped roof wearing an ornament at the peak (momolo). The main entrance is in the form of an arch and is to the right and left of the window. It considers the location in the Cultural Conservation Core Area.

Figure 1. The building was built in 1913.
Source: KITLV, 2006.

Figure 2. The building in 1926 (left) and present condition (right).
Source: KITLV 2006 & Purnomo 2022.

One of the solutions to answer urban problems is to empower old buildings. Old buildings are maintained but also have an economic impact. The old building is used as a tourism support facility. The addition of new buildings in the preservation area must be in accordance with the steps of building preservation (Hayati 2014). The merger of new and old buildings while retaining authentic character while providing appropriate new functions (Gewirtzman 2017). Likewise, with Maison Teraskita, the owner built a new hotel by empowering old buildings and adding new buildings.

The new building is presented in one location with the old building (infill design). The architectural amalgamation is made in contrast between the old building and the new building. The new building looks modern with a glass facade, while the old building has an Art Deco style. The new building with seven floors serves as an enrichment in the preservation area. And its practical function is to be used as hotel rooms and other supporting rooms, such as meeting rooms, offices, social clubs, restaurants and kitchens, swimming pools, and gymnasiums. While the old building only functioned for the lobby, coffee shop, and bar.

The Art Deco style is used in processing interior design. The style is applied to the facade of the old building in the form of a square shape (Geometric Deco) that is repeated. Repetition of a square shape into a decorative trim under the eaves of the facade. While the Art Deco style in the interior design is presented in the furniture. In the lobby room, there is a reception counter table and also teak wood chairs with woven rattan. The other furniture is in the form of a bar table with an Art Deco style. Other equipment such as chandeliers, standing lamps, and wall lamps with the concept of the Art Deco style. The color of the walls

Figure 3. The old building (front) and the new building (back).
Source: Beton Elemenindo Perkasa 2021.

Figure 4. Interior design as a 'link' between old and new buildings.
Source: Purnomo 2022.

Figure 5. Tosca blue as the color of Paris van Java.
Source: https://maisonteraskita.com/gallery/ 2022.

255

and ceiling is dominated by white, and this color becomes the color character of colonial buildings. While turquoise green, turquoise blue, and gold are used as accents. Tosca green and turquoise blue become the colors of Paris van Java. The gold color supports the Art Deco style.

Interior design acts as a 'link' between old and new buildings. The combination of two contrast buildings can be seen as harmony because there is a link that unites the two buildings. Heritage, Luxury, and Tropical interior design concepts become the links. Not only harmony in the building, but also in accordance with the historical area. The design concept is also an additional narrative so that the old building is more meaningful.

4 CONCLUSION

Adaptive reuse is a building preservation strategy at a time responding to the current development of the city of Bandung. Strategies in responding to global problems and creating a sustainable environment as contained in the SDGs. Adaptive reuse has also become an interesting and creative challenge in architecture and interior design. Preservation is not just maintaining the old buildings but also empowering them so that they have an impact on the economy, society, and culture. The construction of a new hotel doesn't have to replace the old building. The initial building was the office of Waskita Karya, a construction company. And now its function has changed as a four-star hotel in the city of Bandung. Changes in function are still considering the new building can be combined with the old building. Therefore, we need creative and innovative designs.

The existence of style is very important in the preservation of buildings. Style is one of the significant elements in historical buildings. The initial building has an Art Deco style, which can be used as a reference in processing the interior design concept. The interior design of Maison Teraskita has a practical function as well as a 'link' between new and old buildings.

REFERENCES

Gewirtzman, D. F. (2017). Adaptive Reuse Architecture Documentation and Analysis. *Journal of Architectural Engineering Technology*, *05*(03). https://doi.org/10.4172/2168-9717.1000172

Hayati, R. (2014). Pemanfaatan Bangunan Bersejarah Sebagai Wisata Warisan Budaya Di Kota Makassar. *Jurnal Master Pariwisata (JUMPA)*, *01*, 1–42. https://doi.org/10.24843/jumpa.2014.v01.i01.p01

Lewis, R. H. (2016). Re-Architecture: Adaptive Reuse of Buildings with focus on Interiors. *Academia.Du*, 1–6.

Martokusumo, W. dan A. S. W. (2021). *Pelestarian Arsitektur dan Lingkungan Bersejarah* (2nd ed.). ITB Press.

Plevoets, B., Cleempoel, K. Van, Šijaković, M., Perić, A., Plevoets, B., Cleempoel, K. Van, Moreno, F. S., Perilla, M. P., Costa, C., & Shahrbanoo, G. (2018). Adaptive Reuse As a Strategy Towards Conservation of Cultural Heritage: a Survey of 19Th and 20Th Century Theories. *Structural Repairs and Maintenance of Heritage Architecture*, *118*(1), 73–92.

Rahadian, R. G. N. (2019). Kajian Konservasi Bangunan Melalui Unsur Pembentuk Arsitektur Dalam Upaya Pelestarian Bangunan tua di Kota Bandung Studi Kasus: Gedung Panti Karya, Jalan Merdeka no. 39 Bandung, Jawa Barat. *Idealog: Ide Dan Dialog Desain Indonesia*, *4*(1), 40. https://doi.org/10.25124/idealog.v4i1.1628

Soewarno, N., Hidjaz, T., & Virdianti, E. (2017). Adaptive Reuse as an Effort to Preserve an Historical District: A Case Study of the Braga Corridor in the City Centre of Bandung, Indonesia. *WIT Transactions on Ecology and the Environment*, *223*, 89–100. https://doi.org/10.2495/SC170081

Sutopo, A., Arthati, D. F., & Rahmi, U. A. (2014). Kajian Indikator Sustainable Development Goals (SDGs). *Kajian Indikator Lintas Sektor*, 1–162.

Process on public service advertisement of COVID-19

S. Nurbani & Y.A. Barlian
Telkom University, Bandung, Indonesia

ABSTRACT: COVID-19 brings misery to people around the world, including Indonesia. One of the strategies carried out by the government to make the public aware of COVID-19 is through advertising campaigns. Most of the media are filled with public service advertisement (PSA) campaigns related to COVID-19. This phenomenon has attracted the author to examine the *processes* presented in these PSAs to find out what on earth occurs through this type of *process*. The purpose of this research is to find out the type of process through the SFL approach in COVID-19 PSA, which will later reveal the phenomena related to COVID-19. Furthermore, this finding can be used as a reference in making creative strategies for advertisers related to PSA for COVID-19. This study uses a qualitative method to reveal social conditions occurring in the community analyzed based on natural data collection.

Keywords: creative strategy, process, PSA, SFL

1 INTRODUCTION

COVID-19 is troubling the world. The beginning of the COVID-19 case was in Wuhan, China at the end of 2019. This was quite a shock to the world because, in 2021, almost the whole world was affected by COVID-19, both health-wise and economically. Many countries have implemented strategies for dealing with COVID-19. Developed countries are competing to create vaccines that are able to reduce the wider impact of COVID-19. From an economic point of view, many countries have implemented strategies, such as lowering interest rates, to ease the financial problems experienced by their citizens. It is hoped that the economy can run even though the reduction in interest rates is very slow. The world of advertising is no exception. The government, in communicating messages conveyed to the public, uses advertising services, which are called Public Service Advertisements or commonly known as public service advertisements (PSAs). Currently, there are many PSAs made by the government to make people aware of the dangers and prevention of COVID-19 as one of the strategies to deal with COVID-19 through the dissemination of information through media campaigns.

Public service advertising campaigns during this pandemic are intensively carried out by the government as one of the steps to solve the problem of the COVID-19 pandemic such that it does not prolong. Talking about advertising campaigns will reveal the message that the makers of this public service advertising campaign want to convey, which, in this case, is the government. The advertising message is something that advertisers want to communicate to the public so that the advertising objectives are achieved (Ilhamsyah 2021). This public service advertising campaign is a tool or medium used by the government to convey information, persuade, or invite the public to do something as desired by the advertiser, in this case, the government. The application of this public service advertising message is one of them, through headlines other than images. Headlines are part of the anatomy of advertisements that function to attract interest that makes readers and the target audience fall in love at first sight (Ariyadi 2017). By making this headline, it is hoped that the reader or target audience will pay attention to the existence of this advertisement, hence they are willing to read the contents of the advertisement and even go further to follow what is expected by the advertiser.

Seeing the above phenomenon, it is necessary to conduct research on PSA in terms of the use of process or simple verbs in a clause. In this case, the headline is the first funnel of information from public service advertisements related to COVID-19. With this research, it is hoped that it will reveal phenomena that occur in the community or the world regarding COVID-19 so that it can be a reference in making public service advertisements related to the pandemic, especially COVID-19 for future Public Service advertisers. In other words, it can indirectly solve one of the problems faced during the COVID-19 pandemic related to the delivery of information through Public Service Advertisements.

Systemic Functional Linguistics (SFL) is a branch of linguistics that is more focused on the study of semantics. SFL can reveal language phenomena occurring in society; therefore, SFL is included in the study of applied linguistics or applied language (macro linguistics). An ideational language is a function where language is used to express the speaker's experience reflected through words about the world, both external and internal, based on the speaker's awareness (Halliday & Matthiessen 2014). This ideational illustrates that speaking or using verbal language is about an experience, also known as a clause as representation. The ideational language can provide what is done, experienced, or what events occur in this universe (Suardana 2021). Language practitioners see or describe the world as consisting of a sequence encoded by a verb that involves something (noun), which has attributes (adjective), and which takes place in the background (circumstance). The process is one of the elements in the major clause that plays an important role. This process is represented in verbs. When viewed from the criteria inherent in the type of action, the process is divided into several types. The main process consists of three types, namely material processes, mental processes, and relational processes, while the combined processes include verbal processes, behavioral processes, and material processes (Halliday & Matthiessen 2004) (Suardana 2021) (Sujatna 2013). Material process, material process the verbs used are verbs related to physical actions that are carried out, whether intentionally or not. Verbs that are grouped into intentional types are verbs usually done by living things, such as eating, drinking, running, washing, and so on. While verbs that are done accidentally are verbs related to events or phenomena, this type of verb is called a 'happening' verb (Halliday & Matthiessen 2014); for example, (1) Rainwater (actor) falls (material), (2) The car (actor) falls (material) into a river (circumstance). In example sentence (1), it is a happening type of verb, while the example sentence (2) is a doing type of verb. Mental processes are reflected in verbs that are mental or non-physical. This process involves feeling, thinking, understanding, and having an emotional reaction to something (Suardana 2021). For example, we (senser) believe (mental) it (phenomena). This process is different from the previous two processes because it emphasizes more "being" or "having" and "becoming". Therefore, this process often does not present a verb but uses a copula. For example, that cake (carrier) belongs to (relational) I (attributive), that he (carrier) is there (relational) in Jakarta (attributive).

Advertising is a way of communication through media containing messages from advertisers to target audiences (Moriarti 2011). Public service advertising is one type of advertising whose purpose is to communicate messages. However, the messages communicated are not in the form of commercial goods or services. This public service advertisement is usually in the form of a campaign carried out by interested parties to communicate information, hence it is conveyed to the target audience (Ilhamsyah 2021). The parties who use this advertising service are usually the government, NGOs, and other interested parties, yet not commercial. In communicating the message, public service advertisements are not much different from commercial advertisements; the only difference is in the creative strategy of the message. In terms of how to convey through public service advertising media, it consists of visual elements including words, color, music, pictures, words in motion, and movement. Creative strategy is a very important thing in the world of advertising because this creative strategy will distinguish an advertisement from other advertisements. Thus, a creative advertising strategy can be said to be a method used to produce attractive advertisements that can be accepted by the target audience and tend to be different from those of their competitors. According to a study (Lee & Johnson 2011), the creative strategy focuses on the message to be communicated and guides all messages that will be used in an advertising campaign. The existence of verbal language and visual language is the main thing in advertising, which is the result of creative strategies used by advertisers.

Through the disclosure of the type of process presented in the public service advertisement, it is hoped that it can provide recommendations for advertising to send a message effectively. By this point, the goal of sustainability in health will be further increased because the public will be more aware of COVID-19, both aware of the dangers of COVID-19 and ways to prevent and live healthier.

2 RESEARCH METHODS

This study uses a descriptive method where the goal is to describe or explain a phenomenon or event in a systematic, factual, and accurate manner regarding the data (Djajasudarma 2006). Other experts say that this method aims to decipher existing data (Ratna 2010). Based on the above, the author will describe the existing data about the types of processes contained in the headlines of public service advertisements to find out the phenomena that occur through the process of the COVID-19 phenomenon in public service advertisements (PSA).

The data were obtained through literature study and direct observation of the object of the COVID-19 prevention PSA as the verbal language of message distribution. The data analysis technique used by the author refers to the theory of Herdiansyah (Herdiansyah 2012), namely that the data obtained are then reduced into certain patterns, then categorization is carried out, and finally, based on the schemas obtained, they can be taken to conclusion. In other words, research data are presented (data display) using a matrix method to see the verbal aspects of the COVID-19 prevention PSAs before all visible phenomena are concluded (conclusion drawing/verification) or verified (Miles & Huberman 2005).

To analyze the process and its types, the author uses the theory of Halliday (Halliday & Matthiessen 2004). The objects or data in this study are public service advertisements taken from the official website covid19.go.id published in July 2021. The object analyzed is in the form of Headlines. While secondary data are a literature study in the form of reference books, data are obtained from various supporting literature, such as papers, and research articles. Secondary data also come from the internet and other sources of information.

3 RESULT AND DISCUSSION

Based on the data obtained, the expression in the form of a clause from the headline of a public service advertisement obtained on the covid19.co.id website, which was published in July 2021, obtained 20 data. These data found two types of processes, namely the material process and relational process. Meanwhile, the process appearing in this data is the relational process. From 20 data found, 10 data will be explained below:

Table 1. Data and results.

Data Number	Data	Type of Process
(1)	Pemerintah siapkan fasilitas pengelolaan limbah medis (31-7-2021)	Material
(2)	3 langkah strtaegis pemerintah hindari korupsi bansos (28-7-2021)	
(3)	Gencarkan tes dan lacak dimasa PPKM (27-7-2021)	
(4)	Indonesia terima 5 juta bahan baku vaksin Covid-19 (22-7-2021)	
(5)	PPKM darurat: Mall ditutup pasar dibuka (2-7-2021)	
(6)	Protokol Kesehatan tanggung jawab semua pihak (30-7-2021)	Relational
(7)	Vaksin Covid-19 terbukti efektif mengurangi tingkat kematian (23-7-2021)	
(8)	208.265.720 orang target sasaran vaksinasi Covid-19 di Indonesia (22-7-2021)	
(9)	Vaksinasi Covid-19 moderna diprioriaskan jadi dosis ke tiga (14-7-2021)	
(10)	Asrama haji di Pondok Gede jadi rumah sakit Covid-19 (12-7-2021)	

Material process

Based on the data there are only 2 processes, namely the material process as much as 45% and the relational process as much as 55%. Thus, it can be concluded that the relational process is the most widely used in the COVID-19 PSA headline published in July 2021.

The data found 9 clauses containing material processes, 5 of which are as follows:

Data (1) show the material process. This is realized using verbs containing motion. In this clause, the word *"siapkan"* is used. The word *"siapkan"* comes from the word *"siap"* as the basic word, which means holding. In this case, a physical process is carried out. As for those who hold it, the government acts as an actor, while the goal is to facilitate medical waste management. Through this clause, the author states that the government is moving to conduct medical waste management facilities to avoid unwanted things because this medical waste will endanger the environment and society.

Data (2) are a material process, this is indicated using verbs that contain physical movements, namely the word *"hindari"*. This word has the meaning of abstaining. The actor or perpetrator who evades is the government and the goal of this evasion is *"korupsi bansos"*. Social assistance is social assistance from the government for people affected by Covid-19.

Data (3) are included in the material process; this is indicated using the verbs *"tes"* and *"lacak"*. This word contains physical movement, namely the physical process of tracking or searching very quickly. The government is the perpetrator or the actor who tracks, but in this clause, it is not shown, because it is assumed that the reader already knows that the government is doing the test and not another party. Meanwhile, the goal of material processing on these data is the COVID-19 test.

Data (4) are included in the material process because the verb used is a word that contains physical movement, namely the word *"terima"*. The perpetrator or actor of this recipient is Indonesia or the Indonesian government. While the goal is *"5 juta bahan baku vaksin Covid-19"*.

Data (5) show the material process realized through the closed verb. This word is a type of verb that involves a physical process. The word *"ditutup"* means "a happening process or an event," because the actor or actors are not shown in this clause. Although the actor or perpetrator is not explicitly stated, the reader already knows that the government has the authority to close the mall, so even if it is not shown in the clause, the reader will not be confused in digesting this message.

Relational process

The relational process in this data has 11 clauses, and this type of process is most often displayed by this public service advertiser, 5 of them are as follows:

Data (6) are a relational process; this is expressed by the word *"tanggung jawab"*, which in the initial sentence contains the word "is," but this word is an ellipsis. The word *"protokol"* is a carrier while *"semua pihak"* is an attribute. This process is also called the process of being, in other words, this process can indicate a fact or information. This clause informs that the health protocol is the responsibility of all parties; here, there is no physical involvement process, only in the form of information or a situation.

Data (7) show the relational process or process of being. The word *"terbukti"* is a word that indicates a relational process. The carrier in this clause is the phrase *"vaksin Covid-19"*, while the attribute is found in the phrase *"efektif mengurangi tingkat kematian"*. This clause is information and not a job or doing.

Data (8) show a relational process, and this is realized through the verb *"adalah target sasaran"*. The carrier in this data is found in the words *"208,265,720 orang"* while the attribute is *"vaksinasi Covid-19 di Indonesia"*.

Data (9) are included in the relational process; this is indicated using the word *"diprioriaskan"*. This word is a process of being because it does not involve physical movement. The word *"Vaksinasi Covid-19 moderna"* acts as a carrier while *"dosis ke tiga"* acts as an attribute.

Data (10) show a relational process. This is stated in the use of the word *"jadi/menjadi"*, which is included in the process of being, not the process of doing, the carrier in this clause is *"asrama haji Pondok Gede"* and the attribute is *"rumah sakit Covid-19"*. In this clause, it is clear that this clause does not represent a physical action but an event in the form of information.

4 CONCLUSION

The conclusion that can be drawn from this research is that in the sentence or clause presented by the maker of public service advertisements on the covid19.co.id website, which was published in July 2021 in the form of a headline, there are only two types of processes, namely material process, and relational process. The relational process is the most common in this PSA headline. Hence, it can be concluded that this PSA maker who is a representative of the government conveys more information or conveys a situation than the action or physical movement. The transmitter of this information is the government. Thus, the process of being dominated more than the process of doing. So that the characteristics of the PSA headline on the covid19.co.id website published in July when this month is the peak of the increase in COVID-19 in Indonesia, is the government still inviting the Indonesian people to know or understand the condition of this COVID-19 rather than do the action. This finding is expected to be an input for advertisers, especially public service advertisements, to create attractive creative strategies that can be easily accepted by the target audience or readers thus the goals of the government or the goals of PSAs can be achieved.

REFERENCES

Ariyadi, W. (2017). *No Title*. Quadrant.
Djajasudarma, T. F. (2006). *Metode Linguistik: Ancangan Metode Penelitian dan Kajian*. PT. Refika Aditama.
Halliday, M., & Matthiessen, C. (2004). *An Introduction to Functional Grammar*. Edward Arnold.

Halliday, M., & Matthiessen, C. (2014). *Introduction to Functional Grammar* (Fourth Edi). Routledge.
Herdiansyah, H. (2012). *Metodologi Penelitian Kualitatif Untuk Ilmu-ilmu Sosial*. Salemba Humanika.
Ilhamsyah. (2021). *Advertising Era Digital* (1st ed.). ANDI.
Lee, M., & Johnson, C. (2011). *Prinsip Pokok Periklanan Periklanan Dalam Perspektif Global*. Kencana.
Miles, M., & Huberman, M. (2005). *Qualitative Data Analysis (Terjemahan)*. UI Press.
Moriarti, S. (2011). *Advertising*. Prenanda Media Grup.
Ratna, N. K. (2010). *Metode Penelitian, Kajian Budaya dan Ilmu Sosial Humaniora Pada Umumnya*. Pustaka Pelajar.
Suardana, I. K. (2021). *Sudut Pandang Systemic Functional Linguistics*. Mitra Cendikia Media.
Sujatna, E. (2013). *Understanding Systemic Functional Linguistics*. Unpad Press.

Implementation of the AISAS method in Waroeng Soejo's frozen food promotion strategy

E. Mentari, I. Wirasari & Ilhamsyah
Telkom University, Bandung, Indonesia

ORCID ID: 0000-0002-1421-7003

ABSTRACT: The shift in consumer behavior in fulfilling food needs is marked by a large number of requests for frozen food products. Waroeng Soejo, one of the small and medium-sized enterprises established in Bandung, captures this opportunity by processing some of its menus into frozen food. However, these products are not widely known by new consumers due to intense competition in this digital era. It is assumed that there is a lack of promotional activities to raise the company's brand awareness. This study uses a qualitative research method with descriptive analysis and the results are used in the AISAS method to be implemented in a promotional strategy for the target audience aged 20–30 years. The result of this study is a proposed promotional strategy through digital channels that can help the brand to be known by new consumers.

Keywords: AISAS method, frozen food, promotion strategy

1 INTRODUCTION

Currently, Indonesian people are adapting to the consumption of frozen food because the industry is experiencing rapid changes. Consuming frozen food has become a lifestyle and one of the most demanded products. Efficiency and effectiveness in meeting food needs have made frozen food a popular choice. This happens because of changes in lifestyle in today's society where they need convenience foods, which are easy to carry everywhere and ready to be served at any time. Those people tend to be too busy with their work and personal matters. *Asosiasi Rantai Pendingin Indonesia* (ARPI) said frozen food products increased by 17 percent in 2020 (Fitri 2021). Even now, in 2022, the frozen food trend is increasingly promising in the future because of the support from many MSMEs and cold-chain delivery that can send these products widely to remote areas. This discussion refers to the eighth Sustainable Development Goals (SDGs) where there are points with the target of achieving a higher level of economic productivity through improving technology and innovation because frozen food has become one of the mainstays in meeting food needs on a busy schedule.

Frozen food becomes a promising business opportunity; however, this opportunity has also been seized by many sellers who have started to make their own frozen food products (Chianardi & Permatasari 2020). It makes the industry face a red ocean market as almost every brand competes for the same trait of the target market (Fiona *et al.* 2019). Not only is the competition getting tougher but also the information environment that surrounds the society has changed in recent years that makes them put up an information barrier. The development of digital information technology, the growth of the internet, and the proliferation of mobile phone users have exposed society to enormous media diversification (Sugiyama & Andree 2011). As a result, according to research by Ira Wirasari said that many businesses stop operating because reluctant to adapt to the times, are too comfortable with the past heyday, can't deal with digital transformation, and are surprised to see new start-ups are just popping up and take their customers (Wirasari 2021).

People's daily activities become much more complicated and will likely to increase as our societies become more connected and technologically advanced. Therefore, now is the time to accelerate the digital promotion strategy as the digital transformation environment becomes fiercer. Waroeng Soejo is one of the small and medium-sized enterprise established in Bandung which provides frozen foods online since 2020. The types of frozen foods offered include beef lungs and chicken meat which are processed with their own special spices. Initially, the business went well and managed to retain 30 loyal customers, but so far it has not shown a significant increase in the number of new customers and not expanding well. Therefore, Waroeng Soejo needs to introduce its frozen food products to new consumers using certain digital media and a specific target audience.

1.1 *Purchase intention towards frozen food*

Consumer conceptions of frozen foods are evolving as people's diets change and buy according to what they perceive to more convenient option. Many factors can influence consumers to purchase frozen food in this information era. Starting from the consumer who sorts out information about products, features, categories, quality, and brands offered before buying (Bughin *et al.* 2010). Others say that taste is the main reason for buying intentions followed by availability, packaging, and price (Saifullah *et al.* 2014). Consumer lifestyle is strongly related to consumer purchasing decisions (Raof *et al.* 2017). Consumers are people who identify and determine their needs, make wishlists, make purchases, and dispose the products (Solomon 2006). By studying consumer preferences, lifestyles, behaviors, and using the Internet as a powerful information and sales channel, sellers can make it easier to promote frozen foods.

1.2 *Promotion strategy*

Strategy is a process of evaluation, planning, and implementation of design to maintain and improve competitive advantage (Sammut-Bonnici 2015). Promotional objectives are divided into providing customer with information about a new product, influencing customers to buy a brand, and reminding customers about the brand (Kuncoro 2010). Today consumers of all ages get information about a product or brand they want by selecting, combining, and linking various media as the situation demands. before buying (Bughin *et al.* 2010). Consumer decisions are strongly influenced by core values and belief systems that form the basis of attitudes and behavior. Sellers that target consumer values are likely to influence their buying behavior. By looking at consumers' knowledge of frozen food, searching for information and lifestyle seller can predict the purchase intention of frozen food directly or through other intermediaries. By understanding, those traits will be a helpful reference for a small and medium frozen food sellers to design and manage their brand to attract new consumers and increase the loyalty of existing customers.

2 RESEARCH METHODS

2.1 *Methods of data collecting*

This study used qualitative research, an approach to investigate and understand the meaning of individuals or groups about social or human problems and focuses on individual meaning (Creswell & David Creswell 2018). The research process was carried out through in-depth interviews, as primary data, based on the theory of human motivation proposed by Sigmund

Freud which assumes that the psychological shaping forces of behavior are largely unconscious and that one cannot fully understand one's own motivations. Therefore, the laddering technique is used which allows tracking a person's motivation from the instrumental to the more terminal by conducting semi-structured interviews using various projective techniques in the form of sentence completion, word association, image interpretation, and role-playing (Kotler & Keller 1956). In this study, interviews and observation were conducted on Waroeng Soejo as an internal source and the sample taken is a husband and wife, as an external source, with age of more than 20 years who are the millennial generation, just got married and have a job, both worked as private employees and are a middle-class group that actively using digital technology. Then, secondary data are taken from journals, books, government publications, websites, and other related sources.

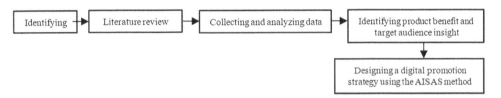

Chart 1. Research method chart.

2.2 *Methods of data analysis*

The purpose of this study is to describe, explain, and analyze the data to determine the factors to create Waroeng Soejo frozen food promotion strategy in order to attract new consumers. The analytical method used is qualitative descriptive analysis, a comprehensive summarization of specific events experienced by individuals or groups of individuals (Lambert & Lambert 2012). The data is analyzed to find out the benefits of the product and insight from the target audience, the results of which are then implemented into a marketing communication model where the emergence of a new social media-based technology called the AISAS model is the formulation of this interaction flow between products/services and consumers that can be used as material for analysis as well as the strategy in advertising planning (Sugiyama & Andree 2011). The process of this model is not linear and unpredictable but is comprehensive to shape consumer responses to advertising messages (Ilhamsyah 2021).

3 RESULT AND DISCUSSION

The discussion starts with identifying the product benefit, Waroeng Soejo's frozen food products have been described and keywords appear, namely Easy, Simple, Fast, and Homey which defines the character of a product that is easy and fast to serve without being complicated and the taste also the quality as good as home food. This product character is needed by most people in this fast-paced era nowadays. Next, identifying target audience insight, a husband and wife aged over 20 who both work as office employees that just got married and started a career, and are technology literate stated that don't have enough time and ability to cook from fresh ingredients, but want to still enjoy home cooking, stay up to date, and be productive. Therefore, they want food products that can be used as energy to continue to be productive, work hard, and do not hinder their activities to keep updating but still feel the pleasure of home cooking. Then concluded Waroeng Soejo frozen food, frozen food with home-style quality and taste that is easy, simple, and fulfill energy needs for hectic schedules, which was later simplified to Waroeng Soejo's frozen food for your busy days.

Those highlights are implemented to the AISAS model, one of the steps of the communication and consumer journey that plays a role in understanding consumer needs and providing effective solutions for consumers to shop for their needs. The food search begins with the awareness of biogenic needs, namely hunger, and thirst. After identifying the needs, the target audience tries to find information about what food products meet their needs but do not hinder their activities. Then a promotion strategy for Waroeng Soejo frozen food was made with the AISAS model based on the trait of the predefined target audience, which is explained as follows:

Attention, the first thing that needs to be done is to get attention. When buying food, the target audience really prioritizes the visual aspects, reviews, ratings, and recommendations advertised through the marketplace and social media. In this case, attractive visual strategies can be used, clear product descriptions, packaging designs that are appropriate to the type of food, and attractive promotional offers on various marketplaces and available social media. Then, Interest, the target audience is getting more affected to buy and consume the brand's products by looking deeper into the brand's account and details. These two categories are communication with low involvement (passive) so they are made as interesting and clear as possible.

Then, the next category is communication with high involvement (active). Search: the target audience, their curiosity is aroused so they want to know more about advertisements and notice them further through various sources on the internet, such as social media and Google Maps. In this case, the brand needs to create its own website to further involve consumers and create a column for direct questions and answers because the millennial generation tends to visit the website to ensure information related to the products offered and their place of business. Next category is Action: the target audience is in direct contact with the advertising message and then contacts the brand to ask questions or ask friends or search on the internet until finally making a purchase. In this category, brands must be ready and quick to respond to product-related questions as well as offer attractive promotions by getting a discount if you place an order on the same day.

Furthermore, Share, after buying and being satisfied with products and services, usually the target audience likes to share their experiences on social media and with people in the surrounding environment. In this category, brands can bring their target audience to write reviews on marketplaces, social media, websites, and other sources by offering reviews in the form of best writing, photos, or videos to get best-seller frozen food products or secret menus for free for subsequent purchases. For this category, brands need to provide shareable content because when a great story or something is interesting to be told in it, then more impactful it can get more attention, encourage conversations, trigger virality, and affect customers to buy.

Overall, the use of AISAS in the frozen food business concept in the digital era plus the power of the internet in expanding information is expected to reach consumers in various dimensions. In this study, Share is the lowest element because most consumers are not willing to share their experiences with others. This could be due to the gap between brand and audience, and this issue is often overlooked by brands, even though the conversation is an important part of the experience. For further research, it can be added with audience engagement strategies in interactions and activities in order to create a holistic customer experience. In the process, various communication channels are involved so that relationships with the audience are maintained, satisfaction is increased and relevance is maintained. The success of the strategy can allow the audience to want to share their experiences both before and after using the brand.

4 CONCLUSION

The frozen food industry is growing rapidly which brings companies into an intense competition to meet the lifestyle needs of people who are busy and fast-paced. The digital era has exposed all individuals to the explosion of information and made consumers put up information barriers, which makes MSMEs become more difficult to market their products in the midst of a vast market sea. MSMEs need to be equipped with the ability to be able to identify the benefits of their products and insight from a predetermined target audience to be able to understand consumer motivation in buying a product. Developing digital literacy will also make it easier for sellers to determine the right solution for every problem. In this case, Waroeng Soejo needed the AISAS method to answer the problem of the difficulty of getting new customers having previously identified product benefits and target audience insights. However, this method cannot be applied to the whole frozen food industry because it requires careful observation, planning, and strategy, and not all SMEs have the same problem. Suggestions for its development are to use this research as a reference in developing a more flexible marketing strategy, it is hoped that this research can help sellers to further develop their business in this digital era.

REFERENCES

Bughin, J., Doogan, J., & Vetvik, O. J. (2010). *A New Way to Measure Word-of-Mouth Marketing.*

Chianardi, D. H., & Permatasari, A. (2020). *Factors Affecting Frozen Food Purchase Intention During The Covid-19 Outbreakin Indonesia (Case of Greater Jakarta).*

Creswell, J. W., & David Creswell, J. (2018). *Research Design: Qualitative, Quantitative, and Mixed Methods Approaches.*

Fiona, M., Utami, D., & Chaldun, E. R. (2019). The Influence of Product Experience on Customer Loyalty of Frozen Food Product. *The Asian Journal of Technology Management*, 12(3), 177–190. https://doi.org/10.12695/ajtm.2019.12.3.2

Fitri, A. N. (2021, October 22). *ARPI: Tahun 2025, Nilai Pasar Frozen Food Bisa Mencapai Rp 200 Triliun.* Kontan.Co.Id.

Ilhamsyah. (2021). *Pengantar Strategi Kreatif Advertising Era Digital* (D. Arum, Ed.). Penerbit Andi.

Kotler, P., & Keller, K. L. (1956). *Marketing Management.*

Kuncoro, M. (2010). *Masalah, Kebijakan, dan Politik, Ekonomika Pembangunan.* Penerbit Erlangga.

Lambert, V. A., & Lambert, C. E. (2012). Pacific Rim International Journal of Nursing Research. In *Pacific Rim Int J Nurs Res.*

Raof, R. A., Othman, N., Noor, M., & Marmaya, N. H. (2017). Consumer's Purchase Decision towards Canned Pineapple Products in Malaysia. In *International Journal of Business and Social Science* (Vol. 8, Issue 9). www.ijbssnet.com

Saifullah, A., Ahmad, N., Ahmad, R. R., & Khalid, B. (2014). *Frozen Food Revolution: Investigating How Availability Of Frozen Food Affects Consumer Buying Behavior.*

Sammut-Bonnici, T. (2015). Strategic Management. In *Wiley Encyclopedia of Management* (pp. 1–4). John Wiley & Sons, Ltd. https://doi.org/10.1002/9781118785317.weom060194

Solomon, M. R. (2006). *Consumer Behaviour: A European Perspective.* Financial Times/Prentice Hall.

Sugiyama, K., & Andree, T. (2011). *The Dentsu Way: Secrets of Cross Switch Marketing from the Worlds Most Innovative Advertising Agency.*

Wirasari, I. (2021). *Digital Knowledge Improvement for Indonesian Small and Medium Enterprises: Cultural Change in Digital.*

Behavioral patterns and interaction of backpacker visitors in limited areas at snooze Hostel Yogyakarta

N. Laksitarini, A.D. Purnomo & L.C. Lase
Telkom University, Bandung, Indonesia

ABSTRACT: Yogyakarta is often used as a vacation destination. Hostels are accommodation facilities that are often chosen by tourists who travel long distances by prioritizing the vacation experience overspending the budget for lodging or are often referred to as a backpacker. Snooze Hostel, located in Yogyakarta, is one of the accommodation facilities that is often used as an alternative place to stay overnight by backpackers who travel to Yogyakarta. The limited space will affect some of the behavior of the visitors, including the higher level of interaction between visitors. This is a manifestation of the result of a stimulant in the form of a space atmosphere that is formed, which will eventually produce perceptions and behaviors in each individual. The purpose of writing this paper is to examine the reciprocal relationship between the limitations of existing space and the behavior of backpacker visitors at Snooze Hostel Yogyakarta. The results of the study indicate that with the layout arrangement in a limited area, visitors will have more opportunities to interact with other tourists because some of the facilities in it are used together. The research method used is a qualitative method with descriptive analysis.

Keywords: backpacker, hostel, behavior

1 INTRODUCTION

Yogyakarta is a city that has a fairly high ranking on the national tourism map. The predicate is related to the strength of the development of the tourism sector in the Yogyakarta area, both in terms of the diversity of tourism objects and the readiness of supporting infrastructure. This also has an impact on the demand for the availability of supporting facilities that can provide comfort for tourists, one of which is adequate accommodation facilities. Hostels are one of the many types of accommodation facilities. They are budget-oriented accommodations, providing shared rooms (dormitory model rooms) that accept individual travelers (usually backpackers) or groups for short-term stays, and provide common areas and shared facilities (Latasi 2019). Referring to the adjective, backpacking can be said as a travel style that is independent and doesn't depend on travel or travel agents. This, of course, has advantages in the freedom to determine the location and destination and the use of a budget that can be minimized as low as possible depending on the ability (Rahmawati 2014).

For backpackers, hostels are the most favorable choice because it relates to budget availability and usually the backpacker will look for a more intimate and personal atmosphere. Similar to other hostels, Snooze hostel offers low-budget accommodation facilities by providing shared public facilities such as a pantry, gathering room, dining area, bedroom, and toilet. Located at Mangunnegaran Kulon Street Number 9, Panembahan, Kraton

District, Yogyakarta City, it is not far from famous destinations in Jogja, namely Malioboro, and Keraton.

The problem that often arises in a hostel is limited space. This is because most hostels are built on limited land, but with the hope of accommodating all the visitor's needs. The concept of shared space often creates various behaviors in visitors. Shared facilities such as kitchens, gathering rooms, and even beds and toilets become spaces that are no longer private because the intensity of meetings and interactions between visitors will be higher. This can be a decent thing to be raised in a study because human behavior will depend on the atmosphere and conditions they experience.

Psychologically, there are two types of responses caused by humans in relation to the process of interaction between space and humans, namely, outward response and inward response. The outward response is the behavior or activity carried out by the human being, and the inward response is the shape of the human form and image of the space they inhabit. Both responses are the output of the personality system of a human being because of stimuli in the form of a spatial atmosphere/behavior setting. The pattern of behavior that exists in humans is a depiction of the interaction of activity factors with a more specific space, namely between activities, space, and time when these activities take place (Setiawan 2010). All perceptions generated by humans on stimuli they receive will proceed through a system of cognition, motivation, and behavior patterns, and can also be influenced by the human concerned's experience background, which will eventually form a response (Hidjaz 2011). Human behavior itself is defined as actions that are consciously carried out by individuals (not reflex actions) in response to the environment because individuals will indirectly choose stimuli before responding.

2 RESEARCH METHODS

This paper examines the behavior patterns and interactions that occur among backpacker tourists in one of the hostels in Yogyakarta. The research location was held on Mangunnegaran Kulon Street Number 9, Panembahan, Kraton District, Yogyakarta City, which is located not far from famous destinations in Jogja, namely Malioboro, and Keraton. This hostel offers a homey atmosphere that comes with the use of a simple and intimate interior. There are facilities such as a mini coffee bar, a pantry, and a gathering room. Like other accommodation facilities, Snooze Hostel has several types of rooms, namely, standard rooms, budget rooms, and dormitories with a room-sharing system. There are two types of dormitories for women only and mixed for men and women. All hostel rooms don't have a private bathroom or air conditioning.

The research method used is descriptive qualitative. Primary data were obtained through field observation techniques and interviews with both managers and respondents (hotel visitors). Secondary data are obtained through printed and electronic literature, such as scientific journals, seminar papers, magazines, and websites. The data are arranged for analysis, and then conclusions are drawn.

3 RESULT AND DISCUSSION

Individual human behavior patterns in space will arise as a response to the space experience they experience. The atmosphere of the room at Snooze Hostel which has a limited size and carries a homey concept, where the placement of furniture is arranged in such a way as at home brings a perception and feeling of comfort to the visitors. Such a space atmosphere affects the individual's response which in the end comes out with a different form of behavior for each individual. As a stimulant factor, the arrangement of the furniture layout will be one of the facilitators until the occurrence of human behavior. For spatial planning that has limited space, the activity or behavior will also have an effect, namely the increasingly limited behavior itself (Hidjaz 2014).

To support a homey atmosphere for visitors, the inner common room is likened to a living room in a house. The use of simple furniture with natural wood materials, and a coffee table that is not too big with the placement of several indoor plants supports the atmosphere like your own home. The scorching temperature of the city of Jogja seems to be muffled by green trees that are almost in most corners of the room. The Snooze Hostel building is a house building like a residence in general and doesn't provide parking space for cars. This is because the existing land is not adequate to provide car parking space, but Snooze Hostel provides motorbike and bicycle parking lots, so hostel visitors can take advantage of their time to travel in Yogyakarta City by using motorbikes or bicycles rented by the hostel. It is closely related to the fact that most of the visitors are backpackers, who didn't bring too much stuff when traveling and didn't even bring a car.

Figure 1. The atmosphere and layout of the common room furniture at snooze Hostel.
Source: https://travelingyuk.com/snooze-hotel-jogja/280683.

In the dormitory type of bedroom area, they use a bunk bed that is arranged vertically. This type of bed can be a solution when the available land has a limited area. To maintain privacy, each booth is equipped with a curtain that can be opened and closed manually by visitors who occupy the booth. The pattern of activity and behavior that occurs in the dormitory-type sleeping area, activities tend to be personal, and are not free because of the location of the beds that are close to each other. This is because people will instinctively maintain their personal space by keeping their distance from other people, to avoid the entry of other people into the individual's personal area.

For the dormitory type of bedroom area, interaction occurs only during activities related to sleep activities. Bathing and changing clothes are done in the bathroom which is located outside the bedroom and is used interchangeably/sharing with other visitors. There are lockers beside the bed to store belongings and necessities that the visitors bring.

Figure 2. Snooze Hostel dormitory-type room layout and atmosphere.
Source: https://instagram.com.

The common room is filled with chair and table facilities which are placed in several strategic areas for gathering. There is a common room inside and outside the room. Visitors are free to use these facilities as long as there are facilities to sit and mingle. There are patterns of behavior and interactions that occur in the shared area, namely, there is no awkward impression even though some of them don't know each other and have never met before. Another common room is the pantry. In the pantry room, there is a kitchen set that can be used by all hostel visitors to prepare food. Cooking utensils have been prepared by the hostel, to be used interchangeably. Here, it takes an element of empathy and responsibility from all visitors who use it, which is expected to wash the cooking utensils that have been used so that they can be reused by other visitors. The interaction process in the pantry occurs during breakfast and dinner hours. Lunch is mostly done outside of the hostel because it coincides with the time to visit tourist attractions. There is a process of interaction outside the personal space, unlike in the bedroom area.

Figure 3. Visitor activities in the common areas of the dining room, gathering room, kitchen, and garden.
Source: https://instagram.com

4 CONCLUSION

Snooze Hostel, located in Yogyakarta, is one of the accommodation facilities that is often used as an alternative place to stay overnight by backpackers who travel to Yogyakarta. In addition to relatively much lower prices, they have more opportunities to interact with other travelers because some of the facilities in it are shared. This is because most hostels have a relatively small space. The use of shared facilities increases the intensity of interaction between visitors, they chat and carry out conversations without feeling shy and awkward, even though some of them do not know each other and have never met before. The interior layout and furniture resulting from the limited space tear down the personal space between visitors, where which rarely happens to hotel visitors in general.

REFERENCES

Hidjaz, T. (2011). *Interaksi Psiko-Sosial di Ruang Interior*. Itenas dan HDII.

Hidjaz, T. (2014). Interaksi Perilaku dalam Suasana Ruang Terbatas Studi Kasus Hotel Kapsul The Pod Singapura. *Jurnal Itenas Rekarupa © FSRD Itenas |*, 2(2), 74–86. http://ejurnal.itenas.ac.id/index.php/rekarupa/article/view/706

Latasi, A. D. L. (2019). *Youth Hostel di Baciro, Yogyakarta Penekanan Desain Hostel Rendah Biaya Pada Bangunan Indis Dengan Menggunakan Metode Blue Ocean-Strategy*. https://dspace.uii.ac.id/

Rahmawati, R. A. (2014). Backpacking Ala Mahasiswa Studi Deskriptif Tentang Gaya Hidup Pada Mahasiswa Universitas Airlangga Surabaya. *Journal Unair*, 3(1), 1–19. https://journal.unair.ac.id/

Setiawan, H. & B. (2010). *Arsitektur Lingkungan dan Perilaku: Pengantar ke Teori Metodologi dan Aplikasi*. Gadjah Mada University Press.

Form of dynamic identity in restaurant interior design

E.A. Wismoyo, T.M. Raja & V. Haristianti
Telkom University, Bandung, Indonesia

ORCID ID: 0000-0002-1736-165X

ABSTRACT: The interior design of the restaurant business has similarities in the stages of the design process, so that the differentiation elements identity as intellectual property, and as a medium of communication to consumers becomes a big variable in the interior design process to create an identity that will affect the differentiation between competitors. A strategy is needed in implementing identity in interior design so that there can be harmony between the characters in an entity and its interior design. Dynamic identity acts as a strategy, which aims to bring out visual variations in the identity and character of an entity in the context of interior design. Interior design elements are used as identity variables that will create a character in an entity. This study uses a descriptive qualitative method, with the analytical method used to compile data from observations and discussions, to identify dynamic identity variables that arise from the object of research. The results obtained are that there are applications of several dynamic identity variables in the interior design of the research object.

Keywords: Dynamic Identity, Interior Design Elements, Restaurant, Visual Identity

1 INTRODUCTION

Interior design is a science that can provide an experience for the users of the space which is one of the important elements that can translate the goals of a company or brand. A tool that serves to communicate the essence of a company's brand entity through the design of the physical environment (Putsure 2011). Along with the development of the world of graphic and interior design, the field of branding design has experienced interesting changes, where the current brand identity is no longer consistent, but allows random variations. According to Siswanto and Dolah (2019), the approach that allows for variation in terms of developing visual identity is called dynamic identity. This is also seen in the interior design of a brand that allows it to implement variations in its visual concept, which no longer uses the same design as other branches. Considering the phenomenon of this need, the business world, both in the field of services and products, has been using interior design as a branding medium for their business. One of them in this case is Holywings which is engaged in food and beverages. The interior design of each Holywings outlet has its uniqueness and is interesting in terms of its visual identity.

Based on observations made by researchers and conducting FGDs with designers from Holywings outlets, namely NADI Architect, primary data related to store documentation photos, as well as brief designs carried out by architects, were obtained. Holywings was designed without a very specific design brief from the owner, only referring to a past project from NADI Architect with the Irish bar concept. Based on the discussion, several keywords regarding the visual identity of the research object were obtained.

This study focuses on the dynamic identity approach in terms of interior design science, by identifying and evaluating interior elements and Holywings' visual identity. The research parameters are only limited to the identification of visual identity, as well as the identification of dynamic identities that appear in interior design elements through several variables.

2 RESEARCH METHODS

This study uses a descriptive qualitative method, the analytical method used is to collect data in the form of photos of the interior documentation of Holywings Paskal Bandung. The data is analyzed and then identified from each interior element to look for visual identity in the interior of Holywings and identify the implementation of the variables of dynamics identity in the interior design of Holywings. Primary data was collected using observation and FGD methods with the interior designer consultant Holywings, while secondary data was collected through references to related journals.

2.1 *Methods of data collecting*

Observations were made at the object location, namely Holywings Pasir Kaliki, Bandung. Observations and documentation of the interior of the object of research were carried out. This observation is carried out to obtain data and information that will be used for the identification and analysis of visual identity and implementation of the dynamics identity approach.

The Focus Group Discussion (FGD) was conducted directly with the designer from Holywings Indonesia, namely Yagi Sastrawidjaja the principal designer of NADI Architect that was held on February 9, 2022, and was conducted online by researchers. From the FGD, data related to interior design elements became the highlight of the identity of the design, technical explanations for the production of interior design elements, and how interior design elements were used as the identity of the design.

The literature used in this study refers to several kinds of literature related to interior design elements, and literature related to visual identity, as well as the dynamics identity approach. Interior elements are objects that have physical properties (visible) that aim to limit an area between the inside and outside as well as the filler inside. According to Wicaksono and Tisnawati (2014), several interior elements in buildings that can be implemented in the application of design are elements of flooring, ceilings, walls and windows and doors, stairs, lighting, and decoration elements.

Imani and Shisebori (2014), stated that forming a certain mentality of a brand through interior design, will have more influence when compared to other methods. At this time, logos and signs, have less influence in the world of branding. His research further explained the conclusion that explains the correlation of branding characteristics with elements in interior design. Space, color, material, shape, lighting, furniture, needs, and human factors are components in interior design that can influence branding activities.

In the field of visual communication expertise, a logo is a graphic symbol that has been perfectly designed and must be consistent, which should not be changed. However, changes and developments in the field of graphics or visual communication will cause excessive design complexity and will result in the discovery of many new challenges in terms of branding design activities. Dynamic Identity is a term popularized by Irine Van Nes (2012) in her book "Dynamic Identities how to create a living brand". In their research, Martin & Cuba (2019) devised a mechanism to achieve a dynamic visual identity. The mechanism is divided into 8 indicators of mechanism variation, which are mentioned in Table 1.

Table 1. Indicators of variation in dynamic identities mechanisms.

Color variation	Color-changing graphic elements
Combination	Combination of graphic elements
Content variation	Some areas or spaces use different depictions, both as background and foreground
Positioning	Graphic elements layout in a different way
Repetition	Repetition of the same graphic element
Rotation	Same graphic element with only rotation
Scaling	Graphic elements that change in scale
Shape Transformation	Graphic elements that change in shape

3 RESULT AND DISCUSSION

Brand aesthetics tend to define visual characteristics that contribute to characterize a brand. This visualization starts with a logo and extends to packaging, website, billboards, advertisements, and interior design (Izadpanah 2021). Data analysis was carried out by combining three primary data, focus group discussions (FGD) conducted with designers from Holywings Indonesia on behalf of Yagi Sastrawidjaja as resource persons, and visual data collection from design objects, and data presentation of Holywings planning designs. These data were combined to adapt the data to the research context to identify the interior design elements of Holywings Pascal as its interior identity. Data analysis is done by grouping space areas based on the orientation of the interior facilities.

At this stage, the identification carried out is to detail the treatment of interior design elements in each area from visual factors, patterns, and shapes. This identification is carried out to determine the visual identity of the interior design of the holywings from each area of the space.

Table 2. Identification of the brand's visual identity in holywings interior design.

Interior Design Treatment at Bar Counter Area	
Backdrop on the bar counter	Uses curved shape printed gypsum with repetitive configurations along the bar counter area, these curves are used as aesthetic elements and as product storage displays. Finishing it uses white wall paint and copper brass finishing pipe metal accents as a frame.
Ceiling on bar counter	The treatment is related to the backdrop area, in the form of printed gypsum with a repetitive curve shape following the dimensions on the backdrop. Finishing on this material is white paint with pendant light linear uplight elements to highlight the curve shape.
Bar counter table	The bar counter table top uses black marble. The body of the counter table is dominated by copper brass color using angled iron plate material with a repetitive configuration along the area of the bar counter table, and use lighting to highlight it.
Bar Table Area	
Column	Four structural columns are symmetrically aligned. The treatment was carried out in the form of a laminated plywood cover with a walnut color finish, and accents using copper brass finishing elbow plate material with uplight lighting on the column cover to highlight it.
Bar table	The bar table set is the main part of this area, each table has a capacity of four seats with a total unit of eight tables. The tabletop material uses a laminated teak wood finish with a natural color finish that brings out the character of the teak wood.

(*continued*)

Table 2. Continued

	Sofa set area
Backdrop wall	The treatment is carried out using teak wood panel walls with walnut color finishing, and the frame as a form of confirmation with copper brass finishing iron material. There is also an uplight lighting treatment on each segment of the panel wall.
	Circulation area
Wall panel	Using a segmented teak wood panel wall with repetitive dimensions extending to one side of the wall. Iron material finishing copper brass as a frame to strengthen the segment which is also repetitive following the dimensions of the teak wood panel segment.

Table 3. Identification of dynamics identities in Holywings interior design.

Color Variation

There are repetitive and varied color combinations in each area of the room. This is intended to create a common thread of spatial identity so that there is continuity.

Figure 1. Color combinations in holywings interior design elements (personal documentation).

Figure 1 shows the walnut color in the teak wood fiber as the dominant color in the design, then the copper brass color is consistently found in walnut and white elements as accents. The white color is found in the interior design elements of the ceiling and wall backdrop of the bar counter area which are repeatedly displayed throughout the area.

Combination

Graphic elements are found in the shape, direction of material fibers, and configuration between the two elements.

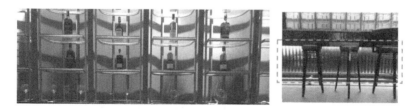

Figure 2. Graphic element combinations in holywings bar area (personal documentation).

Table 3. Continued

The material used is printed gypsum which is adjusted to the planned shape and dimensions. The next material is a copper brass color finishing iron pipe with a shape that follows the shape of printed gypsum. The combination of the two materials between molded gypsum and similarly shaped iron pipes is an interior design element that represents the bar counter area. Wall panel area with a base of teak plywood board material with walnut color finishing and angle iron with copper brass color finishing as an accent of the wall panel.
Content Variation
Treatment of iron material with copper brass finishing is consistently present as supporting material for each existing interior design element. In Figure 2, the interior design elements of the backdrop in the bar counter area, made of iron pipe with copper brass finishing, come with a treatment that adjusts the printed gypsum material as the main material. In Figure 2, the elbow plate with a copper brass color finish is also part of the teak wood panel wall and also repeated in the interior design element of the column as the wrapper.
Repetition
The repetition of the use and treatment of Holywings interior design is also found in the interior design elements. Repetition is used as an area identity and becomes a barrier between one area and another. As shown in Figure 2, there are graphic elements created from the angled plate material that is printed according to certain dimensions, then configured repeatedly and with the same dimensions along the bar counter area.

Evaluation of the dynamics identities approach to the interior of Holywings reveals that there are consistent interior design elements present,such as the use of iron material with copper brass color finishing, and teak plywood material with walnut color finishing. However, the two design elements have different design uses, the copper color finishing iron element consistently has a role as an emphasis factor in a design field; for example, in the circulation wall panel area where walnut color finishing dominates, there are lines elements with iron material to emphasize this area. Then, the role of teak plywood material with walnut color finishing comes with the use of being a consistent element present to provide familiarity to the area in the lounge room. Yagi Sastrawidjaja the designer of Holywings himself stated that there are two large areas, namely the lounge and bar area. The two areas have different identities, where the lounge area has a dominant walnut color identity, and the club area has a gray identity on the GRC finishing concrete texture wall material application.

It can be concluded that the dominant identity in the interior design of Hoywings Pasir Kaliki is the processing of teak plywood with walnut color finishing. The processing of these materials can be found in each lounge area. Then the dynamic identity in this design is the processing of iron material with copper brass finishing. Material processing is not only consistently present with the role of supporting factors or visual reinforcement of interior design elements but this material processing also acts as a visual element that can equate the visual perception of space users to the identity of their space. For example, the interior design element of the counter bar backdrop wall that uses gypsum printed in the form of a curve uses pipe iron material with copper brass finishing to highlight the curve area. This is also repeated in the use of copper brass finishing iron material on the wall elements of the circulation area, so that this can be used as a recommendation for interior design for the next interior design process, there are several materials processing that has different roles related to their role as part of the identity in the space.

4 CONCLUSION

It was identified that there are parts of interior design elements that can be applied to become part of the identity of the interior design. This shows that it can be planned to propose interior design elements as elements of space identity in the interior design work process for certain spaces so that interior designs can be well planned from the start of the work.

The next research potential is the analysis of dynamic identity in one Holywings unit with other Holywings units, but previously it is necessary to identify the material processing used as a mature identity element in one unit. After obtaining the identification data, the identification of the identity of the next Holywings unit will be carried out in a more measured and systematic way.

Subsequent research leads to the analysis of one Holywings unit with other Holywings units related to the use of the dynamic identity approach. Where it can be concluded that the reinforcing elements and factors of interior design elements are the identities of the two business units under one company, namely Holywings itself.

Based on the results of the research that has been done, it can be concluded into the following points:

1. Interior design elements are used in an effort to form a space identity.
2. The dominant interior design element, namely the processing of teak plywood with walnut color finishing has a role as an interior design identity in the lounge room.
3. Interior design elements are reinforcing or supporting interior design elements, namely the processing of iron material with copper brass color finishing has a role as a dynamic identity in the lounge room.
4. Eight indicators of the variation mechanism are implemented in the interior design elements of Holywings Pasir Kaliki.

REFERENCES

Imani, N., Shishebori, V. 2014. *Branding With The Help of Interior Design*. Indian Journal of Scientifik Research

Izadpanah, S. (2021). *Interior Space Brand Identity: Strategies That Matter*. Academic Research and Reviews in Architecture, Planning and Design Sciences, 55.

Martins et al. 2019. *Dynamics Visual Identities: from a Survey of the State-of-the-art to a Model of Features and Mechanism*. Portugal: CISUC, Department of Informatics Engineering University of Coimbra.

Putsure, S. (2011). Power of Interior Branding. *Seyie Design*.

Rashid, S. M., Ghose, K., & Cohen, D. A. (2015). *Brand Identity: Introducing Renewed Concept for Coffee Shops*. People: International Journal of Social Sciences, 1(1). Retrieved from https://grdpublishing.org/index/php/people/article/view/208

Siswanto, R.A., Dolah, J.B. 2018. Dynamic Identity *pada Identitas Toko Buku Gramedia terdapat Prinsip* Controlled Randomness. Bandung: Jurnal Desain Komunikasi Visual, Manajemen Desain dan Periklanan DEMANDIA.

Siswanto, R.A., Dolah. J. B. 2019. *Exploration to the Most Fundamental Form of Dynamic Visual Identity*. Bandung: *5th Bandung Creative Movement International Conference on Creative Industries*.

Van Nes, I. 2012. *Dynamic Identities: How to Create a Living Brand*. Amsterdam: BIS Publisher.

Wheeler, A. 2017. *Designing Brand Identity: An Essential Guide for the Whole Branding Team*. New Jersey: John Wiley & Sons.

Wicaksono, A.A., Tisnawati, E. (2014). *Teori Interior*. Griya Kreasi.

Investigating consumers' attitude toward food souvenir packaging color

W. Swasty
Telkom University, Bandung, Indonesia
Universiti Sains Malaysia, Pulau Pinang, Malaysia

M. Mustafa
Universiti Sains Malaysia, Pulau Pinang, Malaysia

ABSTRACT: Food souvenirs can be perceived as valuable goods; purchasing them gives prestige and social status. In Bandung, most local food products become a part of the tourism industry. This study aims to investigate the consumers' attitude toward the food souvenir package color across two groups of tourists (local and international groups). The study used a qualitative approach, along with an interview for data collection. The analysis was descriptive and the crosstab analysis using NVivo12. This study reveals that packaging design is the fourth factor that is important following typicality or popularity, price, and taste. Background color on the packaging is the most impressive aspect followed by color on logo, on image, and on text. This study shows slight differences between local and international group's attitude toward food souvenir packaging color. The study is beneficial to give insight into food souvenir packaging in promoting tourism sustainability through the creative industry.

Keywords: color, creative industry, food souvenir, packaging, tourism sustainability

1 INTRODUCTION

A souvenir is a product usually carried during travel that involves movement by region and country. Foods and beverages are widely sold and purchased as souvenirs at tourist destinations, which are considered as local identity and tourist experience, and expanding emotions after returning from travel (Horodyski et al. 2014). Food souvenirs are the main tourist items and define tourist attractions in many countries which can be differentiated from other tourist souvenirs, such as mugs, key chains, and T-shirts (Lin & Mao 2015). Specialty food at a destination for them is related to the location's culture and characteristics, so they buy and eat it not just to experience the taste, but also to engage with the food psychologically (Suhartanto et al. 2018).

The common reason consumers want to buy local products is due to price, quality, and ease of access (Trifiyanto 2018). Specialty food promotes local cultures and economic growth, also the area as a tourist destination over the last 20 years as its ability to meet the tourists' needs. Tourists believe it is important to buy food as part of their visit experience and can also serve as souvenirs (Suhartanto et al. 2018). They find that the tourists' understanding of local specialty food is based on promotion and recommendation.

The United Nations sets the 17 sustainable development goals (SDGs) and the SMEs play the role in achieving those goals, mainly the aspects of social impacts and related economic impacts. However, the COVID-19 pandemic has hit the economic performance in many sectors including the tourism industry. There is a need to modify the business models of the tourism industry to be more sustainable (Vărzaru et al. 2021). Many SME food souvenir products do not meet packaging standardization. Based on empirical studies as well as previous studies (Julianti 2014, Trifiyanto 2018, Wahyudi & Satriyono 2017, Yamin et al. 2018), SME food souvenir products are commonly packaged in clear plastic with simple labels only (photocopied or single-color screen printed) or even without the label, less hygienic packaging so that products perishable.

Local cultures can be assimilated by food tourism as they have been one of the most dynamic and creative areas of tourism (UNWTO 2017 as cited in Rachão et al. 2019). Creating a typical taste, scent, look, and texture of a destination, food products are considered as superior to 'traditional' souvenirs when tourists come back to their hometown (Hazman-Wong & Sumarjan 2016, as cited in Ho et al. 2020). Food souvenirs can be perceived as valuable goods; purchasing them gives prestige and social status from sociological aspects (Paraskevaidis & Andriotis 2015). Bandung's creative industry consists of SMEs and most food products tend to be bought locally (Muftiadi & Raharja 2018) and become a part of tourism (Hidayat & Asmara 2017).

Color becomes an important element in designing packages that enable to stimulate consumers' attention and emotions that lead to purchase decisions (Beneke et al. 2015; Singh 2018; Jin et al. 2019; Schuch et al. 2019). Extant studies address the consumers' attitude toward the local product or specialty foods (Trifiyanto 2018) (Suhartanto et al. 2018) but their research did not involve two types of consumers (local and international tourists). In addition, the prior literatures use the quantitative approach which deals with numbers and quantities without considering the meaning behind them. Therefore, this paper aims to investigate the consumers' attitude toward the food souvenir package color across local and international groups of tourists using a qualitative approach.

2 RESEARCH METHODS

This study used a qualitative approach by undertaking individual interviews. Purposive sampling was employed as this study focuses on consumers who purchase food souvenirs. The consumer population consists of local and international tourists. Different groups strive to express diverse views because the manner in which they choose a product or the reasons for purchasing may vary.

This study dealt with small samples of 32 people (16 local tourists and 16 international travelers) because the earlier study suggests that the first 12 interviews established about 92% of concepts (Guest et al. 2017). Google form was distributed online via instant messaging, i.e., WhatsApp, Instagram, and Skype to invite participants. When participants filled out a google form, they initially were provided the ethical consent (No. USM/JEPeM/21060493). The participants also were requested to provide demographic information in Google Form. The in-depth interviews were conducted online and recorded via Zoom platform due to the limitation of movement in the pandemic situation. Semi-structured interviews with local participants were performed in the local language (Bahasa Indonesia), while for international participants, English and/or Bahasa were used. The interview questions were adopted from Huang et al. (2019).

All participants in this qualitative study were given a unique case number to protect their privacy and confidentiality. The interview data collected was coded with descriptive coding (open coding) and crosstab analysis using NVivo12. The descriptive data was expected to reveal information from two perspectives (international travellers and local tourists) regarding their experiences in purchasing food souvenirs in Bandung/Indonesia.

3 RESULT AND DISCUSSION

Below are descriptive data regarding food souvenirs (Q01-Q02) and packaging color (Q03-Q04). For the initial question, the participants were asked about the purpose of purchasing the food souvenir and the reason why they choose food as a souvenir instead of fashion or craft. Table 1 describes the four top answers about the purpose for purchasing the food souvenir and the reason across two consumer types. The table was generated by a crosstab query using NVivo12.

Table 1. Purpose for purchasing food souvenir.

Q01	Loc	Int	Significant Statement
Typical food	9	7	• Souvenirs have their own characteristics for each region. So, if you go to Bandung, for example, there are souvenirs, which are different from other cities (04-LMX). • I think it's better to get the food originally from the place where the food originated. The taste will be different (19-IMY).
Sharing	8	7	• Souvenirs for family, so we bring something from the place we visit for family or relatives (19-IMY).
Acceptable	7	1	• It's also more practical, you don't have to think about what model people like. Or where the craft will be displayed. Food seems to be more practical, and more economical (07-LFY). • The foods run out fastest and can be consumed. Whereas clothes, we have to consider the size and people don't necessarily like (13-LMZ).
Memorable	3	3	• Food makes you miss a place (02-LFX). • So later in Germany, we can eat that food again. Indonesian food is very tasty (22-IFY). • The food shows what we feel in the place (16-LFY). • Buying (food) souvenir is for memorizing our moment at that place (19-IMY).

*Loc = Local; Int = International.

Typical food is essential meal to try when people travel to a place or country. It becomes the main purpose for most participants both local and international travelers for purchasing food souvenirs. From the data, another purpose for purchasing food souvenirs is for sharing with their families, neighbors, and friends to let them know that those foods are originally from that place. And for most local travelers, acceptable is the main reason. Acceptable means that people tend to be able to accept and consume all kinds of food, rather than clothes/shoes which need to consider the size, style, or preference. Besides being practical, food is more affordable than clothes or crafts. Memorable becomes the purpose and the reason for the majority of international tourists. Memorable is worth remembering positively, including experience, place, event, or other things. One of the things that make us miss a place is the food.

Many factors were considered when purchasing the food souvenir that reveal in the interview (see Table 2). Packaging design becomes the main factor considered when purchasing food souvenir, followed by price, typicality or popularity, and taste factors. This finding supports the extant literature (Hartanti et al. 2020) that suggests the souvenirs definitely need a proper packaging design.

Table 2. Factors when purchase food souvenirs.

Q02	Loc	Int	Significant Statements
Packaging design	4	7	• If you never try, if you never taste it before, then the only way to judge is based on the appearance of the product (18-IMY). • The look is good, you can give someone, you can bring it easy to handle in your hand or keep it in the bag (26-IMX).
Price	6	5	• The first is the price, considering the current conditions (01-LFX). • But if you're choosing between different products on the shelf, I would say like the price (18-IMY).
Typical or popular	6	4	• Yes, when we are in one place, the most important thing is buying local food, or those that are famous in that place (03-LMX). • That's the iconic one. Yes, it is most likely we buy it because it represents the area (30-IMY).
Taste	5	3	• The important thing is the taste, suit to appetite. (13-LMZ). • What matters is the taste. The taste must be good (15-IFY).
Acceptability	2	3	• Is it edible or edible for my friends or my family? (20-IFZ)
Condition of Food	0	5	• The weight also because the baggage has limit, right? (24-IFY) • Food that is not easily perishable that can be stored much longer (25-IMY)

Loc = Local; Int = International.

The participants then were asked whether they have been influenced by packaging color when they buy food souvenirs. From data, it is revealed that the majority of local travelers have been influenced by packaging color when they bought food souvenirs as the eye catchy and good appearance of packaging influenced them to buy, mainly when the packaging is displayed in the store rack. The next question is about the most impressive aspect of packaging. Table 3 shows the most impressive aspect of packaging across two consumer types. Dara reveal that background color is the most impressive aspect followed by color on image, on logo, and on the text. This finding reflects those of Spence (2018) who also found that background color influences consumers' food perception, preferences, and eventually their behavior.

Table 3. The most impressive aspect of packaging.

Q02	Loc	Int	Significant Statements
Background color	8	7	• More dominant in the packaging (03-LMX). • Because it's usually the largest thing (07-LFY). • The background is usually the largest thing, so the color is more visible (15-IFY). • The color on the background is like the basic of everything. So, if the background itself like amusing, so whatever color next such as for the logo, the text, and image will fit correctly (20-IFZ).

(continued)

Table 3. Continued

Q02	Loc	Int	Significant Statements
Image color	3	6	• The illustration is the catchiest compared to the others (11-LFZ). • Image is important, I mean when you see a photo of the food, you will think whether the food is delicious or not. if the photo is interesting, you will buy the food (19-IMY).
Logo color	5	3	• The brand is the first to be seen (06-LFY). • In my opinion, the logo is a representation of the business (12-LFZ). • The logo is important, it means the food is legal or not (17-IMY).
Text color	1	0	• From the text, from the font, for me, it has its own uniqueness (22-IMZ)

Loc = Local; Int = International.

4 CONCLUSION

This present study was undertaken to investigate the consumers' attitudes toward the food souvenir package color across two groups of tourists. This study has identified that package design was the main factor that was considered by buyers followed by the price, typicality, and taste of food souvenirs. The study is beneficial to give insight into food souvenir packaging in promoting tourism sustainability through the creative industry by utilizing color in the package design. This insight can strengthen the employment of the Global Partnership for Sustainable Development as formulated in SDGs (United Nations 2019). An issue that was not addressed in this present study was whether there are differences among local and international travelers in evaluating color applied to food souvenir package design. Future studies could evaluate the food souvenir package color from the perspective of two types of travelers.

REFERENCES

Beneke, J., Mathews, O., Munthree, T., and Pillay, K., 2015. The Role of Package color in Influencing Purchase Intent of Bottled Water Implications for SMEs and Entrepreneurs. *Journal of Research in Marketing and Entrepreneurship*, 17 (2), 165–192.

Guest, G., Namey, E., and McKenna, K., 2017. How Many Focus Groups Are Enough? Building an Evidence Base for Nonprobability Sample Sizes. *Field Methods*, 29 (1), 3–22.

Hartanti, M., Nurviana, N., and Lukman, C.C., 2020. The Development of Tools for Designing the Local Characteristic Food Packaging Based on Digital Applications as an Attempt to Accelerate Education. *In: 3rd International Conference on Learning Innovation and Quality Education (ICLIQE 2019)*. Atlantis Press, 517–529.

Hidayat, A.R.T. and Asmara, a Y., 2017. Creative Industry in Supporting Economy Growth in Indonesia: Perspective of Regional Innovation System. *IOP Conference Series: Earth and Environmental Science*, 70.

Ho, C., Liu, L., Yuan, Y., and Liao, H., 2020. Perceived Food Souvenir Quality as a Formative Second-order Construct: How do Tourists Evaluate the Quality of Food Souvenirs? *Current Issues in Tourism*, 1–24.

Horodyski, G.S., Manosso, F.C., Bizinelli, C., and Gândara, J.M., 2014. Gastronomic Souvenirs as Travel Souvenirs: A Case Study in Curitiba, Brazil. *Via Tourism Review*, 6, 1–19.

Huang, L., Mou, J., See-to, E.W.K., and Kim, J., 2019. Consumer Perceived Value Preferences for Mobile Marketing in China: A Mixed Method Approach. *Journal of Retailing and Consumer Services*, 48, 70–86.

Jin, C.H., Yoon, M.S., and Lee, J.Y., 2019. The Influence of Brand Color Identity on Brand Association and Loyalty. *Journal of Product and Brand Management*, 28 (1), 50–62.

Julianti, S., 2014. *The Art of Packaging: Mengenal Metode, Teknik, dan Strategi Pengemasan Produk untuk Branding dengan Hasil Maksimal.* Jakarta: Gramedia.

Lin, L. and Mao, P.C., 2015. Food for Memories and Culture - A Content Analysis Study of Food Specialties and Souvenirs. *Journal of Hospitality and Tourism Management*, 22, 19–29.

Muftiadi, A. and Raharja, S.J., 2018. Pattern of Production Operation in the Creative Industry: A Study in Bandung Creative City. *Review of Integrative Business and Economics Research*, 7 (4), 137–148.

Rachão, S., Breda, Z., Fernandes, C., and Joukes, V., 2019. Food Tourism and Regional Development: A Systematic Literature Review. *European Journal of Tourism Research*, 21, 33–49.

Schuch, A.F., da Silva, A.C., Kalschne, D.L., da Silva-Buzanello, R.A., Corso, M.P., and Canan, C., 2019. Chicken Nuggets Packaging Attributes Impact on Consumer Purchase Intention. *Food Science and Technology*, 39 (Suppl.1), 152–158.

Singh, R.K., 2018. The Effect of Packaging Attributes on Consumer Perception. *International Journal for Innovative Research in Multidisciplinary Field*, 4 (5), 340–346.

Spence, C., 2018. Background color & its impact on Food Perception & Behaviour. *Food Quality and Preference*, 68 (December 2017), 156–166.

Suhartanto, D., Chen, B.T., Mohi, Z., and Sosianika, A., 2018. Exploring Loyalty to Specialty Foods Among Tourists and Residents. *British Food Journal*, 120 (5), 1120–1131.

Trifiyanto, K., 2018. Masa Depan Produk Lokal: Analisis Pengaruh Etnosentris Konsumen, Disain Kemasan, dan Persepsi Labelisasi Halal terhadap Minat Pembelian Produk Lokal (Studi pada Produk Lokal Lanting khas Kebumen). *Jurnal Fokus Bisnis*, 17 (02), 15–24.

United Nations, 2019. *The Sustainable Development Goals Report 2022.* United Nations publication issued by the Department of Economic and Social Affairs (DESA).

Vărzaru, A.A., Bocean, C.G., and Cazacu, M., 2021. Rethinking Tourism Industry in Pandemic COVID-19 Period. *Sustainability*, 13 (12), 6956.

Wahyudi, N. and Satriyono, S., 2017. *Mantra Kemasan Juara.* Jakarta: Elex Media Komputindo.

Yamin, M.M., Abidin, E.E., and Sulaeman, 2018. Sosialisasi Pengemasan Kue Tradisional Di Desa Sepabatu, Kec. Tinambung, Kab. Polewali Mandar. *RESONA Jurnal Ilmiah Pengabdian Masyarakat*, 2 (1), 31–41.

Folklore-based art performances in digital art space as a form of revitalizing local wisdom in the metaverse era

R. Rachmawanti* & C.R. Yuningsih
Telkom University, Bandung, Indonesia

*ORCID ID: 0000-0003-1500-7267

ABSTRACT: Indonesia is an archipelagic country with ethnic and cultural diversity, where there is a lot of uniqueness, especially in traditional art in the form of folklore. Folklore is a potential asset in an art performance because it can present interesting displays of representations from the traditional culture of the community. Information and communication technology that develops in this digital era demands innovation in various fields including art. This article discusses the study of folklore-based performing arts using digital art spaces, as an effort to revitalize art into digital performances. This study is qualitative research with an ethnographic approach and experimentation at the final stage of the research. The results of the study show that the innovation of folklore art performances into the digital art space is a powerful way to maintain its existence and be more widely known by future generations in the metaverse era.

Keywords: Folklore, Local Wisdom, performance art, digital art space, metaverse

1 INTRODUCTION

The fourth industrial revolution, the industrial era of 4.0—which is currently taking place—has major influences on every aspect of life. The development of many technologies of physical and digital, combined with analytical and artificial intelligence, cognitive technologies, and the Internet of Things (IOTs) are the main characteristics of this era. The emergence of interrelated "digital company" is also a current phenomenon that enables people to generate speed and accuracy in transaction. One aspect that is affected by technological developments is art and culture, especially in the world of performing arts.

Performing arts, including dance, theater, and music, are usually lived in the arena itself, although now these performances have been packaged through new media, such as film, television, and digital formats, in addition to being staged live. Indonesian performing arts that contain this diversity have undergone various processes of change, including various engineering, reinvention, re-actualization, reproduction, and art creation. This development resulted in a series of ideas about the meaning of national culture, especially regarding the arts. Technological media makes performing arts activities easier to present and use as needed. This ease of access can indirectly increase the level of public appreciation and have an impact on the increasingly rapid competition for performance production in Indonesia. One of the cultural assets owned by the Indonesian people is the diversity of folklore.

Folklore is part of the culture and traditions of a collective society, which is spread and passed down from generation to generation in various versions, both in oral and exemplary forms, movements, or a series of activities as outlined in the framework of art. Regarding the type of culture, Yadnya (1981: 2528) states that folklore is part of a traditional, informal, and national culture. This view implies that folklore is not only ethnic but also national. On the other hand, Potter argues that folklore is "a living fossil that refuses to die" (Leach 1994: 401).

Folklore is also related to technology. Currently, many folklore-based art performances are uploaded on the internet, with attractive packaging made by creative technology, making this nation richer because of its folklore. Through the internet, people can reply to folklore, using rhymes. Folklore has become a medium of intercultural communication. These conditions have an impact on production in the performing arts in Indonesia should be able to compete with products coming from abroad (Minati *et al.* 2015).

The transformation of the performing arts as a cultural product is coupled with technological developments (Hardjana 2004). Digital art performances are growing, with the support of various website and smartphone-based application platforms. This indicates that the visualization of performing arts will be more developed in the virtual world. The virtual space provides and provides facilities for users to find new ways to participate in an art performance, including enjoying folklore performances. The presence of the digital art space application as a virtual space in producing a performance is an opportunity for folklore-based performing arts to develop and continue to exist, with a contemporary appearance that attracts the current generation.

2 RESEARCH METHODS

This study is qualitative research with an ethnographic approach and experimentation at the final stage of the research. In this case, the study of folklore is analyzed using the science of performance arts, culture, and technology which, in the end, will cover the entire realm of cultural studies consisting of the Manti fact, socio-fact, and artifact areas. The process of research that has been carried out begins with observing various phenomena that occur in the existence of folklore, especially those that are often adopted in the performance arts. After the study on folklore art performances is carried out, this study proceeds to the experimentation stage by making a simulation of folklore performances in the digital art space, using the Digital Kultur application.

2.1 *Methods of data collection*

The sample of this research data was taken from the implementation of the folklore music festival for group vocals and solo vocals conducted in 2019 and 2020 using a digital culture application. The data became the basis for technology development for the application, then a simulation experiment was made in the form of a competition. The ethnographic method in this research is used to analyze the types of folklore that are in accordance with the concept of performing arts in the digital space, which in this case are folk songs. The experimental method is used during the application development process from Digital Kultur to a media and virtual space for organizing folklore performances.

2.2 *Methods of data analysis*

Data analysis was carried out through several stages including curation, reduction, and analysis using an art and cultural studies approach so that conclusions were obtained regarding the menti-fact, soci-fact, and artifact aspects of the folklore-based performance.

3 RESULT AND DISCUSSION

3.1 *Folklore as a local wisdom*

Local Wisdom is part of a culture that is in a society that cannot be separated from the community itself. Local wisdom becomes basic knowledge of life, obtained from experience or the truth of life, and can be abstract or concrete, balanced with nature and culture belonging to a certain community group (Mungmachon 2012: 174). The local wisdom that is best known by the public is folklore.

The diversity of folklore in Indonesia, covers various fields of art, especially in the performing arts. Folklore is a potential asset in performing arts because it is able to present interesting displays, which are the result of representations of the traditional culture of Indonesian society. This can be seen from the increasing number of performing arts that adopt and collaborate performance materials with traditional themes, most of which are taken from folklore.

Folklore is increasingly spreading, known by many people through electronic media. Folklore has become a medium of intercultural communication. According to Brunvand (in Danandjaja 1997: 21), folklore can be divided into three major groups: oral folklore, partly oral folklore, and non-verbal folklore. Oral folklore is defined as folklore whose form is purely oral. The forms of this type of folklore include (a) folk speech such as accent, nicknames, traditional ranks, and titles of nobility; (b) traditional expressions, such as proverbs, proverbs, and memes; (c) traditional questions, such as puzzles; (d) folk poetry, such as pantun, gurindam, and syair; (e) folk prose stories, such as myths, legends, and fairy tales; (f) folk songs. Furthermore, Dundes (Endaswara 2016) added that folklore has the following functions: (1) to strengthen the feeling of collective solidarity, (2) to provide justification for a society, (3) to give direction to society in order to allow it to criticize others, (4) to be a tool for protesting injustice, and (5) to be a source of fun and entertainment..

Partial oral folklore is defined as folklore whose form is a mixture of oral and non-verbal elements. Meanwhile, non-verbal folklore is defined as folklore whose form is not verbal, even though the way of making it is taught orally. The form of this type of folklore is broadly divided into two namely material and non-material. Materials include folk architecture, handicrafts, food and beverages, and traditional medicines. On the other hand, the non-material ones include traditional gestures, sound signals for folk communication, and folk music.

The digitization of folklore is an effort to preserve culture. In this digital era, many things need to be adjusted, including cultural preservation. Folklore, which was originally delivered in oral form, also requires adjustment using digital media such as films, animation, games, digital books, digital comics, and web and mobile-based applications. Just like the purpose of the emergence of folklore which is to provide education both in terms of character and other education, folklore in digital form also has the aim of providing education to children or readers. Digital media that is more interactive is expected to provide optimal learning outcomes (Widhiyanti & Gunando 2021).

3.2 *Folklore-based performance in digital kultur*

Many digital media have been developed to convey folklore. The delivery of folklore in the digital form actually has the same goal as the traditional or oral delivery of folk tales. Folklore spread throughout the archipelago is a cultural wealth that deserves to be popular through digital media (Rosita 2019). One research has been done and has produced very useful applications of folklore. The form of digital media that has been developed is multimedia applications in the form of text, illustrations, images, videos, and sound in the form of web and mobile applications. Dharsana et al. (2016) conducted a process of digitizing Balinese folklore based on an Android application.

All types of folklore that have been mentioned above can be made in the form of performing arts. One type of folklore that is often displayed in performances is the type of folk music/songs in group and solo vocal formats. Based on the results of the study of the data that has been obtained, folk music is one of the folklores that can be performed with various collaborative movements.

The world of performing arts is synonymous with live performances that show works and are attended by the audience. Performing arts innovators have developed a new method for organizing performances, namely collaborating live performances with digital technology. One of the digital platforms created to accommodate performing arts is Digital Kultur.

Digital Kultur is an invention in the form of an application made for organizing various cultural festivals in Indonesia. Using digital live performance technology with recording and live streaming systems, Digital Kultur makes it easy for participants to take part in a series of festivals or performance events. It is designed to be one of the solutions to address technological challenges in the world of music and performing arts. This digital platform is used as an application for organizing music and cultural festivals. Digital Kultur has made it easy for participants to compete, perform, and express their abilities in music not only within a small event, but also in a large-scale cultural festival, involving dance, folklore, and other art fields.

Figure 1. Folklor performance in digital kultur.

Table 1. Folklore performance in music.

Year	Location	Performer	Folklore Music
2018	Bandung	SMA & SMK	1st Song: Manuk Dadali
			2nd Song: Other Sundanese Folk
2019	Jawa Barat	SMA & SMK	Sabilulungan & Other folksongs;
			1) Bungong Jeumpa
			2) Sigolempong
			3) Cublak Cublak Suweng
			4) Janger
			5) Yamko Rambe Yamko
2020	Kalimantan Barat (Daerah Perbatasan)	SD, SMP, & SMA	Paris barantai

The existence of Digital Kultur is not only an intermediary for the implementation of folklore music performances in the digital era, but can be a vehicle for revitalizing local wisdom, which is conveyed through folk songs/folklore. Thus, the concept of inheritance and transfer of knowledge rooted in local culture as a national identity can be implemented quickly, precisely, effectively, and efficiently, and is easily accessible on various digital media platforms. Through this performance, the values of local wisdom can be integrated into every organization of folklore-based art performances with wider community involvement.

4 CONCLUSION

Folklore is a potential asset in an art performance because it can present interesting displays of representations from the traditional culture of the community. Advances in technology make folklore-based performing arts become easier to be presented and utilized according to the needs. The innovation of folklore art performances into the digital art space is a powerful way to maintain its existence and be more widely known by future generations in the metaverse era. The folklore-based performance in digital art space can be a vehicle for revitalizing local wisdom, which is conveyed through the values of local wisdom that are integrated into wider community involvement.

REFERENCES

Danandjaja, James. 1997. *Folklor Indonesia*: Ilmu Gosip, Dongeng, dan lain lain. Cetakan V. Jakarta: PT Pustaka Utama Grafiti.

Dharsana, dkk. 2016. *Digitalisasi Cerita Rakyat dalam Rangka Pelestarian Budaya Berbasis Aplikasi Mobile.*

Dundes, Alan (ed). 1965. *The Study of Folklore*. New Jersey. Prentice-Hall, Inc.

Endawarman, Suwardi (ed.). 2013. *Folklor Nusantara: Hakikat, Bentuk dan Fungsi*. Yogyakarta: Penerbit Ombak Dua.

Hardjana, Suka. (2004). *Musik Antara Kritik dan Apresiasi*. Jakarta: Kompas.

Leach, Steve; Stewart, John and Kieron Walsh, 1994. *The Changing Organization and Management of Local Government*, McMillan Press Ltd.

Minarti, Heli; Tajudin, Yudi A, Gesuri, Eka. (2015). *Rencana Pengembangan Seni Pertunjukan Nasional. Badan Ekonomi Kreatif Kementrian Pariwisata.*

Mungmachon, Mmiss Roikhwanput. 2012. Knowledge and Local Wisdom: Community Treasure. *International Journal of Humanities and Social Science*, 2 (13), 174–181.

Rosita, Erlinda. 2019. Potensi Cerita Rakyat sebagai Objek Material Bersastra dalam Media Digital. *Prosiding Sembadra. Universitas Sriwijaya.*

Widhiyanti, Kathryn and Gunanto, Samuel Gandang. 2021. *Nusantara Folklore in the Digital Age Proceedings of the 2nd International Conference on Interdisciplinary Arts & Humanities (ICONARTIES) 2021.* http://dx.doi.org/10.2139/ssrn.3800616

Yadnya, Ida Bagus Putra. 1984. *"Folklor Esoterik dan Eksoterik,"* Widya Pustaka, Th II No 1 Agustus, Denpasar: FS Udayana.

Nighttime attraction based on of Gedung Sate area

R.H.W. Abdulhadi, H. Anwar & I.Z. Budiono
Telkom University, Bandung, Indonesia

ABSTRACT: Gedung Sate, Gazebo Sport Field, and Monument of Perjuangan Rakyat Jawa Barat as part of public space have significant value in the development of culture, history, architecture, facilities, and its infrastructure in Bandung. Its significant part in Bandung has earned its reputation as a monumental area, and its visual images should attract visitors. Apart from its daytime visual appearance, the night time appearance of this area should also be an important aspect in attracting visitors. This research is aimed at the perception of users towards this area. The analysis is done using quantitative methods in the form of a questionnaire for the respondents. The results of this research show that the application of illumination to shape night time visual appearance in this area attracts visitors and piques the interest of the visitors to the area. This research can be used as a ground for the development of lighting in this area.

Keywords: attraction, city lighting, public space, area illumination, night time visual image

1 INTRODUCTION

City branding is something that emerges and develops from tourism-related marketing strategies and appears in various media, (Zhou *et al.*, n.d.). Images that appear from various media are visual images that can be a selling point for the city. This is, of course, closely related to urban planning and city management in creating an attractive city for tourists (Riza et al. 2012). To create an interesting image of the city in relation to city branding, there should be an urban experience that consists of several aspects such as aesthetics, landscaping, facilities, and functions of the environment (Servillo et al. 2011). The basic character in the public space can be divided into four parts: 1. Buildings (monuments, landmarks, etc.); 2. Landscape (street lights, bus stops, etc.) 3. Infrastructure (trees, street furniture, etc.); 4. Usage (events, gatherings, markets, etc.). Points one to three represent physical forms that make up public spaces, while point 4 represent activities that make up public spaces (Carmona et al. 2008).

As in the case of city branding of Bandung, several interesting places emerge in relation to city branding and urban experience. One of which is very iconic is the Gedung Sate area which plays an important role and has deep historical value. The placement of Gedung Sate, which originally functioned as a Dutch East Indies government office, has a north-south axis with a natural marker on the north in form of Mount Tangkuban Perahu. In line with the times, this area then grew and developed into a multifunctional area with public spaces such as sports fields, and parks Perjuangan Rakyat Jawa Barat monument is a marker in the north and the last addition is the COVID-19 hero monument.

The arrangement of the area and public space around the Gedung Sate area provides a new physical form in terms of landscape arrangement, construction of new hotels and monuments, as well as arrangement of lighting. Meanwhile, in terms of activities, the area still has the same value and function as before, that act as a space for socializing, exercising, and economic activities such as a food stall.

The importance of city branding does not stop at the daytime, and the importance of the visual image of the city at night, brighten by lighting design becomes an important message for city promotion (Li et al. 2017). A strategy should be carried out in creating identity at night. Building images and street corridors become important along with their physical forms such as shape, color, arrangements, ornaments, or any special façade characteristics (Nikoudel & Mahdavinejad, n.d.)

In terms of its image at night, the Gedung Sate area has been updated with the placement of new luminaires along the Gasibu sports field, new lighting arrangements at the fasade of Gedung Sate and Pullman hotel, colored and luminous fixtures at the COVID-19 monument. What is then interesting is whether the new arrangement of luminaires and illumination Gedung Sate Area gain attraction for passersby to enjoy this public space longer, with the lighting attributes that occur.

1.1 *Illumination in public spaces*

Based on the theory put forward by Lynch (2020), in city's image, the inner part of a city can be divided into several things: 1. Landmarks; 2 Districts; 3. Nodes; 4. Edge; and 5. Road. From the perspective of lighting in urban planning, as stated by Carmona et al. (2008) that humans possess a perception of a place—in this case, a public space—will differ in changing lighting conditions during the day and night. At night the perception of objects and places depends a lot on the effects caused by artificial lighting. For example, objects or places with the highest lighting are considered landmarks. Districts in a bright public space can be identified by the presence of dark areas and lit areas, while the role of a pathway can be seen as a road corridor that has sufficient levels of brightness. In a lighting design, nodes and edges can be seen by the intensity of the lighting in the area.

1.2 *Attributes of light in public spaces*

Illumination in public spaces has different designs related to the functions and needs carried out in these public spaces. With the variety of lighting systems that occur in many public spaces that have been designed, there are lighting attributes that are the focus of meetings with visitors passing through public spaces, namely, (1) brightness, (2) contrast, (3) light distribution, (4) legibility, (5) color (Ünver 2009).

2 RESEARCH METHODS

The method used in this study is a quantitative method to assess the interest that arises from the lighting made in the Gedung Sate area. Data collection in the study was carried out in two ways. The first is by making direct observations in the field by observing the conditions of the application of artificial lighting in the Gedung Sate area, both in the pedestrian area and on the facade of the surrounding building. Observation produces data in the form of video and photo documentation. The second is data collection through online questionnaires. A total of 100 from age of 18 until 67 were recorded and have criterion as one who has visited the Gedung Sate area at night.

2.1 *Methods of data collecting*

The data collected in this study was conducted through a questionnaire which was then processed using a Likert scale. According to Rohmad & Supriyanto (2015), the Likert scale is a scale that can be used to measure attitudes, opinions, and perceptions of a person or group of people about a particular symptom or phenomenon. When answering questions on a Likert scale, determine their agreement with a statement by choosing one of the available options. The form of the Likert scale answer consists of strongly agree, agree, undecided, disagree, and strongly disagree.

2.2 Methods of data analysis

In this study, the analysis was carried out by first compiling the existing lighting attributes, namely, brightness, contrast, light distribution, legibility, and color which then dissolve with the elements that make up the area in the Gedung Sate area. The element that forms the area is the main data of documentation obtained through field observations in the Gedung Sate area, which is then selected into several parts. This is done because in this area the existing lighting can be based on literature studies into buildings, pathways, landmarks, and squares.

The buildings proposed in the questionnaire list are Gedung Sate and the Pullman hotel, while the pathway is a pedestrian area around the area. The landmarks in this area are the Gasibu field and the COVID-19 monument. The points are then made into a questionnaire which is then calculated using Likert. The Score is divided into Strongly Interested (SI), Interested (I), Neutral (N), Not Interested (NI), and Strongly Not Interest (SNI). Then, results of the data is collected to determine the percentage of respondents for further analysis. From the results of this analysis, conclusions regarding interest in the Gedung Sate area being made.

3 RESULT AND DISCUSSION

The question model used in the questionnaire is descriptive type. The aim is to provide a definite calculation of the lighting variable being asked. The variables used are visitor attractions regarding brightness, contrast lighting, light distribution, legibility, and color in the Gedung Sate area. Questions posed to respondents were categorized based on lighting attributes consisting of (1) brightness, (2) contrast, (3) light distribution, (4) legibility, and (5) color (Ünver 2009). The results obtained from 55 respondents with an age range of 18 to 67 years, gave satisfactory overall results. This is shown through an interest in brightness, contrast lighting, light distribution, and legibility in the Gedung Sate area, respectively, by 75.15%, 73.3%, 78.2%, 67.3%, and 73.3%.

Based on these variables, respondents' interest was assessed using a Likert scale. The score is divided into Strongly Interested (SI), with a score of 5; Interested (I) with a value of 4, Neutral (N) with a value of 3, Not Interested (NI) with a value of 2; Strongly Not Interest (SNI) with a value of 1. The results of all respondents are then added up. To provide an interpretation of the assessment, then the sum results are assessed based on the highest and lowest total scores. The highest value of each variable is 825, and the lowest score is 165.

Brightness plays an important role in the perception of human interest in a place, as has been studied by Boyce (2019) that by the clarity of place, a feeling of security and alertness in a location will be obtained. In the relationship between lighting contrast and perception of attractiveness, which is based on research (Saad Ghalib & Saadi Lafta 2022), attraction is obtained through high contrast in lighting, especially with the use of color. This is because, with the high contrast of lighting, the clarity between the design elements and the clarity of the form will be obtained. The distribution of lighting either through the light emitted by the armature or by a layout that has harmony, rhythm, and regularity is one of the factors of user interest in a place due to clarity in judging (Saad Ghalib & Saadi Lafta 2022).

The following are the results obtained from respondents regarding the assessment of lighting with variables of brightness, contrast, light distribution, legibility, and color.

Table 1. Attraction of city elements through brightness.

No	Questions	SNI	NI	N	I	SI
1	Does the level of brightness on the facades of the Gedung Sate and Pullman hotels attract your attention?	5	0	10	25	15
2	Does the level of brightness on the Gasibu Field and the COVID-19 Monument attract your attention?	5	0	15	15	20
3	Does the level of brightness on the pedestrian attract your attention?	0	10	10	25	10
	Amount	10	10	35	65	45
	Total Score	10	20	105	260	225
	Σ Score			620		
	Maximum Score			825		
	Percentage (%)			75,15%		

Table 2. Attraction of city element through contrast.

No	Questions	SNI	NI	N	I	SI
1	Does the level of contrast on the facades of the Gedung Sate and Pullman hotels attract your attention?	0	10	10	20	15
2	Does the level of contrast between the Gasibu Field and the COVID-19 Monument attract your attention?	0	5	20	15	15
3	Does the level of contrast on the pedestrian attract your attention?	0	10	15	20	10
	Amount	0	25	45	55	40
	Total Score	0	50	135	220	200
	Σ Score			605		
	Maximum Score			825		
	Percentage (%)			73,3%		

Table 3. Attraction of city elements through light distribution.

No	Questions	SNI	NI	N	I	SI
1	Does the light distribution on the facades of the Gedung Sate and Pullman hotels attract your attention?	0	5	20	15	15
2	Does the light distribution on the Gasibu Field and the COVID-19 Monument attract your attention?	0	5	15	20	20
3	Does the light distribution on the pedestrian attract your attention?	0	5	20	15	15
	Amount	0	15	55	50	50
	Total Score	0	30	165	200	250
	Σ Score			645		
	Maximum Score			825		
	Percentage (%)			78,2%		

Table 4. Attraction of city elements through legibility.

No	Questions	SNI	NI	N	I	SI
1	Does the legibility of the facades of the Gedung Sate and Pullman hotels attract your attention?	5	5	10	15	10
2	Does the legibility on the Gasibu Field and the COVID-19 Monument attract your attention?	5	0	20	20	10
3	Does the legibility of the pedestrian attract your attention?	5	5	10	15	20
	Amount	15	10	40	50	40
	Total Score	15	20	120	200	200
	Σ Score			555		
	Maximum Score			825		
	Percentage (%)			67,3%		

Table 5. Attraction of city elements through color.

No	Questions	SNI	NI	N	I	SI
1	Does the color of light on the facades of the Gedung Sate and Pullman hotels attract your attention?	10	0	5	15	25
2	Does the color of light on the Gasibu Field and the COVID-19 monument attract your attention?	5	5	15	10	20
3	Does the color of the light on the pedestrian attract your attention?	5	5	20	5	20
	Amount	20	10	40	30	65
	Total Score	20	20	120	120	325
	Σ Score			605		
	Maximum Score			825		
	Percentage (%)			73,3%		

4 CONCLUSION

From this study, it was concluded that the lighting occurred in the Gedung Sate area gave a high interest to users, this was due to clear environmental legibility, bright light levels, the use of light color, high lighting contrast, good lighting design and its arrangement at the Gedug State area. Through this research, it is hoped that the existing lighting in the area can be applied to other public spaces that still have shortcomings in attraction.

REFERENCES

Boyce, P. R. (2019). The Benefits of Light at Night. *Building and Environment*, *151*, 356–367. https://doi.org/10.1016/J.BUILDENV.2019.01.020

Carmona, M., de Magalhães, C., & Hammond, L. (2008). *Public Space: The Management Dimension*, 1–232. https://doi.org/10.4324/9780203927229/Public-Space-Matthew-Carmona-Claudio-De-Magalh

Li, P., Wang, K., Chen, T., & Wang, F. (2017). Negative Perceptions of Urban Tourism Community in Beijing: Based on Online Comments. *International Review for Spatial Planning and Sustainable Development*, *5*(2), 93–103. https://doi.org/10.14246/IRSPSD.5.2_93

Lynch, K. (2020). "The city image and its elements": From the Image of the City (1960). In *The City Reader* (pp. 570–580). Taylor and Francis. https://doi.org/10.4324/9780429261732-67

Nikoudel, Mahdavinejad, and V. (n.d.). *EBSCOhost | 129442845 | Nocturnal Architecture OF Buildings: Interaction OF Exterior Lighting and Visual Beauty*. Retrieved October 23, 2022, from https://web.p.ebscohost.com/abstract?direct=true&profile=ehost&scope=site&authtype=crawler&jrnl=02362945&AN=129442845&h=zYjQirfvc3jM%2BB5iPysGWl5zfPjHLtP6R1ix3yOwKjX1py%2BQUVduomhfHzJYUnYZODHIKafsqYNvNmCbIcnMug%3D%3D&crl=c&resultNs=AdminWebAuth&resultLoca

Riza, M., Doratli, N., & Fasli, M. (2012). City Branding and Identity. *Procedia – Social and Behavioral Sciences, 35*, 293–300. https://doi.org/10.1016/J.SBSPRO.2012.02.091

Rohmad, H., & Supriyanto. (2015). Pengantar Statistika Panduan Praktis Bagi Pelajar dan Mahasiswa. *Pengantar Statistika Panduan Praktis Bagi Pelajar Dan Mahasiswa*, 131.

Servillo, L., Atkinson, R., & Russo, A. P. (2011). Territorial Attractiveness in EU urban and Spatial Policy: A Critical Review and Future Research Agenda. Http://Dx.Doi.Org/10.1177/0969776411430289, *19*(4), 349–365. https://doi.org/10.1177/0969776411430289

Ünver, A. (2009). *People's Experience Of Urban Lighting İn Public Space*. https://open.metu.edu.tr/handle/11511/18433

Zhou, L., Cities, T. W.-, & 2014, undefined. (n.d.). Social media: A New Vehicle for City Marketing in China. *Elsevier*. Retrieved October 23, 2022, from https://www.sciencedirect.com/science/article/pii/S0264275113001807?casa_token=BJMuu_Ej02wAAAAA:uVz1V3rv29orJT6J4NL-pj2nWzfd–DCZ1sGTNsv_mSix0t7YiCTu6k5P2qVuoEdgC0v8i-EsvfDCQ

The development of the virtual environment and its impact on interior designers and architects. Case study: Zaha Hadid Architects

V. Haristianti & D. Murdowo
Telkom University, Bandung, Indonesia

*ORCID ID: 0000-0001-9566-4691

ABSTRACT: This study aims to discuss the development of interior design and architecture in the realm of the virtual environment. Nowadays, virtual environments are becoming more common. Thus, many architects and interior designers have spotted new opportunities to extend their design frontiers into the virtual world. This study uses a qualitative method. Data were collected by means of a literature review. The analysis proceeds hand-in-hand with other parts of developing the qualitative study, namely the data collection, and the writing of findings. The results show that today's digital acceleration brings many possibilities for interior designers and architects. Some of the impacts that occur due to the development of the virtual environment, in general, include the rise of the digital economy and the rise of digital assets. In addition, this also causes a change in the business model where interior designers and architects in the virtual environment are no longer consultants but content creators.

Keywords: metaverse, spatial transformation, technology, virtual environment

1 INTRODUCTION

The metaverse is a hypothetical future version of the internet built around immersive virtual worlds, virtual reality, and augmented reality (Laukkonen 2022). Metaverse is not virtual reality but it will be accessible in VR. With metaverse, users can own spaces where they design their own projects (Nicola 2022). The development of the digital world is actually not new. However, the pandemic propelled the online world to a new level, catching the attention of private buyers and brokerages alike who began to funnel tier money into this augmented reality (Overstreet 2021). Virtual environments are becoming more common and ever more sophisticated as companies race to build platforms that will draw people to their respective corners of the metaverse (Matoso 2022). Currently, the existence and development of the metaverse is a hot topic of discussion and is very relevant to be discussed; one of which is in the world of architecture and interior design. All the recent hype forces us to evaluate our role as architects and interior designers and whether or not we want to consider servicing the digital realm. Schumacher in Kolata, 2022 said the design of the metaverse falls within the remit of the discipline of architecture and the wider design disciplines, not video game artists.

Places, like things and buildings, were primarily understood through use and experience (Heideggers in Sharr 2007). All design is about framing social interactions. Architecture is a key to comprehending reality (Tuan 2008). To design architectural projects, real or virtual, implies the development of a spatial-visual language (Schumacher in Kolata 2022). The constructed form has the power to heighten the awareness and accentuate between 'inside' and 'outside' (Tuan 2008). In the virtual environment, the metaverse has a similar concept to a city in the real world. It can be accessed clearly in terms of navigation. In addition, it can be explored in a detailed three-dimensional form so that the experience of space can be created properly for the person who explores it (Schumacher in Kolata 2022). In general, this paper will discuss the development of the virtual environment and its impact on architects and interior designers. The case study of this paper is Zaha Hadid Architects which believes the metaverse is where much of the architectural action and innovation will be happening in the coming period. It is hoped that this paper will explain that today's digital acceleration brings many possibilities for interior designers and architects. The position of this research is the initial research to be used as a basis for the next stage of research.

2 RESEARCH METHODS

2.1 *Methods of data collecting*

This research is qualitative research (Creswell 2018) with a one-time setting of cross-sectional studies and the number of contacts (Kumar 2005). Qualitative methods rely on image and text data so the literature study method was used as a way of collecting data. References about the development of the virtual environment from various sources including scientific journals, reference books, articles, and excerpts from interviews on the internet are used as data sources by the author. This technique is considered quite in accordance with the position of the research which is in the early stages, where this method allows data to be generated in a short time. After the literature that is considered suitable is selected, the author then extracts the writings related to these architectural projects and metaverse designs from Zaha Hadid Architects. Other data as a compliment is a collection of other people's writings about their thoughts on metaverse projects from Zaha Hadid Architects. It is hoped that with the many views of various authors, the data can be more varied and broad, so that at the time of processing obtained good validity.

2.2 *Methods of data analysis*

This research is categorized as descriptive research (Kumar 2005). The data analysis method used is qualitative data analysis (Creswell 2018). The description of interpretation about the meaning of the research theoretically is obtained by reviewing the literature study through the stages of 'collecting data' applied by reading the selected writing samples, trying to understand them by interpreting them, and rewriting all the ideas contained in each writing. Data analysis will proceed hand-in-hand with other parts of developing the qualitative study, namely the data collection, and the write of findings (Creswell 2018). The first analysis stage of this research is 'sorting', which is the stage of grouping ideas that are considered to have the same topic or purpose. Next is the labeling stage, completed by stating the topic of the group of ideas. Finally, the last is the stage of interpretation of the data based on sorting and labeling that has been done. All these procedures are conducted by winnowing the data which is focusing on some of the data and disregarding other parts of it (Creswell 2018).

3 RESULT AND DISCUSSION

3.1 *Results*

This section will discuss two works by Zaha Hadid Architect on a virtual environment portal, namely, The Liberland Metaverse, which was created on the Mytaverse platform, and NFT-ism at Art Basel Miami which was used as a case study. After reading various sources of data and analyzing them, the following results are obtained.

3.1.1 *The liberland metaverse*

This project is an NFT version of the Free Republic of Liberland which the libertarian Czech politician Vít Jedlicka proposed in 2015. Launched by Zaha Hadid Architects in early 2022 under the name The Liberland Metaverse, the real-world design area is three square miles of uninhabited land. The dispute between Croatia and Serbia does not have the infrastructure, diplomatic recognition, and the entry of a neighboring country that had gone viral and was predicted to be the third smallest sovereign state in the world, preceded by the Vatican City and Monaco (Figure 1). In that area on the Metaverse platform—a cloud-based platform that creates 3D environments, a dramatic virtual scheme has now been designed which is currently being used as a futuristic oasis sporting a buzzy NFT trading room and sweeping office towers (Waddoups 2022). This design is planned to be a template for the micronation's eventual physical presence. Those interested in joining can purchase an empty plot centered around a curated city core, and access it as an avatar (Stouhi 2022). The Liberland Metaverse can be seen in Figure 1.

Figure 1. The image of real-world Liberland and Digital replica of the physical micronation of the Republic of Liberland. (Source: https://www.dailymail.co.uk 2022)

Currently, more than 7,000 e-citizens have signed up with 780,000 applications in the backlog. Zaha Hadid Architects designed all of the buildings in their typical style with curvaceous, sinuous forms, and rounded corners (Finney 2022). The interesting thing is that this project, according to its website one of the buildings called a virtual campus, will be used as a network hub for crypto projects, crypto companies, and crypto events. Also, the people who bought a plot of land and set up a business in this virtual city would also have a stake in physical Liberland. Another uniqueness is that many of the buildings have elements not supported from the ground—something that is not possible with gravity in the real world. This is also said by Zaha Hadid Architects as a potential for the development of virtual environments, namely, boost parametricism. The infrastructure that Zaha Hadid Architects has built in The Liberland Metaverse includes the city hall, plaza, and exhibition center.

3.1.2 *NFTism*

NFTism at Art Basel Miami is a virtual gallery exploring architecture and social interaction in the metaverse. This gallery features a spatial design created by Zaha Hadid Architects that focus on user experience, social interaction, and dramaturgical composition. This virtual architecture is powered by performance-aligned, field-tested parametric design technologies (Zaha Hadid Architects 2021). The project focuses on cyberspaces that enable human-to-human communication via computer networks. The event was to introduce how the metaverse supports new forms of creative cultural production like digital art and virtual art museums (Kit 2022). Some visualizations of NFTism can be seen in Figures 2 and 3):

Figure 2. The Virtual gallery (Left) vs Figure 3. The Z-car, a concept vehicle. Source: https://www.zaha-hadid.com/design/nftism-at-art-basel-miami-beach/ 2021

The virtual gallery features, among other items, designs previously commissioned by Kenny Schachter, namely the 'Z-boat', the 'Z-Car One', a sculptural bench-table 'Belu', and a stool 'Orchis'. These designs are collaborations between Zaha Hadid Architects and Zaha Hadid Design. This NFT-ism project is a joint project between ZHA and JOURNEE. JOURNEE is a leading technology company specialized in connecting brands with their audiences in the metaverse by providing the world's most advanced metaverse technology platform. In addition, the exhibition NFTism is curated by art market impresario Kenny Schachter. Schachter hopes that NFTism will inspire other architects and that people will be more open-minded and embrace innovation in the metaverse with as much quality as there is in the physical world. Also, NFTism will exist as a kind of benchmark.

3.2 *Discussion*

Reflecting on the two works by Zaha Hadid Architects above, several things were obtained to be discussed as points of discussion regarding the extent to which the development of the virtual environment had an influence on architects and interior designers. For centuries, architects, interior designers, engineers, and builders largely dictated the shape of the built environment and its regulations. However, with the current development of the metaverse, anyone with a pioneering spirit and a little bit of cryptography can plant their flag and build their own slice of the virtual world however they wish (Matoso 2022). More people have the opportunity to do business and make a profit. So when viewed from an economic perspective, many think that the development of this virtual environment will lead to the rise of the digital economy and the rise of digital assets.

In addition, the progress of computing-based design is also considered to be growing rapidly. Because in a virtual environment, rules such as construction, materiality, and budget can be overridden. This may be contrary to the thoughts of some architects such as Pallasmaa who think that technology is a factor that makes the world turn into an unerotic thing. Pallasmaa is one of the figures who rejected the modernist movement in architecture. According to him, architecture is a medium that connects designers with building users. To achieve this, the designer must clearly understand, in his mind, the spatial form of the building as a whole and only then manifest it in the built work (Sabatini et al 2018). So the metaverse may also cause a lot of controversies, especially for architects who share the same views as Pallasmaa.

The next thing that might happen from the development of the metaverse and virtual environment is a change in the business model in the world of architecture and interior design. In a physical environment, a consultant is a place where an architect and interior designer work, as well as transact with clients. This is different from what happens in a virtual environment. The role of architects and interior designers is turning into content creators. To be able to sell a design in the metaverse, the design results from an architect and interior designer are packaged into content in the form of visualization so that potential clients can see it. In the real world, this is usually done by a visualizer. The function of the visualization is only one of the parts of the design stages because, after that, there are still construction and post-construction stages that usually take more time and energy. This is one of the things that is very significant when compared to the state in the metaverse. The indication is that the work of an architect and interior designer will end in a shorter cycle because there is no need for a development stage.

4 CONCLUSION

The development of web 3.0 and the metaverse is currently still at an early stage. It is difficult to understand the exact day when the metaverse will be fully launched. However, it is believed that this virtual environment will continue to develop in the future. Not only as a tool for interacting between individuals and other individuals in new ways but the metaverse is also allegedly able to penetrate many things, including education. This possibility can be seen in the two case studies in this study. We can see that currently, the barrier between physical and virtual environments has been getting thinner. As architects and interior designers, we must be able to respond to change with an open mind and also prepare ourselves to face the changes that occur. All the information collected is expected to increase knowledge related to architectural technology science. The author is aware that there are many shortcomings in this study. Because this is still initial research using the literature study technique only. The next research plan is to try to develop the design of planning projects on the metaverse platform and compare the design methodology with the design methodology in the physical world so that before-after studies can be produced as clear and more valid participants.

REFERENCES

Architecs, Zaha Hadid, 2021. *NFTism at Art Basel Miami Beach Designed by Zaha Hadid Architects, Powered by JOURNEE*. https://amazingarchitecture.com/news/nftism-at-art-basel-miami-beach-designed-by-zaha-hadid-architects-powered-by-journee. Accessed on 20th May 2022.

Creswell, J. W., Creswell J.D. 2018: *Research Design: Qualitative, Quantitative, and Mixed Methods Approaches* Fifth Edition. London: Sage Publications, International Education and Professional Pubisher.

Hall, J. 2015. *Czech Politician Who 'Founded' a New Country Named Liberland in no Man's Land Between Croatia and Serbia is Arrested for Trespassing*. https://www.dailymail.co.uk/news/article-3076499/Czech-politician-arrested-trying-enter-3-square-mile-patch-land-Croatia-Serbia-claimed-founded-new-country.html. Accessed on 20th August 2022.

Finney, A. 2022. *Zaha Hadid Architects Designs Virtual Liberland Metaverse City*. https://www.dezeen.com/2022/03/11/liberland-metaverse-city-zaha-hadid-architects/. Accessed on 14th May 2022.

Kit, K.T. 2022. Sustainable Engineering Paradigm Shift in Digital Architecture, Engineering and Construction Ecology within Metaverse. *World Academy of Science, Engineering and Technology International Journal of Computer and Information Engineering* Vol. 16, No. 4, 2022.

Kolata, S. 2022. *The Metaverse as Opportunity for Architects: An Interview with Patrik Schumacher*. https://www.archdaily.com/980196/the-metaverse-as-opportunity-for-architects-an-interview-with-patrik-schumacher?ad_source=myarchdaily&ad_medium=bookmark-show&ad_content=current-user. Accessed on 14th May 2022.

Kumar, R. 2005. *Research Methodology: A Step by Step Guide for Beginner*. London: Sage Publication.

Laukkonen, J. 2022. *What Is the Metaverse? All About Metaverse, its Meaning, and What You Can Do There*. https://www.lifewire.com/what-is-the-metaverse-6260309. Accessed on 30th Auugust 2022.

Matoso, M. 2022. *Metaverse: A Fertile Ground for Architects?*. https://www.archdaily.com/979614/metaverse-a-fertile-ground-for-architects?ad_source=myarchdaily&ad_medium=bookmark-show&ad_content=current-user. Accessed on 20th May 2022.

Nicola, G. 2022. *How Metaverse Might Change Interior & Architecture Design?*. htttps://www.tallboxdesign.com/metaverse-interior-architecture/. Accessed on 14th May 2022.

Overstreet, K. 2021. *Bitcoin, NFT's, and the Metaverse: Reflecting on the Year of Digital Real Estate and Design*. https://www.archdaily.com/973814/bitcoin-nfts-and-the-metaverse-reflecting-on-the-year-of-digital-real-estate-and-design?ad_source=myarchdaily&ad_medium=bookmark-show&ad_content=current-use. Accessed on 14th May 2022.

Sabatini, S.N., Kurniati, F., Haristianti, V. and Sudrajat, I., 2018. Sumbangsih Juhani Pallasmaa dalam Teori Arsitektur. *RUAS (Review of Urbanism and Architectural Studies)*, 15(2), pp.49–60.

Sharr, A. 2007. *Thinkers for Architects: Heidegger for Architects*. New York: Routledge.

Stouhi, D. 2021. *Zaha Hadid Architects Presents Virtual Gallery Exploring Architecture, NFT's, and the Metaverse*. https://www.archdaily.com/972886/zaha-hadid-architects-presents-virtual-gallery-exploring-architecture-nfts-and-the-metaverse?ad_medium=widget&ad_name=related-article&ad_content=978522. Accessed on 20th May 2022.

Stouhi, D. 2022. *Zaha Hadid Architects Designs "Cyber-Urban" Metaverse City*. https://www.archdaily.com/978522/zaha-hadid-architects-designs-cyber-urban-metaverse-city?ad_source=myarchdaily&ad_medium=bookmark-show&ad_content=current-user. Accessed on 20th May 2022.

Tuan, Y.F. 2008. *Space and Place The Perspective of Experience Sixth Printing*. Minneapolis: the University of Minnesota Press.

Waddoups, R. 2022. *A Micronation by Zaha Hadid Architects Is Forming in the Metaverse*. https://www.surfacemag.com/articles/liberland-zaha-hadid-architects-metaverse/. Accessed on 20th May 2022.

Recommendations for restructuring urban development in Kemang area

N.A. Hapsoro* & K.P. Amelia
Telkom University, Bandung, Indonesia

ORCID ID: 0000-0001-6036-0145

ABSTRACT: Kemang is one of the fastest growing areas in South Jakarta. Along with the development of Jakarta, transformations in the land use in Kemang have changed a lot. Dominated by settlements, currently these functions are mixed with other functions with a modern tendency. In this regard, the DKI Jakarta Provincial Order adopted a policy with the Governor's Instruction Number 140 of 1999 concerning Guidelines for the Granting of Permits in Kemang Modern Village, where trading activities are allowed on the condition that business actors must register their businesses. This policy is expected to control regional development and maintain the characteristics of Kemang. Based on this, it is necessary to prepare a recommendation guide for the planning of Kemang so that it becomes the basis for the Technical Proposal and Urban Design Guidelines for Kemang to be implemented in the field. The preparation of this recommendation guide uses a descriptive qualitative method, and the output is a proposed concept of land use, building mass intensity, linkage, and open space.

Keywords: Jakarta, Kemang, urban design, guidelines, land use

1 INTRODUCTION

The narrow space with high density, and the individualistic nature of city dwellers who tend to be selfish and prioritize their own groups are the characteristics of unhealthy areas that are considered as time bombs in encouraging crime, especially if there is a combination of these characteristics. This condition becomes an obstacle to sustainable city development because there are at least three indicators that must be developed in a balanced way to create a sustainable city (Sutiyarsi *et al.* 2019). The Kemang area is one of the areas in South Jakarta that is developing towards this condition. Known as a commercial area, based on the City Master Plan (RIK) of 1965–1985, the Kemang area has an urban land allotment (Urban Type) and rural land designation, with a typology of Kampung Type building use of 5% to 20%. In line with the development of Jakarta city, the function of the land in the Kemang area has changed. Kemang has become an area that seems exclusive to both expatriates and the upper class, and has left behind Betawi culture and lower class society (Prayogo & Fuad 2020). Dominated by settlements, today's functions mixed with other functions with a modern tendency. Since the 1970s, according to DKI Jakarta Provincial Regulation No. 1 of 2014 land use in the Kemang area, especially along Kemang Raya, has changed from a private function to 90% commercial (Pinendita & Fuad 2020). Changes in land use unconsciously have an impact on changes in urban design elements in an area. These two things are cause-and-effect relationships that influence the quality of urban design (Harjasa *et al.* 2016).

In this regard, the DKI Jakarta Provincial Order adopted a policy with the Governor's Instruction Number 140 of 1999 concerning Guidelines for the Granting of Permits in

---

*Corresponding Author: ariefhapsoro@telkomuniversity.ac.id

Kemang Modern Village, where trading activities are allowed on the condition that business actors must register their businesses. The policy refers to several laws and regulations: Law Number 26 of 2007 concerning Spatial Planning, Law Number 32 of 2009 concerning Environmental Protection and Management, Law Number 2 of 2012 concerning Land Acquisition, Presidential Regulation of the Republic of Indonesia, and Law Number 71 of 2012 concerning the Implementation of Land Procurement for Development in the Public Interest. With the issuance of a regional regulation that regulates regional spatial planning, DKI Jakarta has a legal basis for carrying out development. All forms of development in DKI Jakarta, which concretely utilize space must be in accordance with the spatial plan for the DKI Jakarta area as determined in the DKI Jakarta RTRW (dan Raihan *et al.* 2017).

The Governor's Instruction Policy Number 140 of 1999 is expected to be able to control the development of the area and maintain the characteristics of the Kemang Area, so that the Kemang Area can grow in harmony with the surrounding area. New York can be used as a precedent to create a better urban space for its people. New York is seen as having shown that giving people more space can bring benefits to the city and its people (Prawata 2015). The design strategy adopted by the New York City Department of Transportation can be applied to Kemang. The Jakarta government, city planners and developers have a role in planning and shaping the city and can use creative strategies to improve the quality of life for city residents. Based on this, it is necessary to prepare a recommendation guide for the planning of the Kemang Area so that it becomes the basis for the Technical Proposal and Urban Design Guide Lines for the Kemang Area to be implemented in the field.

2 RESEARCH METHODS

This study uses a qualitative method. Data collection techniques were carried out by direct observation and literature study. The analytical method used in this research is content analysis and qualitative descriptive analysis.

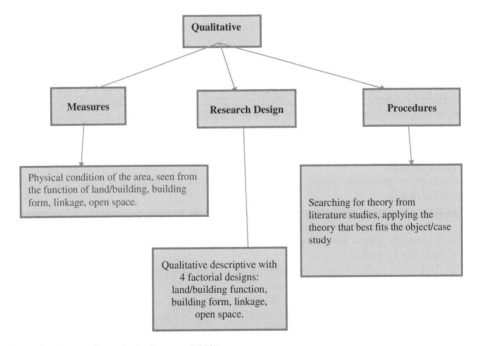

Figure 1. Research methods diagram (2022).

3 RESULT AND DISCUSSION

3.1 *Land use concept*

Diverse functions are centered on West and South Kemang, especially along the main road corridors. A shared space concept is needed to shift the meaning of exclusivity to be more inclusive (Pinendita & Fuad 2020). Meanwhile, north and east Kemang, as much as possible, remains a residential area only. If there is a commercial function, it is only a supporting function. The commercial areas that had been large-scale in the north and east Kemang were moved to the west and south Kemang, and instead, part of the residential areas in the west and south Kemang were moved to north and east Kemang.

3.2 *Building mass intensity*

Arrangement of orientation and building mass is to create order and efficiency, especially for the functions that want to be accommodated in the building, in addition to producing a pattern of mass order that is synergistic with the outdoor space.

The concept of orientation and mass of the building on this site includes:

a) Building mass blocks still allows open spaces to enlarge air and pedestrian circulation.
b) There is a harmony of building styles along the corridors of the street with a slight accent at certain points to break the monotony.
c) Buildings must be able to provide clear directions both for pedestrians and motor vehicles.
d) Zones at the ground floor level in mixed-use areas are used as public areas, trade and service areas at the next level, while parking as much as possible in the basement.
e) The mass placement should also adapt to the orientation of sunlight
f) In the residential zone, part of the settlement in the north and east Kemang typology was converted into apartments and townhouses to save the land and increase the green area.

3.3 *Linkage*

West Kemang corridor to south Kemang is made in the same direction to reduce congestion. The circulation paths served in this area generally consist of pedestrian paths, disabled lanes, and motor vehicle lanes. It is hoped that pedestrian/vehicle problems can be minimized with the presence of raised sidewalks and signage. Wayfinding devices can be used to guide pedestrians to the area to be addressed (Anwar 2020). Land acquisition for road widening, 2–5m considered necessary to widen the highway and add pedestrians, especially in the western to southern Kemang corridor. This circulation design aims to make this area more accessible, especially for pedestrians and motorized vehicles.

1. **Pedestrian path**

    Concept of providing pedestrian network facilities for pedestrians and open courts, among others:

    a) Arrangement of pedestrians in the environment to create comfort for pedestrians. uninterrupted by vehicle lanes
    b) Pedestrian paths flow continuously and connect to every activity and function between regions.
    c) Pedestrian makes it easy for people with disabilities, the elderly, or children, such as ramps and visually impaired lanes.
    d) The floor pattern application follows the concept of space, for example, moving or stationary patterns
    e) Motor vehicle access is minimal but still provides affordable parking facilities to the object you want to go to.

2. **Parking system**
   Off street Parking Provision Plan
   a) The existence of basements for parking or parking buildings in activity centers.
   b) Require all trade activities, services, and public facilities to provide parking by the scale of service of such activities.

   Plan to provide parking for people with disabilities
   People with physical disabilities who use mechanical assistance such as wheelchairs and canes need a more comprehensive than standard parking space. A minimum of two parking spaces per parking lot must be designed for people with physical disabilities or at least one parking space per 20 vehicles. It should be placed as close as possible to the driveway from the building and, if potential, no more than 30 meters. Parking spaces designed for use by people with physical disabilities assisted by mechanical assistive equipment must have a minimum width of 3 meters plus an intermediate length of 1-meter-wide provision of ramps for the curb. A 1:6 (17 percent) ramp is entirely appropriate for short distances.

3.4 *Open space*

Open space has the function of providing light and air circulation in buildings, especially in the city center. Presents the impression of perspective and vista on the urban landscape, especially in the dense city center area. Provide a recreational arena with special forms of activity. In this case study, the following is a list of recommended open space design guidelines:

a) Improve the quality of the outdoor space on the site
b) Serves as a means of activity and socializing close to nature.
c) Increase the city's green open space
d) Protecting the city's ecosystem
e) Green Open Space is in almost every zone of the region and becomes the main core space of the zone.
f) Green belt functions as a buffer that can be placed in the path of vehicles and pedestrians.
g) The planting of protective trees and directional trees is adjusted to the function of the area they shade.
h) Determining plant types that are easy to care for, such as trembesi trees, palms, etc.
i) Determining the type of local plant that is easy to get, cheap, and still shows aesthetics.
j) Implementation of the streetscape.
k) The garden roof concept, especially in commercial areas.

4 CONCLUSION

With all its problems, the Kemang area needs to be regulated and controlled for its development as soon as possible, so the site can still grow harmony with the surrounding. Proposed techniques can be used as a guide for the design and control the aspect of land use by switching the function of the Kemang area. West and south Kemang became mixed-function, and north and east Kemang became residential functions only. The existing west and south Kemang dwellings should be moved to north and east Kemang. From the aspect of the mass intensity of the building, in the residential zone, the typology of the building is changed to a townhouse and apartments to add more green space on site; it could be passive or active. It is also encouraged to add a parking building, so there is no more on-street parking. Circulation aspect, a need for land acquisition that will later use to make the road wider, pedestrian paths, and also ensuring that pedestrian lanes are made flowing, continuous, and able to connect every activity and function between their areas. And in the aspect of open space, the need to create green space as the core of each function requires the existence of a garden roof in each building, especially in commercial areas, and the addition of greenbelts in addition to vehicle lanes and pedestrian lanes. There are technical proposals to be applied as a guide for the design and control of Kemang area, so it can continue to grow into a harmonious and healthy space for its city residents.

REFERENCES

Anwar, H. (2020). Transformasi Kawasan Hunian Menjadi Kawasan Komersil Studi Kasus Kawasan kemang. *Jurnal Teknologi Dan Desain*, *1*(2), 36–44. https://doi.org/10.51170/jtd.v1i2.6

dan Raihan, U., Balai Rakyat Kelurahan Utan Kayu Utara, J., & Jakarta Timur, M. (2017). *Ruang Terbuka Hijau Dalam Pembangunan Berkelanjutan Di Daerah Khusus Ibukota Jakarta* (Vol. 14, Issue 1).

Harjasa, P., Zulkaidi, D., Ekomadyo, A. S., Bandung, T., Kebijakan, P., & Itb, S. (2016). Pengaruh Perubahan Guna Lahan dan Intensitas Guna Lahan terhadap Kualitas Ruang Kota. *Pengaruh Perubahan Guna Lahan Dan Intensitas Guna Lahan Terhadap Kualitas Ruang Kota*, *1*, 105–110.

Pinendita, T., & Fuad, A. H. (2020). *Shared Spaces by Placemaking to Create Inclusive Kemang Raya*. 475 (Idwell), 200–211. https://doi.org/10.2991/assehr.k.201009.021

Prawata, A. (2015). Creative User Generated Urbanism. *Procedia – Social and Behavioral Sciences*, *184* (August 2014), 232–239. https://doi.org/10.1016/j.sbspro.2015.05.084

Prayogo, J. R., & Fuad, A. H. (2020). *Pluralizing Method for Generating Heterogeneous Public Space: A Case Study of Palang Pintu Festival Kemang*. 475(Idwell), 212–220. https://doi.org/10.2991/assehr.k.201009.022

Sutiyarsi, T., Koestoer, R. H., & Susiloningtyas, D. (2019). The challenge of slums toward a sustainable city. *IOP Conference Series: Earth and Environmental Science*, *338*(1). https://doi.org/10.1088/1755-1315/338/1/012014

Extended reality: How digital technology transformed in film festivals?

D.A.W. Sintowoko & H. Azhar
Telkom University, Bandung, Indonesia

H. Humaira
University of Exeter, Exeter, United Kingdom
ORCID ID: 0000-0003-4702-8409

ABSTRACT: The overall economic impact, social role and status of cultural professionals during COVID-19 remain relatively understudied in the film Industries. The professional had to decide on an emergence alternative to go entirely virtual activity with no physical event. The present study highlighted technological improvements; it is about how spectators involved more humanities using extended reality tools in film screening festivals. Purposive sampling was used in the current study. A qualitative method with a case study approach was applied as well to see how the Amsterdam and Venice film festivals are driven by extended reality. Also, this study wants to see how extended reality has a potential system for Indonesia film festivals as well. Results show that film has shifted into the industrial revolution technology in a rich country, where the film festival has transformed into new environments and visualizations, especially in engaging multiple senses such as sight sound, smell, and touch. Second, physical and digital objects can exist and interact in real time. However, extended reality might still become a challenging issue related to fields in developing country.

Keywords: extended reality, film festivals, film industry

1 INTRODUCTION

Indonesia has a rich culture. The local culture has been implemented using various digital technologies, such as film. The film is a medium where one can reflect culture mixed with storytelling (Okanovic V. *et al.* 2022). Showing cultural identity through visual language (film) is an improvement. Although the situation of pandemics has become a new normal condition, however, visualization of a film screening in Indonesia is quite the same. Hernandez (2020) strongly criticized that the national film festival has a missed chance for the agents to investigate the huge gap in representing the crew to the public. He claimed that the Indonesian film industry, especially the umbrella of film festivals has not detained the consideration of the common public as a festival. However, instead of pointing out the film ecosystem, there is a more challenging issue on film and its technology.

In the last decade, filmmakers have seen exceptional investment from technology giants. For instance, Apple, Metaverse, Google, and Adobe have into Augmented Reality (AR), Mixed Reality (MR), and Virtual Reality (VR). All of them are now consolidated under the umbrella of the Extended Reality (XR) (Schleser (2022). In short, film technology has integrated within the mediascape, turn boosting new manners and creative work passes for the film festival. Film technology has now been positioned as offering the potential to transform smartphones with complex exploration such as mobile cinema (Atkinson 2017), creative mobile media (Prasad 2017), and mobile filmmaking (Berry 2017). Thus, these devices are potentially demonstrating XR for film curation.

Previous research found that XR has continued to grow and gain credibility in interactive animated documentaries (Ehrlich 2022 March); social science (Tassinari et al. 2021); and observing cultural heritage (Rizvić et al. 2021). The board's definition of interactive documentary positions as "any project that starts with the intention to engage with the real, and that uses digital interactive technology to realize this intention" (Aston & Gaudenzi 2012). Traditionally, studies on film are focused on cinematic structure instead of its technology (Sintowoko 2022); visual language in the hybrid culture (Sari & Sintowoko 2022), and film narrative (Tan 2013). None of the highlighted XR as a potential film festival technology in Indonesia. Therefore, the present study is at an early stage and will be focusing on the chance of XR as a potential technology for film festivals in Indonesia with applied case studies from Amsterdam and Venice film festivals. Thus, the present study wants to show the innovation between the potential of the physical and digital environment in Indonesian film festivals as well, since none of the previous studies focus on this issue.

2 RESEARCH METHODS

A case study as a qualitative research approach was used in the current study. The principal objective of the present article is to portray the main characteristics of some film festivals, especially in Extended Reality (XR) held in some countries, including Amsterdam (2016), the United States (2015 & 2019), Thailand (2017), and Venice (2018). These are called purposive sampling due to the advanced system of film screening during a pandemic. They used extended reality as an innovation of film screening festivals which is representing a new atmosphere of how extended reality become potentially applied in other countries, like Indonesia as well. Even though extended reality sounds quite new in Indonesia, however, it might be challenging due to the number of Indonesian film festivals being varied. Regional is the typical origin of Indonesian film festivals that grow from the grassroots. As such, as a first step, we applied a qualitative method with a case study approach to see how the Amsterdam and Venice film festivals as a reflection of how extended reality becomes the potentiality for Indonesia. Thus, the Indonesian film festival is collected through the Indonesian Film Board in the present study.

3 RESULT AND DISCUSSION

3.1 *Extended Reality (XR)*

Extended Reality (XR) technologies in the Film Industry that simulate an innovative movie screening can tremendously help all film festivals. Virtual Reality (VR) traditionally refers to computer technologies constructed with realistic images, sounds, mood cues, and sensations that potentially reflect in an immersive environment. Standard VR systems have a mere possibility of users' interface in virtual experiences. It used a multi-projected environment to generate realistic visuals and sounds unlike in the conventional 2D.

Formerly, VR has been adopted in various fields, such as design, architecture, art, entertainment, communication, tourism, business, and education (Alizadehsalehi et al. 2020). Finally, VR is the use of computer technology to create a simulated experience that can be similar to the real world's atmosphere due to its advanced 3D environment. Historically, VR has been founded in the 1860s, when the new 360 degrees of art through panoramic murals began to appear. Further, the film industry has predicted by some researchers that XR is a tool for sensory experience. The design was purposed to engage spectators through visual appearance. This new technology was created for artistic and creative experimentation that continuously raising (Elstner 2020). Historically, the extended reality was initiated by the cinematographer, Morton Heilig, who produced the *Sensorama*. *Sensorama* is a unique machine, known as a big VR box, with complect sensor system for human senses (Mann *et al.* (2018). Previous researchers also mentioned that Morton Heilig has three basics idea combining color video, 3D including audio, and smells where the spectators allow sitting in a vibrating chair to feel an immersive experience (Matahari 2022).

Figure 1. VR cinema opens in Amsterdam 2016.

3.2 *The Indonesian film industry and festival*

The official number of the Indonesia Film Festival is about 77 film festivals. The festivals are spread over 12 provinces. These range from grassroots-level festivals, like at schools and universities. At the national level, there are *Festival Film Indonesia* (FFI), *Jogja-Netpac* (JAFF*), and Festival Film Dokumenter* (FFD). Others are *Arkiperl* for experimental film and *Minikino Film Week*, an international short film festival. The festivals are quite different from one another because Indonesia is a big country that consists of various islands. Indonesia is an archipelago with 260 million people spread over 15.500 islands and currently has only 1,412 screens (Indonesian Film Board 2022). Due to the lack of screens and the concentration of cinemas in some big cities, there is still a huge demand that is currently not served by commercial movie theatres. Indonesian Film Board (BPI) reports that from 15,500 islands, there are only 10 regions with the most movie theatres by December 2017, with a number of screens, followed by 19 movie theatres from Central Java, 11 movie theatres from North Sumatra, 8 movie theatres from South Sumatra, 8 movie theatres from Jogjakarta, 7 movie theatres from North Sulawesi, and 7 movie theatres from Bali, respectively. The number of Indonesian film production has increased annually from 2011 to 2018, where the lowest case was 82 movies. Since 2018, more than 150 movies were made and screened in alternative screening venues throughout the nation. The government has created regulations to support the development of the Indonesian film industry. One of them is presidential regulations number 44 of 2016 for the creative industry where the cinema must also comply with the regulation which set 60% local content/movie in the cinema. However, to encourage the Indonesian film ecosystem, there are a number of film festivals outside of commercial cinemas. Thus, the next sub-topic will discuss Indonesia Film Festival.

3.3 *XR as a future*

Extended reality (XR) has a big potency for the Indonesian Film Industry. XR encompasses a conglomeration of virtual reality and mixed reality, where the technology engages multiple senses such as sight sound, smell, and touch. XR can be defined as a system that fulfills 3 basic features a combination of the real world and virtual world in real-time interaction and accurate 3D. There is sensory information that can be destructive meaning the natural environment shows virtual objects in the way augmented reality alters the perception of the real world, whereas VR completely replaces the user's field world environment with simulated camera-based apps.

Figure 2. Venice VR expanded 2021.

4 CONCLUSION

The present study indicates an analysis of potential Extended Reality film festivals specializing in VR, AR, XR, and 360 degrees in Amsterdam and Venice, identifying common and specific features of each one. Based on the data, the current study performed an initial approximation regarding the sample of the study. First, film festivals are such for exhibiting non-commercial proposals where the public is allowed to participate. Second, the Indonesian film festival has a strong chance to use advanced technology for film festivals from the grassroots. Film Festivals are traditionally aimed at the non-commercial, and those interested in Extended Reality aim precisely to obtain the maximum benefit for the public.

REFERENCES

Aston, J., & Gaudenzi, S. (2012). Interactive Documentary: Setting the Field. *Studies in Documentary Film*, 6 (2), 125–139.
Alizadehsalehi, S., Hadavi, A., & Huang, J. C. (2020). From BIM to extended reality in AEC industry. *Automation in Construction*, 116, 103254.
Elstner, M. (2020). *Use Cases of Extended Reality in the Construction Industry.*
Ehrlich, N. (2022, March). Virtual Veracity: Animated Documentaries and Mixed Realities. *In 2022 IEEE Conference on Virtual Reality and 3D User Interfaces Abstracts and Workshops (VRW)* (pp. 7–14). IEEE.
Mann, S., Furness, T., Yuan, Y., Iorio, J., & Wang, Z. (2018). All Reality: Virtual, Augmented, Mixed (x), Mediated (x, y), and Multimediated Reality. *arXiv preprint arXiv*, 1804.08386.
Matahari, T. (2022). WebXR Asset Management in Developing Virtual Reality Learning Media. *Indonesian Journal of Computing, Engineering and Design (IJoCED)*, 4(1), 38–46.
Okanovic, V., Ivkovic-Kihic, I., Boskovic, D., Mijatovic, B., Prazina, I., Skaljo, E., & Rizvic, S. (2022). Interaction in Extended Reality Applications for Cultural Heritage. *Applied Sciences*, 12(3), 1241.
Rizvić, S., Bošković, D., Okanović, V., Kihić, I. I., Prazina, I., & Mijatović, B. (2021). Time Travel to the Past of Bosnia and Herzegovina Through Virtual and Augmented Reality. *Applied Sciences*, 11(8), 3711.
Schleser, M. (2022). Studies in Documentary Film. *Smart Storytelling*, 1–17.
Sari, A.S., & Sintowoko D.A.W. (2022). Costume and Feminism: Character in Film Kartini. *Capture: Jurnal Seni Media Rekam*, 13(2), 148–157.
Sintowoko, D.A.W. (2022). Mood Cues dalam Film Kartini: Hubungan antara Pergerakan Kamera dan Emosi. *Rekam: Jurnal Fotografi, Televisi, Animasi*, 18(1), 1–16.
Tan, E. S. (2013). *Emotion and the Structure of Narrative Film: Film as an Emotion Machine*. Routledge.
Tassinari, M. E., Marcuccio, M., & Marfia, G. (2021). Extended Reality in Social Science: A Conceptual Clarification. *In Proceedings of Edulearn21 Conference*, (Vol. 5, p. 6th).

Destination branding strategy *Bitombang* old village as alternate tourism destination in Selayar, Sulawesi Selatan

M. Andhyka Satria Putra & F. Ciptandi
Telkom University, Bandung, Indonesia

ABSTRACT: Bitombang Old Village is located in Bontoharu, Selayar Islands Regency, having various tourism potentials, including architecture, culture, history, and natural scenery. However, the image of the tourist village is not well known to the wider community due to the lack of promotions carried out. The research method used is the design thinking method with the following stages: 1) Empathy to explore all information related to basic knowledge regarding all information about Kampung Tua Bitombang by conducting literature studies, interviews, and ethnography; 2) Identification to identify problems found and proposes the urgency of the problem through a data confirmation process; 3) Ideas in the form of a problem analysis phase and idea generation in the form of the solutions offered; 4) Implementation in the form of strategy design analysis through the elaboration of relevant theories; and 5) Validation, namely evaluating effectiveness using differential semantics. From the analysis carried out, this potential can be used as a tourist attraction, becomes an educational value for tourists, and maintains the balance of ecosystems and historical heritage.

Keywords: Bitombang Old Village, destination branding, tourism destination, Selayar Island

1 INTRODUCTION

The potential and uniqueness of each region in Indonesia if it can be optimized properly can become a profitable tourist destination because, through this tourism industry, income for an area can increase from tourist visits that come while at the same time increasing the workforce of the local population itself [1]. The potential and uniqueness of the old village of Bitombang have a great opportunity to be developed into a tourist destination area based on culture and history, in addition to other tourist destinations that already existed in the Selayar Islands. Some of the tourism potentials owned by the Selayar Islands Regency that are very prominent are the potential for historical tourism, cultural tourism, and marine tourism. This tourism potential resource is according to the Selayar Regency Regional Regulation Number 20 of 2002 concerning the determination of tourism objects in the Selayar Regency area which is translated into the Selayar Islands Regent's Decree Number 453/VII/YEAR 2017. It covers the destination names and locations of Tourism Destinations in the Selayar Islands Regency. Based on the strategic plan of the Office of Culture and Tourism of South Sulawesi Province in 2021, the Province of South Sulawesi develops strategies and policy directions for developing Tourism and Culture as follows:

1. In increasing tourist attraction, optimization of marketing promotions and destination development is carried out which is directed at strengthening digital marketing facilities and the use of digital branding and e-package tours. In addition, the development of

destinations is more inclined toward halal destinations which refer to halal tourism products, as well as the development of ecotourism destinations in South Sulawesi.
2. Increasing tourism competitiveness, South Sulawesi Province is taking a path through optimizing the quality of tourism, and optimizing cooperation with tourism partners at the center and districts/cities, especially to support halal tourism destinations and ecotourism destinations.
3. Regarding cultural development, the province of South Sulawesi has a target of increasing the development of local arts and culture through facilitating activities for understanding the values of the cultural character of South Sulawesi and facilitating arts and cultural activities through organizing arts and cultural festivals with the characteristics of South Sulawesi and strengthening the cultural character of South Sulawesi.
4. In addition, other targets are to increase the management of arts and culture, cultural heritage through improving the management of art studios, improving the quality of management of preservation and utilization of cultural heritage, and increasing the quality and quantity of registration of cultural works.

Based on the discussion on the potential of regional tourist destinations and also on the strategic plan of the Tourism Office of the Selayar Islands Regency which has been described above, the historical tourism potential of the old village of Bitombang for branding destinations is expected to be able to provide proposed strategies that can help increase tourist attraction. There are various kinds of tourism in the Selayar Islands: the old village of Bitombang is one of the tourist destinations that has a unique culture and history that has become an icon of the old village of Bitombang.

2 RESEARCH METHODS

The research method uses qualitative description and it is carried out in principle using a design thinking approach [8] which is carried out using five stages, as follows:

1. *Empathy stage*
   This stage is done by determining in advance the object, subject, and research population. The object of research is the tourist destination of Bitombang Old Village, Bontoharu, Selayar Islands Regency, South Sulawesi, and the tourism potential in Bitombang Old Village, with research subjects, including the Head, Manager of Bitombang Old Village and the Selayar Islands Tourism and Culture Office to explore information related to the population village (gender), geographical location, and natural and cultural potential.

   Collecting data through literature review to find information about tourist destinations is based on journals and books, and then through direct observation to understand the symptoms or the community in the village situation of the Old Village of Bitombang. Starting from the potential that exists in the Old Village of Bitombang to the tourism potential around it. Next, he conducted interviews with parties including the head of Kampung Tua Bitombang, the people of Kampung Tua Bitombang, the Department of Tourism and Culture of the Selayar Islands, and tourists from the Selayar Islands. Finally, it is the study of documentation in the form of pictures or photos that record past events.

2. *Define stage*
   This stage is carried out in several steps:
   (a) Data reduction, to select data that is considered relevant and create a clear picture of the problem.
   (b) Data presentation, in the form of presenting data conclusions that have been selected at the previous stage to build a clear problem framework.
   (c) Determination of the urgency of the problem.

3. *Ideate stage*

At this stage, a comprehensive analysis is carried out by elaborating on the support of theories that are relevant to the state of the data in the field, to give birth to an idea of the right solution in the form of an ideal strategy concept.

4. *Prototype Stage*

This prototype stage is not realized in the form of a product, but a written idea in the form of schematics and guidelines in order to realize the destination branding strategy for Kampung Tua Bitombang.

2.1 Methods of data collecting

The initial methods of data collecting are observations that were made to obtain valid data based on direct observations of how the people of the old village of Bitombongan are active. Existing tourism assets in the Old Village of Bitombang and other tourism potentials can be used as assets. In addition, observation is used to obtain accurate and comprehensive data from respondents, which is very much needed by researchers and then semi-structured interviews. In this study, it is aimed to anticipate that the researcher does not get all the answers needed because the interviewee does not have a guide to answer. The parties to be interviewed by the author are the Head of Bitombang Old Village, the villagers of Bitombang Old Village, the Department of Tourism and Culture of the Selayar Islands, and tourists from the Selayar Islands. The last method of data collection is a literature study. which means collecting data from various literatures such as journals, previous research, and articles related to destination branding, potential tourist destinations, and the Bitombang Old Village.

3 RESULT AND DISCUSSION

Based on the data collection that has been done, the researchers found several issues or problems that exist in the old village of Bitombang, the first is the problem of the degradation of cultural inheritance among older people who understand the substance and procedures of a tradition and the youth in the old village of Bitombang. Based on the results of interviews with youths in Bitombang, these youths only understood the procedures but did not understand the substance or philosophy of the tradition, while the indigenous people told them that they hoped that the youth would contribute more to the customs and ceremonial traditions.

Figure 1. Village of Bitombang. Source: (self-documentation 2022).

Another problem is in traditional ecology, namely, problems regarding the environment in the old village of Bitombang. Based on the traditional ecology of the old village of Bitombang, livestock has always been released into the wild and returned home in the evening; however, according to the tourism and culture office of Selayar and based on a strategic plan to advance tourist destinations, it becomes an obstacle for the tourism office because it is lacking or reduces the quality of this tourist destination because livestock throws feces anywhere, including major access roads.

From the results of data collection, researchers found several traditional artifacts. The architecture of the house is the main attraction in the Bitombang Old Village Area. The house in this village has a height of about 10–15 meters at the back and only about 2–3 meters at the front. According to history, there are at least two reasons that houses in this area were designed that high. First, it is adjusted to the geographical conditions of the village which is located at an altitude and with terraces. Second reason is to avoid chaos because in the past there were frequent thefts and civil wars. The houses in this village are known as durable houses because they are hundreds of years old. This is associated with the process of making a house that has a special ritual in building it. The ritual is still preserved by the residents of Bitombang until now. The ritual of building houses in the Bitombang Old Village area is largely determined by the role of the *Mataguri* as the leader of the house builder. Starting from the selection of wood for the *Sapo fort* (house pillars) totaling 25 of *bittil/holasa* wood, the selection of *pallangga* stone (the base of the house in the form of stones chosen directly from the river), the establishment and reading of prayers at the *Lalaki* fort (the fort in the middle of the house) to the wife's body measurements to determine the length of the house frame. Based on an interview with one of the community leaders, *Mataguri* in the Old Village of Bitombang, in building houses and the village of Bitombang, there is still a culture of mutual cooperation. The development process begins by appointing one of the residents as coordinator. In the construction process, all residents were assisted comprising men, women, and children. With this cooperative system, building a frame house only takes 3–4 days. However, the process of building houses in Bitombang then and now has changed. Most of the construction of houses at this time uses experienced carpenters.

4 CONCLUSION

The conclusion shows that the community can teach the tradition of the old village of Bitombang to young people. The most fundamental reason is in order to minimize cultural degradation as well as to maintain the young generation. Regarding ecology, it is recommended that the local government must consider more ways to keep the environment of the old village of Bitombang cleaner. Thus, they might enrich tourist interests, human nature, and customs of the old village of Bitombang to minimize the existing conflicts.

ACKNOWLEDGEMENT

Hibah DRTPM skema Penelitian Tesis Magister (PTM) Tahun 2022 No kontrak 126/SP2H/RT-MONO/LL4/2022; No. Keputusan KWR4.057/PNLT3/PPM-LIT/2022 dan No. Kontrak Turunan Peneliti 358/PNLT3/PPM/2022.

REFERENCES

Aaker, D. A. (1996). Measuring Brand Equity Across Products and Markets. *California Management Review*, 38(3), 102–120.
Aaker, J. L. (1997). Dimensions of Brand Personality. *Journal of Marketing Research*, 34(3), 347–356.

Cai, A. (2002). Cooperative Branding for Rural Destinations. *Annals of Tourism Research*, 29(3), 720–742.

Dinas Perumahan dan Permukiman Kabupaten Kepulauan Selayar. 2017. Identifikasi Kawasan Permukiman Kumuh di 10 Kecamatan Kabupaten Kepulauan Selayar Tahun Anggaran 2017.

García, J. A., Gómez, M., & Molina, A. (2012). A Destination-branding Model: An Empirical Analysis Based on Stakeholders. *Tourism Management*, 33(3), 646–661.

Javier Escalera-Reyes. (2020). Sustainability Article Place Attachment, Feeling of Belonging and Collective Identity in Socio-Ecological Systems. *Journal Sustainability* 12, 3388; doi:10.3390/su12083388

Qu, H., Kim, L. H., & Im, H. H. (2011). A model of destination branding: Integrating the concepts of the branding and destination image. *Tourism Management*, 32(3), 465–476.

Plattner, Hasso, Meinel, Christoph. 2011. *Design thinking: Understand, Improve, Apply – Understanding innovation*. Berlin: Heidelberg

Exploring sustainable fashion market in Indonesia. Case study: *Sukkha Citta, Setali* Indonesia, and *TukarBaju*

R. Febriani
Telkom University, Bandung, Indonesia

ABSTRACT: Today, sustainable practices have become well-known and influential in the attributes of the textile and fashion industry. Many local brands have emerged that apply the principles of sustainable fashion as a focus and added value for their companies. Sustainable methods also affect people's behavior in consuming or buying goods of interest, especially fashion products. Many urban communities in Indonesia are now choosing environmentally friendly clothing or extending clothing life by exchanging and redesign it (upcycle). This paper uses a descriptive qualitative approach through literature study, observation, and random sampling by diversity of the sustainable fashion market in Indonesia from the consumer behavior perspective. This research shows that several market categories carry sustainable fashion practices: markets or brands that produce clothing from raw materials, make upcycling clothing, and rotate wearable clothes to be used as new products for several target markets.

Keywords: consumer behaviour, fashion industry, lifestyle, sustainable fashion

1 INTRODUCTION

Even though there has been a significant shift towards an eco-friendly concept, the reality is that fashion is still one of the most polluting industries. Sustainable fashion emerged as a response to the complications caused by global industrialization. Although the situation is not the same as what happened in Europe and the United States with large production capacities (in Indonesia, fashion development in general is still driven by players on a small or medium scale), but Indonesia still needs to observe the conditions of fashion development and find ways to apply it to become a solution relevant to domestic conditions.

The sustainable fashion industry is logically a pioneer in finding good things (goods/products), not just showing new things continuously as targets. However, we realize that the challenge at this time is that most of Indonesians are already dominated by the behaviour of speed of consumption, seeking what is instant and trending, as well as a lack of concern for environmental impacts. According to Kotler and Keller (2012) [1] the factors influencing consumer behaviour are cultural, social, personal, and psychological. That is the basis for determining someone to decide or make a purchase. Therefore, sustainability in fashion, if appropriately managed and utilized morally responsibly, can be a movement that has a significant impact on this country.

This paper will discuss the diversity of the sustainable fashion market in Indonesia by sampling the practices of different sustainable fashion activists. The above phenomenon is combined with several key concepts, such as the lifestyle of urban communities in Indonesia from the perspective of consumer behaviour, as well as development opportunities in the creative economy in Indonesia.

2 RESEARCH METHODS

This study uses a qualitative method with a marketing approach, using the consumer behaviour module from Kotler and Keller to see the development of sustainable markets in urban communities in Indonesia.

2.1 Methods of data collecting

The research method is carried out using the qualitative approach, with data collection methods as follows. Firstly, literature or review that is used: scientific journals, books, internet media, and others that discuss the merger of the fashion industry, sustainable fashion, sustainable fashion movement, and consumer behavior. Secondly, observations relating to the work that was inspired were carried out by indirect observations on local brands or organizations engaged in sustainable fashion. Third, random sampling on three types of business platforms engaged sustainably, namely: (a) Sukkha Citta, a local brand that is concerned about environmentally friendly products and prioritizes transparency and worker welfare in the production process; (b) Setali Indonesia, a movement and service provider in extending the life of fashion products by reuse, repair & recycle; and (c) #TukarBaju, an organization under the social entrepreneur Zero Waste Indonesia, has the concept of extending the life of fashion products by exchanging used clothes suitable for use as an economical and environmentally friendly solution.

2.2 Methods of data analysis

The process of data analysis, author uses consumer behavior model as a reference for research indicators.

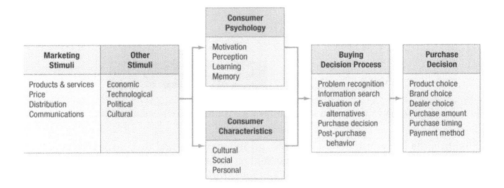

Figure 1. Model of consumer behaviour by Kotler and Keller.

3 RESULT AND DISCUSSION

Increasing consumers' understanding of sustainable fashion today is essential because the leading cause of unsustainable fashion is excessive consumption, which creates high inventories. One effective way to suppress supply is to change demand, as with classical economic theory, namely supply and demand. If you want to make the fashion industry players change how they do business, it is necessary to make efforts to change consumption habits. According to the book The Sustainable Fashion Handbook written by Sandy Black (2012) [2], sustainable fashion requires cooperation from all internal companies; the entire supply chain covers all steps from production to distribution to consumer engagement. Once again, it is emphasized that consumers must participate actively, especially in changing the highly constructed fashion cycle. Consumers aware of sustainability will continue to encourage brands and businesses to produce greener and more responsible clothing throughout the year. Currently, urban communities in Indonesia are offered various solutions that can support a sustainable lifestyle and become one of the choices in purchasing decisions, especially fashion products.

In the model of consumer behaviour, according to Kotler and Keller, we can see that several factors influence purchasing decisions, starting with marketing stimuli. First, these stimuli will affect the psychology and characteristics of consumers; when the consumer psychology and characteristics are affected, then in the buying decision process, consumers will act with what consumers have considered until they make decisions about product choices, brands, purchase amounts to payment methods.

Most people with a middle and upper social class tend to choose new clothes from local brands that carry sustainable fashion practices in their purchasing decisions. Moreover, this consumer class has a high average income, equivalent to a reasonably high purchasing power of products. One local brand that fits this target market is Sukkha Citta.

Figure 2. (a) Sukkha Citta's media social (b) 'Kapas' collection from Sukkha Citta.

Sukkha Citta [3] is an Indonesian brand, an Indonesian brand that carries the concept of sustainability and slow fashion. The name Sukkha Citta is taken from Sanskrit, which means literally, I want everyone involved in this brand, from the makers to the users of the products, to feel joy. Departing as a brand engaged in social business, at first, Sukkha Citta presented handicraft products and now produces clothing using textile materials printed by craftsmen (primarily women) from Java, Bali, and Sumba with decent wages. Using traditional batik and weaving techniques with exclusive motifs inspired by everyday life in Indonesia, the Sukkha Citta brand has added value to the creation of its products.

For people with middle to upper to middle social status, awareness of consuming sustainable fashion products can be seen in the efforts or movements made to reduce the problem of fashion waste. One of them is Setali Indonesia.

Setali Indonesia [4] is an organization engaged in decluttering and thrifting to support the sustainable fashion movement, which is a movement to protect the environment from the dangers of fashion industry waste. Its main activity, namely upcycling, is an effort to make old clothes look new and extend the life of the product. This is a practical solution with a creative design approach that can adapt in Indonesia's urban society lifestyles.

Figure 3. (a) Setali Indonesia's media social (b) Upcycle product from Setali Indonesia.

Figure 4. (a) #TukarBaju's media social (b) #TukarBaju activation program.

Meanwhile, the middle and lower-middle-class people tend to contributing to the sustainable fashion movement by exchanging suitable clothes. One of the communities that promote this is Zero Waste Indonesia – #TukarBaju.

#TukarBaju [5] provides an alternative for those of us who still want to be stylish and change according to fashion but not by consuming new things. #Tukarbaju is a concept where people bring their used clothes and exchange them for other people's clothes. A cost-effective and eco-friendly solution to keep changing fashion styles without spending the expense of buying new clothes.

4 CONCLUSION

The sustainable fashion market in Indonesia seems to be divided into several categories of market classes based on the behavior patterns of consumers who live it. Consumers who prefer local brands to support sustainable fashion practices can simultaneously help the national economic recovery. Consumers who want to stay abreast of trends can turn old products into something new. The result will be the same as reducing fashion wastage and creating a unique limited edition product. Meanwhile, consumers who want to save money and continue to support sustainable fashion can use the clothes swap program as a wise decision-making solution. There are also several brands of sustainable fashion activists from these three class categories that are similar but will be described in more detail in the following research.

REFERENCES

[1] Kotler P. a. K. L. K., 2012. *Marketing Management*, England: Pearson Education.
[2] Black S., 2012. *The Sustainable Fashion Handbook*, London: Thames & Hudson Ltd.
[3] Indonesia S., "*Setali Indonesia, Vermak Jeans Keliling,*" 2019. [Online]. Available: https://www.instagram.com/p/B41-fp9gj6v/. [Accessed 2022].
[4] Citta S., "*Sukkha Citta, Kapas,*" 2022. [Online]. Available: https://www.instagram.com/p/Cc5GtwBvikx/?igshid=YmMyMTA2M2Y=. [Accessed 2022].
[5] Baju T., "*Keriuhan #TukarBaju,*" 2019. [Online]. Available: https://www.instagram.com/p/BxG8TnSAGrf/. [Accessed 2022].

Portable handle push-up bar design: How could plastic waste encourage healthy living?

H. Azhar*, D.A.W. Sintowoko, Akhmadi & P. Viniani
Telkom University, Bandung, Indonesia

M. Bashori
Radboud University, Nijmegen, The Netherlands

*ORCID ID: 0000-0002-8177-482X

ABSTRACT: With the current push to boost plastics recycling rates, the low-hanging fruit of clean mono-streams of plastic garbage has long been picked. If Telkom University's ambitious recycling goal is to be one of Indonesia's Green Campuses, plastic waste streams that have previously been classified as 'problematic' and routinely sent to cremation must also be examined. With the slogan "Save Planet, No Plastic," Yayasan Pendidikan Telkom (Telkom Educational Foundation) increases environmental awareness through a creative campaign. The author was experimenting with plastic rubbish treatment. One such stream is the frying of plastic garbage to create a liquid substance that can be molded into novel products. Following that, the design from recycling technique was effectively applied to developing a new product to promote healthy living, a push-up bar. The plastic-made portable push-up bar design was created efficiently and was beneficial in minimizing stress on the targeted muscle groups and joints.

Keywords: plastic rubbish, push-up bar, healthy living, sustainable design

1 INTRODUCTION

Plastic waste and environmental sustainability are two crucial challenges in large cities. According to the Indonesian Central Statistics Agency (Badan Pusat Statistik 2021), the annual amount of plastic garbage in Indonesia is 66 million tons. 3.2 million metric tons of plastic waste are poured into the ocean. Meanwhile, 10 billion plastic bags, or as much as 85,000 tons, are discarded into the environment each year. Geotimes (2022) estimated that Jakarta's daily waste production was 6,500 tons, of which 13% was plastic waste. On the island of Bali, the amount reached 10,725 tons per day, while in Palembang, it increased dramatically from 700 to 1,200 tons per day. This number places Indonesia in second place, behind China, as the nation with the most plastic trash pollution in the ocean. This number is compounded by imports of plastic garbage from other nations, which reached 320 thousand tons in 2018, a 150% rise from the previous year. The impact will be increased pollution in Indonesia and a decline in environmental quality.

Ecological collapse, air pollution, and deforestation are contributors to global warming, and the solution to these problems is sustainable design. According to Jason F. McLennan (2004), sustainable design is an approach to design that aims to improve our ecosystem while avoiding or eliminating negative impacts. Based on three sustainable design principles, sustainable design refers to human activities that are harmonious with and respectful of nature (Stuart Walker 2006). All of them are in some way connected to sustainable design. As a result, the notion of sustainable design focuses not only on environmentally friendly design but also on the community's social life and the production of economic value or growth in both local and worldwide markets.

Reducing, reusing, and recycling plastic trash is a viable method for reducing environmental pollution caused by plastics. Waste4Change (2021) reports that Indonesia generates 175,000 tons of garbage every day. However, only 7.5% of this waste may be recycled and turned into compost. Up to 10% of garbage is built up, 5% is burned, and 8.5% is not managed. This study will concentrate on the experimental treatment of home plastic garbage, a perennial issue in the environmental sphere. Researchers will cure plastic waste by heating it to transform it into a profitable commodity.

Plastic treatment research is used to produce portable handle push-up bars. Push-ups are an everyday, uncomplicated physical activity. Although push-ups can be performed without equipment, a push-up bar can facilitate the exercise. The benefits include assisting in achieving optimal performance and lowering the risk of wrist damage. The market's abundance of push-up bars may cause confusion among consumers. During this epidemic, individuals are becoming more aware of the need to take better care of their health, which led to the selection of this product (Azhar et al. 2021) (Pambudi et al. 2022). One of the initiatives is to participate in sports at home. The demand for simple exercise equipment, such as push-up bars, is affected by the growing popularity of sports at home. Researchers will attempt to create portable push-up bars from plastic garbage in this investigation.

This research will use PP plastic waste upcycle material to build a portable handle product for the push-up bar in response to the previously discussed problems. The objective is to give environmentally acceptable material options and implement sustainable design principles by decreasing plastic bag waste. This research is also in compliance with the order given by the Telkom Education Foundation (Yayasan Pendidikan Telkom 2019) to hold the socialization of the Save Planet No Plastic program. This assistance is also made available for research funding in the area of applied basic research in the year 2021.

2 RESEARCH METHODS

Qualitative research is used in this study. Finding the roots of social problems in specific communities is one of the goals of qualitative research. For this method to work, several core processes must be in place, such as the ability to ask and gather questions, collect and analyze data from specific individuals, events, or circumstances, and then inductively analyze that data. Based on research goals, this study's results can be different (John W. Creswell 2014).

2.1 *Methods of data collecting*

The data collection methods used by the authors were surveys, observations, and reading archives. The authors first conducted observations and surveys about plastic waste management. This step was held in Bandung, Indonesia, for two months.

2.2 *Methods of data analysis*

The qualitative research method used in this project combines a material analysis and product design. Portable handle push-up bars were heated to a high temperature and then fused to undertake material testing. Because it is the simplest and uses the fewest resources, this approach was selected. Plastic fusing will focus on this study's work to create new materials for upcycled products. Analyses on the following topics will be a study of the method's technical application and the concept of a recycled product design that will be made by fusing pieces of plastic.

3 RESULT AND DISCUSSION

3.1 *Plastic selection*

To prepare for the fusing method of processing rubbish from plastic bags, it is necessary to select the type and quantity of plastic bottle caps to be used. This is the initial step in preparing for the approach. The nature of the product, as well as its appearance and shape, will be affected by the type of plastic bottle caps used, even though the quantity of plastic bags utilized will ultimately affect the product's thickness and overall strength.

Figure 1. Plastic bottle caps selection.

Selecting the type of plastic bottle caps
Avoid using plastic bottle caps that may be disassembled into smaller pieces. It is not advised to use the fusing method with this sort of plastic bottle cap, as the substance within has the potential to crumble more quickly when heated. The PP variety of plastic bottle caps should be selected since it is the most environmentally friendly and because it is the form of plastic bag that is most commonly used and may be found in household garbage. As a result of the material's low melting point, it is simple to manipulate with standard household tools, such as iron.

The number of plastic bottle caps utilized
This project takes somewhere between many distinct plastic bottle caps pieces. When this quantity is utilized, it is possible to obtain satisfactory results regarding the fused plastic bag's strength and thickness. Three to five pieces of thicker plastic bags are anticipated to be required for this project. This amount can generate adequate plastic strength and thickness through the fusing procedure.

3.2 *Plastic fusing procedure*

Heating multiple layers of flexible plastic can make it more rigid. Stiff plastic is created using this process and has a particular feel and coloration reminiscent of vintage items. To begin, gather the tools and materials you'll need: A press, scissors, and a few sheets of trashy plastic are required. Make a rectangle out of the plastic by slicing the top and bottom. After that, use parchment paper or tracing paper to cover the bottom of the base before adding the plastic. Depending on the desired output, the number of plastic layers is adjusted. Between five and seven layers of plastic are ideal. The parchment or tracing paper should then be placed over the top. The plastic should be ironed slowly and uniformly at a high enough temperature to keep it from melting. Fusing produces material that can be used to make new objects.

3.3 *Plastic fusing experiment*

Plastic trash is processed through a heating procedure using a simple instrument heated in an electric microwave. The primary component of the material is plastic bottle tops. This material was chosen due to the quantity of plastic, but it is not easy to recycle into high-quality plastic products. Thus, it does not sell well if sold to plastic trash collectors. This tool was chosen because it is readily available and generates sufficient heat for the material processing procedure. The processing of plastic bottle cap raw material produces attractively textured slabs. As a result of its composition of plastic sheets, the resultant material is also more durable than before.

 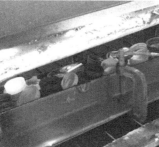 

Figure 2. Plastic bottle caps fusing experiment.

Figure 3. The prototype of portable handle push-up bar.

3.4 *Portable handle push-up bar concept design*

The product design concept must take into account the following criteria:

1. Facilitate the application of plastic fusing procedures so that they are simple to manufacture.
2. Utilize simple, easy-to-obtain, and easy-to-use tools whenever possible to maximize efficiency.

3. Functional product; portable handle push-up bar
4. The product's target market consists of environmentally conscious young individuals who appreciate using repurposed goods.

Figure 4. The concept of a portable handle push-up bar.

4 CONCLUSION

Based on the analysis and design process described previously, it is possible to conclude that research on the design of portable handles for push-up bars made from upcycled PP plastic waste can be conducted. Plastic garbage is problematic for landfills due to its high volume and difficulty in decomposition. Here, the author transforms plastic garbage into a portable push-up bar. Based on the design results, the author constructs a portable push-up bar out of approximately 3 kg of polypropylene (PP) plastic bottle caps. Then there are several other tools, such as iron molds with a thickness of 5 mm and top and bottom covers. The melting point of PP plastic is roughly 220 degrees Celsius; if the temperature is too high, the plastic will burn, and the oil within it will evaporate. During the cooling phase, molten PP plastic will shrink from the size of the mold or bend. Multiple causes can cause PP plastic to contract or deform. The first is the thickness of the molten plastic, and the second is the cooling process that occurs when the plastic is taken from the mold before it has cooled thoroughly. Plastic that has been melted will not be flawless since specific locations contain air and form small holes in the plastic. Several investigations involving plastic seeds, cut bottle caps, and entire bottle caps led to this conclusion.

Based on the design of a portable push-up bar manufactured from plastic waste, the following recommendations can be made for the development of future research: For the creation of molds based on the results of the design, it is best to pay close attention to the size of the product to be developed in order to minimize plastic shrinkage. During the cooling phase, the plastic is left in the mold until it is cooled to prevent it from bending. Last but not least, the enumeration of plastic raw materials is crucial when melting plastic to reduce the amount of incoming air and, therefore, the number of holes in the plastic mold.

ACKNOWLEDGMENT

This article is the result of the Basic and Applied Research 2021, fully funded by PPM Telkom University.

REFERENCES

Azhar, H., Atamtajani, A. S. M., & Andrianto. (2021). Mahogany Fruit Material Exploration for an Essential Oil Nebulizer in the New Normal Adaptation. In *Dynamics of Industrial Revolution 4.0: Digital Technology Transformation and Cultural Evolution* (pp. 193–197). Routledge. https://doi.org/10.1201/9781003193241-36
Badan Pusat Statistik. (2021). *Statistik Lingkungan Hidup Indonesia 2021*. https://www.bps.go.id/publication/2021/11/30/2639657be1e8bd2548469f0f/statistik-lingkungan-hidup-indonesia-2021.html
Geotimes. (2022, April 21). *Indonesia Darurat Sampah*. https://geotimes.id/opini/indonesia-darurat-sampah/
Jason F. McLennan. (2004). *The Philosophy of Sustainable Design: The Future of Architure*. Ecotone Publishing.
John W. Creswell. (2014). *Research Design: Pendekatan Kualitatif, Kuantitatif, dan Mixed*. Pustaka Pelajar.
Pambudi, T. S., Azhar, H., & Andrianto. (2022). Upcycled Design: From Plastic Bag to Bicycle Bag. In *Embracing the Future: Creative Industries for Environment and Advanced Society 5.0 in a Post-Pandemic Era* (pp. 86–90). Routledge. https://doi.org/10.1201/9781003263135-17
Stuart Walker. (2006). *Sustainable by Design: Explorations in Theory and Practice*. Earthscan.
Waste4Change. (2021). *Waste Credit Service by Waste4Change: Why Your Company Should Consider It?* https://waste4change.com/blog/waste-credit-service-by-waste4change-why-your-company-should-consider-it/
Yayasan Pendidikan Telkom. (2019). *Save Planet No Plastic*. https://ypt.or.id/save-planet-no-plastic/

Surround sound system matrix in Reaper for learning module

A.A. Anwar
Telkom University, Bandung, Indonesia

ORCID ID: 0000-0002-4570-0145

ABSTRACT: Surround sound production and processing is limited to the post-production division, to film players such as cinemas and home theaters. We describe the arrangement of the channel flow on the soundtrack in DAW by the matrix method. Surround Sound spatial sound processing is indispensable for the advancement of film production in Indonesia. Previous similar research was in computer programming and electricity at sound physics which only certain circles could understand. Inspired by advances in new media technology such as Dolby Surround, we independently traced the matrix method on the DAW Reaper sample and referenced the tutor to find the simple settings for preparing audio mixing conditions with a 5.1 surround system. We propose a series of contribution sets of channel flow arrangements that lead to the simplification of the audio mixing preparation process. The impact of each contribution on the role of the settings in the study of pedagogy. Qualitative experiments show that our methods significantly outperform learning-based approaches in terms of sound design.

Keywords: learning module, matrix method, Reaper DAW, surround sound system

1 INTRODUCTION

Many studies have explored the benefits of interactive experience technology for society. One of them is the integration of virtual reality and augmented reality technology on various needs in limited fields. However, it is predictable that in the future VR and AR technology will develop and penetrate other fields (Muñoz-Saavedra *et al.* 2020). This immersive technology is getting more and more stretched during the Covid-19 pandemic (Singh *et al.* 2020). The education system is also participating in gradually utilizing this technology (Elmqaddem 2019).

In its representation, several new media technologies in the interaction experience category always try to make themselves feel real. Not only on the element of sight, the element of hearing must also continue to follow and keep pace with technological developments and advances from time to time. The design, reproduction, and application of 3D spatial sound for aural experiences are common when people judge that the media is getting more and more popular. A study shows that managing virtual reality conditions requires good consideration, measurement, and understanding so that users can feel sensory stimuli properly (Huisman *et al.* 2019). This is the basis for our thinking to further research the application of surround sound systems to be understood and developed into science in the form of teaching materials.

Education in Indonesia, even up to university, has not shown much development and deepening in the technical direction of surround sound systems specifically and completely. We took the initiative to explore in research to find a basic guide or learning module that is easy to understand and apply about setting up a surround sound system. Later, this guide can be used by educational institutions or industrial institutions. Of course, research on surround sound systems to reproduce spatial sound requires further research on audio production devices, recording techniques, virtual and augmented reality applications, and audio mixing techniques to create immersive perceptions by representing acoustic environments (soundscapes).

We conducted online observations of the curriculum at several universities in Indonesia to see how far they have deepened their knowledge of the application of surround sound systems, especially for moving image works such as film and animation.

Table 1. The results of a simple survey of the film study program curriculum.

No	University	Study Program	Audio Design Approaching
1	Universitas Negeri Makasar	Film Study Program	filmmaking skills with the general editing process
2	Universitas Padjadjaran	Film and Television study program	content creation and film production in general
3	Universitas Jember	Television and Film Study Program	collecting data for cultural content, and the sound system is specifically for making film music and field recording
4	ISI Yogyakarta	Recording Media Arts Study Program	sound education to sound recordists
5	ISI Denpasar	Film and TV Study Program	knowledge in the film editing process in general
6	ISI Padang Panjang	TV and Film Program	sound systems in films in general
7	Universitas Islam Indonesia	Communication Study Program	general direction of film production
8	Telkom University	Visual Communication Design S1 Study Program	Basic of Audio Design

The sound factor can help the user's presence feel real. The results of the study show that the sensory sound of synchronous footsteps makes listeners feel present in a virtual space (termed Localization) when compared to just listening to the soundscape (Kern & Ellermeier 2020). The development is increasingly widespread in the design of immersive soundscapes by utilizing sound recording equipment (such as a spherical microphone) and 3D audio reproduction (playback) via surround headphones or surround array speakers (Hong et al. 2017). Spatial sound perception has almost the same impact as realistic sound, especially when equipped with VR visual technology (Rajguru et al. 2020). This supports the argument that humans can localize themselves with spatial sound representations.

Spatial sound reproduction can affect perception with localization impacts. The spatial sound reproduction process requires the ability of the sound channel matrix in the DAW software, then passed to the loudspeaker system. The process will be forwarded to the audio mixing stage to prove and measure the "feeling of being in it" or Localization. Thus, research on spatial sound technology requires further research on more advanced technologies. In fact, research on spatial sound has been widely studied. However, in Indonesia, there has not been much research or education on spatial sound recording and reproduction, due to human resource factors. It's not too late to develop that knowledge.

2 RESEARCH METHODS

We investigated a simple set of matrix routing and settings in software. The output of this research is a learning module for setting up a surround sound system on the DAW Reaper.

After finding the setting, we codified the findings of the setting qualitatively through a learning-based approach (Simasiku 2016).

Research on setting and routing matrix in software was carried out in stages based on the hypothesis that setting up a surround sound system is enough to have a multichannel sound card output up to 6 channels and the use of ReaSurround effects on audio source tracks in Reaper. Some of the setting variables of the experiment are stored and collected by archiving. We do a visual study by observing tutorial videos on YouTube (Dabašinskas 2015). The finding of the routing matrix in Reaper is illustrated with a flow chart to facilitate mapping in the design of learning modules which are prepared using a qualitative approach based on sound design learning.

3 RESULT AND DISCUSSION

Today, information and limited access to education are everyone's problems. This causes limited access to broad knowledge to be obtained and studied by young people who are interested in it. It is undeniable that the spearhead of all learning activities in educational institutions, of which is the ability to work and entrepreneurship. Creating a surround sound system requires proficiency in understanding amplification for loudspeaker equipment; installation comprehension skills for connecting computers, sound interfaces, and loudspeakers; and requires an understanding of routing or patching system output channels on DAW software, and the course requires sound manipulation design skills to optimize 3D sound effects.

The surround setup tutorial on video references leads to the audio mixing process for movies. Not only channel patching, sampling rate settings, pan law, default gain, and volume faders on the track control panels are also pre-set, so that surround settings are in decent condition. Sampling rate is a measure of how many audio samples are in bits per second. Pan law is the loudness offset tolerance in decibels (dB). Pan law makes a tolerance for sound accumulation on both stereo channels which if it is in the center position, it will be considered accumulated and pan law will work by reducing the decibels. By Thomas, the pan law is left at the default, which is at +0.0 dB. Even so, the default gain for the process of sending & receiving channels, the default is +0.0 dB. Fader ranges are set at +12, and again, that's Reaper's default project setting. We don't need the master track, so we hide the master track in the mixer view in Reaper.

Surround settings are based on simulating the distribution of sample audio using ReaSurround, then recording motion sound with the automation feature. The steps are:

1. Create a track with the role of the audio track source.
2. Create tracks as master audio channel.
3. Create tracks used for rendering.
4. Create tracks for monitoring when audio mixing, by dividing:
   - Tracks used for high frequency accumulation
   - Tracks used for low frequency accumulation

ReaSurround will be used on the Audio Source Track. The Master Channel Track works to split the audio receive from the Audio Source Track for distribution to other groups of tracks (channel patching). Monitoring Track Grouping is used for the audio mixing process. Before the sound comes out through the audio interface, there is a LFE (Low-Frequency Enhancement) mapping process to balance the mapping on two different types of speakers (full range frequency & low range frequency). This will simplify the matrix routing process on the physical output on the audio interface. While the grouping of rendering tracks is used for the audio mastering process.

The tracking matrix method is the key to the success of mapping audio interface channels in Reaper. The basis for using the matrix is channel patching on the master track. This track is used by receiving three stereo channels of the audio source track and 16 mono sends with the first 6 channels for the rendering track group, 5 channels for the high-frequency range group, and 5 channels for the low-frequency range group. Low-Frequency Range (LFR) tracks will be sent to a track called LFE Total.

The flow becomes simpler with codification, then becomes the basic material for designing learning modules. Codification can increase the ease of understanding and student achievement (Simasiku et al. 2015). So, we conclude that the codification process can make it easier for students or students to understand sound design learning, especially the surround sound system routing matrix.

We codified the routing matrix method for a surround sound system in Figure 1 with a simpler flow in Figure 2. Codification is the process of sorting through to prescribing a standard language usage model (I Gusti Putu Sutarma 2013). The language planning process as proposed by Haugen in this study begins with sorting out several alternative sets of settings and proceeds to codification. The third step is the implementation of the module to learning institutions after going through the standardization process, and the last step is the elaboration of vocabulary for the transformation of foreign terms into general terms that are easy to understand (Haugen 1959).

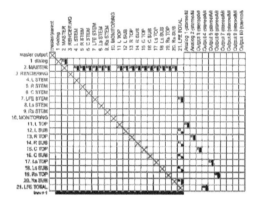

Figure 1. Routing matrix for surround sound system in DAW reaper.

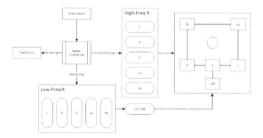

Figure 2. Simplification of the routing matrix after the codification process.

The fourth Sustainable Development Goals ensures inclusive and equitable quality education and promotes lifelong learning opportunities for all. Open access to information does not rule out the possibility for researchers to develop educational science. With the support of the concept of independent learning initiated by the Indonesian Ministry of Education, we take this opportunity to unlock useful knowledge in technical and vocational skills to support decent work and open new fields of entrepreneurship.

4 CONCLUSION

Finding Surround Sound System settings requires a specific understanding of the interface in the DAW itself. It is possible that there will be similarities or differences in matrix routing in other DAWs or in Plugins. This research provides an opportunity to develop sound production readiness for films more deeply. The learning module will be designed in the next stage through several considerations of standardizing the learning module. Some important points that become the main points of designing a learning module for setting up a surround sound system in Reaper are Channel Mastering, Send & Receive Tracks, Grouping of HFR tracks and LFR tracks, and Matrix Routing Channels to the audio interface.

This research produces teaching materials that can be used to provide a 3D sound auditory experience. Science in sound experience is an old subject that becomes a new subject in the presentation of learning media for performances. This module will be able to complement the curriculum or learning in film performance media. It can be a new reference in formulating a curriculum in the field of scientific film performances or other sound shows.

REFERENCES

Dabašinskas, T. (2015). Film Surround Sound Mixing Setup in Reaper (no 3rd party plug-ins). In *YouTube Content Video*. YouTube.

Elmqaddem, N. (2019). Augmented Reality and Virtual Reality in education. Myth or reality? *International Journal of Emerging Technologies in Learning*, *14*(3). https://doi.org/10.3991/ijet.v14i03.9289

Haugen, E. (1959). Planning for a Standard Language in Modern Norway. *Anthropological Linguistics*, *1*(3).

Hong, J. Y., He, J., Lam, B., Gupta, R., & Gan, W. S. (2017). Spatial Audio for Soundscape Design: Recording and Reproduction. *Applied Sciences (Switzerland)*, *7*(6), 1–21. https://doi.org/10.3390/app7060627

Huisman, T., Piechowiak, T., Dau, T., & Macdonald, E. (2019). Audio-visual Sound Localization in Virtual Reality. *Auditory Learning in Biological and Artificial Systems*, 7.

I. Gusti Putu Sutarma, I. K. S. (2013). Penggunaan Bahasa Indonesia di Industri Pariwisata. Studi Kasus Perencanaan Bahasa Pada Industri Pariwisata. *SOSHUM. Jurnal Sosial Dan Humaniora*, *3*(2).

Kern, A. C., & Ellermeier, W. (2020). Audio in VR: Effects of a Soundscape and Movement-Triggered Step Sounds on Presence. *Frontiers in Robotics and AI*, *7*. https://doi.org/10.3389/frobt.2020.00020

Muñoz-Saavedra, L., Miró-Amarante, L., & Domínguez-Morales, M. (2020). Augmented and Virtual Reality Evolution and Future Tendency. *Applied Sciences (Switzerland)*, *10*(1). https://doi.org/10.3390/app10010322

Rajguru, C., Obrist, M., & Memoli, G. (2020). Spatial Soundscapes and Virtual Worlds: Challenges and Opportunities. In *Frontiers in Psychology* (Vol. 11). https://doi.org/10.3389/fpsyg.2020.569056

Simasiku, L. (2016). The Impact of Code Switching on Learners' Participation during Classroom Practice. *Studies in English Language Teaching*, *4*(2). https://doi.org/10.22158/selt.v4n2p157

Simasiku, L., Kasanda, C., & Smit, T. (2015). Can Code Switching Enhance Learners' Academic Achievement? *English Language Teaching*, *8*(2). https://doi.org/10.5539/elt.v8n2p70

Singh, R. P., Javaid, M., Kataria, R., Tyagi, M., Haleem, A., & Suman, R. (2020). Significant Applications of Virtual Reality for COVID-19 Pandemic. *Diabetes and Metabolic Syndrome: Clinical Research and Reviews*, *14*(4). https://doi.org/10.1016/j.dsx.2020.05.011

Succupedia: Augmented reality-driven horticultural book and its potential on becoming a metaverse object

A.P. Budi & A. Harditya
Sampoerna University, Jakarta, Indonesia

ABSTRACT: During the pandemic, the new culture of working from home ignited new interests and hobbies to help entertain people from the monotony of lockdown. Horticultural activity has seen a surge in popularity over the past year. Unsurprisingly, horticultural activity has indeed helped people to stay physically and mentally well. The wave of new horticulturalists has led to a growth in sales of house plants, especially succulents. Unfortunately, many plants died due to a struggle in finding proper care. Therefore, it is necessary to provide education and guidance on succulent plant cultivation through the *Succupedia* book. *Succupedia* is an augmented reality-driven (AR) book that aims to provide knowledge about succulents for beginners. In parallel, the pandemic has left the creative industry devastated to seek new sources of income, as many projects and events are cancelled, and the discovery of Non-Fungible Token (NFT) is becoming popular. Conceptually, this article extends how AR objects inspired by horticultural activity could potentially become an NFT object inside of the Metaverse as meta-species.

Keywords: Augmented Reality, Metaverse, NFT, Horticultural, New Media

1 INTRODUCTION

On January 30, 2020, the World Health Organization (WHO) declared 2019-nCOV to be a Public Health Emergency of International Concern (PHEIC). Since the first case of COVID-19 in Indonesia was announced in early March 2020, policymakers and ministers have made various proposals and suggested regulations for fighting against the pandemic (Indonesia 2020). Due to these physical and social distancing rules, people spend 80–90% of their time at home. All the outdoor activity has been restricted and people have more free time in their house. In these times, many people are returning to do their hobbies or trying some new things to escape from their repetitive routines. Not surprisingly, throughout these challenging times, gardening has indeed helped people to stay physically and mentally healthy as well as to relieve stress. This phenomenon brings up a wave of new gardeners that has led to a growth in sales of house plants, although succulents within the immediate span of time they do not last long. A survey found that $\frac{1}{3}$ of new plant owners struggle to find proper care. It is necessary to provide education and guidance on succulent care because during challenging times since there is an increase in economy in this area. On that account, we must sustain the positive horticultural market reaction (Threes.com 2022). In parallel, the pandemic has left the Indonesian creative industry devastated to seek new sources of income. In an interview with Adi Wibowo better known in the industry as Blak, founder of a multimedia service company based in Jakarta, claims that his company struggles to survive during the pandemic as 80% as he admitted that the "direct network of most projects is offline events where physical distancing is a disadvantage." Four months later he experimented on NFT production to be sold in the Metaverse (Wibowo 2022). There are four types of Metaverse, it consists of Virtual Worlds, Augmented Reality, Mirror Worlds and Lifelogging. According to Blak, "there are differences between NFT and Metaverse, NFT is a trading object regardless of the Metaverse, because NFT belongs in the marketplace." On the other hand, Metaverse cannot be a trading space without an NFT, currently it is explored as a social space where variety of communities socialize. Extensively, this article critically analyses the connection of horticultural plants in the physical world and how AR objects inspired by it could potentially become an NFT object inside of the Metaverse as meta-species. Finally, it will become clearer that the purpose of AR objects may have extended the engagement or immersion of the book, so we have clearer view of the purpose of *mixed reality* production (Grasset *et al.* 2008). The purpose of this is that the article tries to discover the similar specificity of AR and Metaverse object. Thus, the article raises questions as follow:

Research Question	Objectives
Could AR objects transform and improve textual educational content delivery?	Improve textual educational content delivery.
Will printed material shift its role in graphic design production as a delivery medium?	Evaluate printed material in graphic design production as a delivery medium.
Could Mirror World's Metaverse be the most effective simulation for plant education?	Apply Mirror World's Metaverse simulation in plant education.
How can we design real-life plants into an engaging meta-species that meets the purpose of the Metaverse?	The purpose of metaverse object is to be part of a collection. Therefore, plants need to fit that category.

2 RESEARCH METHODS

This project will apply design-based research (DBR) methodology. This project will refer to the DBR process by Siko and Barbour. The process consists of analysis and exploration; design and construction; evaluation and reflection. It is important to analyze the reasons for failure and try another technique to fix it (Armstrong *et al.*, n.d.). Additionally, this article uses qualitative methods in interviewing direct sources from the creative industry to gain insight on actual industry post-pandemic situation as well as their experience on exploring NFT as an alternative revenue and Metaverse as an alternative space. The first step is editorial design, the process of arranging visual and textual elements to catch reader's attention and communicate information. Regardless of the medium, understanding the effect layout has on your audience's visual perception is crucial. As shown in Figure 1, the main layout for

Figure 1. Layout for succupedia.

this book is divided into two sections. The first section on the left is full illustration with contrast color. Later the illustration will be used as a Natural Feature Tracking (AR-NFT); not to be confused with Non-Fungible Token (NFT). To make the illustration become an AR-NFT, the .jpg or .png format illustration needs to be converted into .iset, .fset, and. fset3 formats (Vate-U-Lan 2012). While the layout on the right side is the description about each succulent type.

Secondly, after the book's layout is done, proceed to test print the material for the book. To make the printed AR-NFT easier to read by the programs, the paper selection is based on several criteria such as brightness, whiteness, and finish. Brightness defines the optical properties while Whiteness defines the reflective properties of the paper across the visible light spectrum. The paper finish is the look and feel of its surface and it is achieved during the paper-making process (Wayne *et al.* 2016). *Everyday* and *Splendorgel* paper types meet all the criteria. However, the book proceeds with *Splendorgel* since it has a smoother texture

Figure 2. User interface of 3D assets.

than *Everyday* paper type. Created the 3D assets for augmented reality projects exported as .gltf / .glb format this includes the user interface as shown in Figure 2 below, which consists of instructions of the succulent care, which is the primary interface elements superimposed in the real world (Billinghurst *et al.* 2001). The AR system is created using AR.js, an open-source JavaScript library for website-based augmented reality. The purpose of using this is to make it easier for ease of AR experience. It is an unnecessary experience to use specific devices or using third-party applications when viewing AR. AR.js will define a specific 3D object for a specific marker. So, when a camera recognizes a marker, the website will show a 3D model on the top of it. Therefore, there will be pages for each succulent type on the website.

3 RESULT AND DISCUSSION

3.1 *Improve textual educational content delivery.*

Below is the result of the book *Succupedia* and how it works as shown in Figure 3 and 4. The audience can use their mobile phone to scan the barcode for the website first. Then, point the camera at the illustration. After a few seconds, the 3D model will appear on the surface of

Figure 3.   Succupedia book.   Figure 4.   QR code for video result.   Figure 5.   Succupedia as AR Object

the book. Figure 5 shows that with the AR feature in the Succupedia book, the audience not only reads the text but can also see the 3d shape of each plant and how to care for it. This provides audience to experience immersive experiential learning (Ha *et al.* 2011).

3.2 *Evaluate printed material in graphic design production as a delivery medium*

Based on the experiment, AR on printed material is two times more responsive (1.50 seconds) than using a screen (2.72 seconds). The detected picture on the screen is less accurate since the screen is very reflective. There were many issues found; one of many is that only two AR-NFTs out of eight managed to visualize the 3D object clearly. This is because AR-NFT works well when the image has high complexity and colour saturation, which results a more accurate marker detection. Second, one out of eight 3D models cannot be uploaded into the website due to excessive polygon counts that affects the gltf / glb file size. Thus, it is recommended to simplify the 3D object since the maximum size for uploading 3D model is only 25Mb.

3.3 *Apply Mirror World's Metaverse simulation in plant education*

In parallel with the book, we have expanded the *Succupedia* by creating a Mirror World's Metaverse simulation. The simulation is created by using *Decentraland* SDK, an open source metaverse development kit provided by *Decentraland* Metaverse. Within this mirror world, there is an area where there is every plant in the *Succupedia*, and players can freely walk from one to another. By implementing mirror world's simulation, *Succupedia* is hoped to have education value for new gardeners to learn how to care for houseplants in both physical or virtual world, and in the end can reduce the risk of dead plants. While in the previous section Figure 5 shows how Succupedia plants are previewed as AR objects and Figure 6 and 7 below shows Succupedia plants in the metaverse.

Figure 6 & 7.   Succupedia in Mirror World's Metaverse simulation.

3.4 Metaverse object is to be part of a collection.

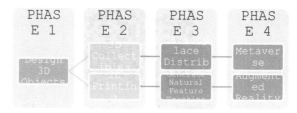

Figure 8. Succupedia in Mirror World's Metaverse simulation.

As seen on Figure 8, the production phase begins with the standard 3D design process, no specific instructions, however, how we distribute the 3D objects, as shown in Phase 2, that needs to be highlighted. The distribution process requires significant number of designers, due to the magnitude of the metaverse space. *Succupedia* can be implemented as NFT collectibles in the marketplace and evidently sufficient as Metaverse objects. Below is the analysis summary for the measured objectives:

Objectives	Effectivity Measurement	Remarks
Improve textual educational content delivery	Yes	AR objects provide immersive experiential learning that can transform textual educational content into object augmentation.
Evaluate printed material in graphic design production as a delivery medium	Yes	The result shows that the implementation of augmented reality features in a printed book is efficient in certain types of paper. One out of four papers are qualified for criteria such as reflectivity, flexibility, color, and texture.
Apply Mirror World's Metaverse simulation in plant education	Yes	Mirror World's Metaverse simulation can be a way for the new gardeners to learn how to care for houseplants, so that later it can reduce the risk of dead plants.
The purpose of metaverse object is to be part of a collection. Therefore, plants need to fit that category	Yes	The effectiveness of this is still arguable because plants have many varieties, and it is entertaining and educating to have plants as collectible items in Metaverse.

4 CONCLUSION

In conclusion, digitalization has brought numerous innovative and sophisticated technologies; this replaces the old technologies, such as the printing industry. Although there were many issues during the process, this *Succupedia* project proved that the implementation of augmented reality features in a printed book is efficient. Augmented reality can make a significant contribution, especially to the education field as well as the Metaverse. Therefore, this method can be applied as a learning tool to deliver teaching materials and for metaverse entertainment. Since the implementation has an extremely high potential that using this method can revive the printing industry in this digital era and develop a greater practical concept of Mirror Worlds' Metaverse.

REFERENCES

Armstrong, M., Dopp, C., & Welsh, J. (n.d.). The Iterative Process of Design-based Research. *Analysis and Exploration*. In *Pressbooks*.

Billinghurst, M., Kato, H., & Poupyrev, I. (2001). The MagicBook: A Transitional AR Interface. *Computers & Graphics*, 25(5), 745–753. https://doi.org/10.1016/S0097-8493(01)00117-0

Grasset, R., Dünser, A., & Billinghurst, M. (2008). Edutainment with a mixed reality book. *Proceedings of the 2008 International Conference in Advances on Computer Entertainment Technology – ACE '08*, 292. https://doi.org/10.1145/1501750.1501819

Ha, T., Lee, Y., & Woo, W. (2011). Digilog Book for Temple Bell Tolling Experience Based on Interactive Augmented Reality. *Virtual Reality*, 15(4), 295–309. https://doi.org/10.1007/s10055-010-0164-8

Indonesia, K. K. R. (2020). *Pedoman Pencegahan dan Pengendalian Coronavirus Disease (COVID-19)*.

Threes.com. (2022). *Gen Z Bought Houseplants to Improve Mental Health During Pandemic; Stress and Dead Plants Result Instead*.

Vate-U-Lan, P. (2012). An Augmented Reality 3D Pop-Up Book: The Development of a Multimedia Project for English Language Teaching. *2012 IEEE International Conference on Multimedia and Expo*, 890–895. https://doi.org/10.1109/ICME.2012.79

Wayne, C., Hass, A., Jeffrey, K., Martin, A., Medeiros, R., & Tomljanovic, S. (2016). *Graphic Design and Print Production Fundamentals*. Graphic Communications Open Textbook Collective.

Wibowo, A. (2022). *How the Creative Industry Shifted Towards NFT and Metaverse During Pandemics*.

The effect of digital marketing through social media is increasing online consumer trust in a pandemic period

Christian & F.V. Kurniawan
Universitas Ciputra, Surabaya, Indonesia

ORCID ID: 0000-0002-8688-6361

ABSTRACT: Technology is transforming the world quickly, making life easier for those living in it. Because of technology, people use social media to connect with others and conduct long-distance communication more swiftly, especially when an outbreak of the COVID-19 virus occurred. They must adapt their way of life to live entirely online and virtually. Social media, which at first served as a venue or platform for socializing only, is today a platform utilized for a variety of purposes, including carrying out marketing. The goal of this research is to determine and examine the role that social media plays in digital marketing's ability to boost customer trust online, especially during the pandemic. A qualitative approach is used in the research procedure, literature review along with case studies. According to the study's findings, social media marketing on the internet during the epidemic affected consumers' perceptions toward brands.

Keywords: consumer trust, digital marketing, pandemic, social media

1 INTRODUCTION

The world has changed due to ongoing advancement and development. When looking at the past few years, there have been a lot of changes, from the growth of the infrastructure, economy, society, and culture to the way of the people who live there. Additionally, since the advancement of technology, communication media, and information swiftly held and accessed by everyone in the world, the world has been moving faster. The development of social media and the internet can benefit people, especially in terms of quick information access, interaction, and distance communication.

Social media refers to some internet-based apps that leverage Web 2.0 technology that can be used to exchange and distribute information; (Untari *et al.* 2018). Users of social media can use and benefit from a variety of features. A Hootsuite analysis shows that 57.5% of individuals use social media to pass the time, followed by 58% of those who use it to interact with friends and family. 50% of people are looking for activities or items to purchase. Because of the opportunities, it has provided, people now use social media regularly. Social media has evolved from its original purpose of being a place for people to engage virtually and share some of their moments to one where people may get as much information as they can. Only by what is presented via social media can people learn about a person, group, agency, and others.

Everyone has access to social media because of its usability and accessibility. Due to this openness, an increasing number of people are developing social media platforms to suit their particular requirements. People's behaviors, lifestyles, and propensity for consumption may alter as a result of the intensifying usage of social media (Frederick *et al.* 2021). Through social media, many businesses provide their brands, goods, and services online. Every company competes to present digital marketing content that can attract as many target audiences as possible. New businesses focus exclusively on developing interesting social media to raise public awareness of their company. They overlooked the requirement that the provided material likewise be intriguing, unique, and leave a good impression on the community.

The COVID-19 virus pandemic struck people all over the world in 2019. All physical interactions are limited, so they cannot carry out activities together or even meet each other. People must also spend most of their time at home in order to decrease the spread of the COVID-19 virus and the number of fatalities it causes, everyday tasks and activities have been replaced by online and virtual ones. People's mobility is reduced as a result of this desperate situation, and they are compelled to adapt to new things and lifestyles. As a result, the already high use of social media is indirectly experiencing a quick increase once more. Both domestically and internationally, these operations are carried out; for instance, purchasing goods from overseas is now easier.

According to a Hootsuite and social analysis from February 2022 (Figure 1), Instagram, Facebook, and TikTok are the most popular social media platforms in Indonesia. Social media is used regularly by 68.9% of Indonesia's overall population, and the number of social media users who are active has climbed by 12.6% since last year, which shows that social media has a lot of potential for usage in connecting brands with their intended audience. A change in these kinds of circumstances can enhance a brand's consumer-facing digital marketing approach.

In light of the economic downturn brought on by COVID-19, this information may present a chance for businesses to gradually find a means to survive and contribute to the achievement of Sustainable Development Goal number 8: decent work and economic growth. Based on sources and case studies that serve as references, this journal will illustrate how the impact of digital marketing via social media can boost consumer trust, particularly during the pandemic.

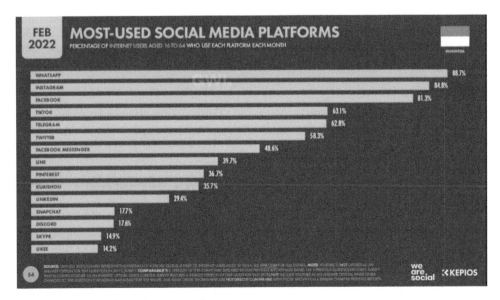

Figure 1. Most-used social media platforms.
Sumber: https://datareportal.com/reports/digital-2022-indonesia

2 RESEARCH METHODS

The research method used in this study is a qualitative approach in the form of literature data from several journals and books, both locally-based and internationally based, that is credible and relevant to the topics discussed. Additionally, research is conducted using case studies and some of the findings from earlier studies that corroborate the journal's proposed contents.

3 RESULT AND DISCUSSION

The COVID-19 pandemic has brought about a complete halt to all human activity, including daily activities, education, and even economic operations. The economic downturn was brought on by people being forced to stay at home; numerous businesses closing as a result of the inability to maintain business activity; the disappearance of brick-and-mortar stores. The one who can embrace new technology as their tool will survive. Social media cannot be dissociated from every person in the modern world. Every person has multiple social media accounts for a variety of purposes.

Every social media platform currently in use has a target demographic, including users of various ages, professions, behaviors, and interests. Today, brands and entrepreneurs use social media as a channel for selling their goods and services. Particularly during a pandemic when individuals spend a lot of time on social media, social media is one of the most effective marketing tools available. One of the most popular social media platforms for this purpose is Instagram. Hootsuite reports that there are approximately 156 new posts from business accounts on Instagram per day, demonstrating that social media is the most effective platform for boosting online consumer trust in a brand's product or service.

Effective branding must be done before engaging in further social media marketing and promotion so that consumers are aware of the brand image they wish to project. A brand needs to be able to establish a special and gratifying connection with its customers. The company's branding strategy can have a big impact on how well-known the brand is and how it is perceived when first enters the market and can create a close bond between the brand and its intended audience (Connolly 2020). One of the key tactics for boosting marketing is branding, which also serves as a differentiator and distinctiveness between one brand and its rivals. Online consumer confidence can also rise with effective social media branding. Customer loyalty to a business is essential for boosting online sales, and customers will continue to follow this loyalty while making purchases.

Digital marketing strategy

In general, UGC (i.e., user-generated content) and FGC (i.e., firm-generated content) are the two types of marketing strategies that are frequently employed in the realm of digital marketing through social media (Heng Wei *et al.* 2022). The digital marketing strategy through social media can boost the trust of online customers if these two items are coupled and have a comprehensive planning procedure. UGC refers to marketing that is done directly or indirectly by consumers, such as when they leave comments, reviews, and opinions about the goods or services a company provides.

With the help of this user-generated content, consumers can feel more confident about the value and number of goods and services being provided in the real world. For instance, many people choose to use or purchase a good or service after reading positive reviews from other customers who have already seen or used it, real goods or services. FGC, on the other hand, is advertising developed by businesses or brands to forge strong bonds with their target market and customers in order to increase consumer familiarity with and interest in their goods and services. FGC can cover company services, promotions, and the introduction of new products or services.

Firm-Generated Content (FGC) marketing

Companies or brands that use FGC as part of their digital marketing strategy through social media must undergo a thorough planning process from scratch before they can begin marketing (Lee & Kotler 2020). This is also consistent with the study by Ardiansah & Maharani on efficient social media marketing tactics (2020). Many things can be done, including researching the priorities of the client or the intended target audience, establishing a clear focus and goals, examining competitors, producing relevant and appealing content, choosing consistent and effective posting times, and utilizing digital advertising.

Businesses or brands must explore and utilize the digital characteristics offered by social media. Different characteristics must be employed correctly and in accordance with the goals to be met. The Instagram Reels tool, for instance, is used to upload little videos that can draw viewers in a matter of seconds. Businesses must think about what features and social media will be utilized to win over customers' trust before implementing the Instagram Reels function, which is similar in appearance and operation to the TikTok application's core capabilities. Therefore, marketers must carefully undertake market research and comprehend the brand persona of their target audience in order to avoid mistakes when employing existing features and social media.

In addition to utilizing features other than those offered by social media platforms themselves, businesses or brands must produce original material to pique the interest of consumers. The material must be compelling, pertinent, and capable of boosting consumer confidence in the associated brand. Information on brands, benefits of goods or services, images of goods or services, customer testimonials, or interactive content involving consumers directly can be considered forms of content. Audiences and consumers may be left with a lasting impression via interactive material.

The growing use of social media will make it more difficult for brands to immediately reach all of their target customers. Almost all social media platforms offer paid digital advertising tools that help brands and businesses more quickly contact their target audiences. Digital advertising is a form of advertising that disseminates information about goods and services made available to the general public online. This digital advertisement serves as a channel for fostering relationships and communication between brands and consumers in addition to showcasing the goods and services the brand offers (Pratomo 2020). The digital advertising tool will rapidly present its audience with relevant brand advertisements or target particular communities directly. For instance, advertising for a company or brand in the food and beverage sector will frequently show up to individuals who are interested in adjacent industries.

User-Generated Content (UGC) marketing

Harahap et al. (2020) claimed that social media is a free platform where numerous individuals come together, interact, and share information. In this instance, UGC is crucial to the marketing strategy's success in addition to FGC. Social media users not only appreciate the content that is shared, but they also unintentionally help a business sell itself (Tuten & Solomon 2018). The audience can be directly involved in participating in interactive activities held on social media, such as taking challenges or just leaving their opinions and thoughts about the products or services offered, by utilizing the freedom of social media combined with exciting and interactive content. This may raise brand recognition, brand trust, and brand image among online consumers. People are more likely to believe what consumers have to say when there are more positive reviews and promotions made on social media by customers.

Case study: Scarlett whitening

According to a case study on Scarlett Whitening on the role of digital marketing on local beauty companies undertaken by Digdowiseiso et al. in 2021, digital marketing through social media can have a large and favorable impact on consumer trust. As is well known, Scarlett Whitening is a local beauty company that not only has high-quality products but also strongly focuses on social media promotion. In order to provide a sophisticated customer experience and customer engagement, Scarlett Whitening engages in numerous online exchanges. As part of its marketing strategy, Scarlett Whitening also works with well-known singers, including K-Pop stars like Song Joong Ki, Twice, and others to serve as their brand ambassadors. This may draw in their customers, who are mainly female. On all the social media platforms they use, Scarlett Whitening combines all the components and facets of effective digital marketing. Online customers become aware of the Scarlett Whitening brand as a result. Customers are relying more and more on the content, reviews, and other materials they share on social media to form their opinions of the products they sell.

4 CONCLUSION

If used properly, social media can allow businesses affected by the COVID-19 pandemic to survive and even develop economically, giving the stakeholder a decent job and income (SDG 8). It is also the most efficient way for brands to sell their goods or services during a pandemic. These two marketing techniques can help brands gain more online customer trust: FGC, which begins with businesses that introduce brands to consumers through engaging and interactive content, collaboration, and use of UGC, where consumers themselves will contribute to the process of brand marketing through sincere reviews, comments, likes, and shares, among other methods. Marketing plans that are implemented through thorough and mature processes can benefit from the social media platforms and features already in place and apply marketing strategy guidelines in a correct FGC and UGC way, hence increasing online consumer confidence.

REFERENCES

Ardiansah, I., & Maharani, A. (2020). *Optimalisasi Instagram Sebagai Media Marketing: Potret Penggunaan Instagram Sebagai Media Pemasaran Online pada Industri UKM*. CV Cendikia Press.

Connolly, B. (2020). *Digital Trust: Social Media Strategies to Increase Trust and Engage Customers*. Bloomsbury Publishing.

Digdowiseiso, K., Lestari, R., & An'nisa, B. (2021). The Effects of Brand Ambassador, Digital Marketing, and Instagram Use on Brand Trust: A Case Study of Scarlett Whitening Product. *Budapest International Research and Critics Institute (BIRCI-Journal): Humanities and Social Sciences, 4*(4), 12027–12033. https://doi.org/10.33258/birci.v4i4.3268

Febriani, N., & Dewi, W. W. A. (2019). Perilaku Konsumen di Era Digital: Beserta Studi Kasus. In *Journal of Chemical Information and Modeling* (Vol. 53, Issue 9). UB Press.

Frederick, B., & Krisna Maharani, A. (2021). EKSISTENSI MEDIA SOSIAL PADA MASA PANDEMI COVID-19. *Jurnal Penelitian Pendidikan Sosial Humaniora, 6*(2), 75–83.

Heng Wei, L., Chuan Huat, O., & Arumugam, P. v. (2022). Social Media Communication with Intensified Pandemic Fears: Evaluating The Relative Impact of User- and Firm-Generated Content on Brand Loyalty. *Asia-Pacific Journal of Business Administration*. https://doi.org/10.1108/APJBA-07-2021-0319

Kemp, S. (2022, February 15). *Digital 2022: Indonesia — DataReportal – Global Digital Insights*. https://datareportal.com/reports/digital-2022-indonesia

Lee, N. R., & Kotler, P. (2020). *Social Marketing: Behavior Change for Social Good* (6th ed.). SAGE Publications, Inc.

Pratomo, E. R. (2020). The use of Advertising and Social Media in Today's Teenage Lifestyle. *Visual Communication Design Journal, 5*(1), 35–45. https://doi.org/https://doi.org/10.37715/vcd.v5i1.2683

Tuten, T. L., & Solomon, M. R. (2018). *Social Media Marketing*. SAGE Publication Ltd.

Untari, D., & Fajariana, D. E. (2018). Strategi Pemasaran Melalui Media Sosial Instagram (Studi Deskriptif Pada Akun @Subur_Batik). *Widya Cipta, 2*(2), 271–278.

The application of contemporary artistic theme in Aloft Jakarta Wahid Hasyim Hotel

T. Sarihati, M.A. Alfarizy, R. Firmansyah, A.N.S. Gunawan & D.G. Dijiwa
Telkom University, Bandung, Indonesia

ORCID ID: 0000-0002-9107-8932

ABSTRACT: As time transforms, hotels that initially provided only lodging facilities continue to innovate and now start to provide many other supporting facilities. Also, today's hotels have different target markets of young millennials who have up-to-date taste that is commonly known in design language as contemporary. In addition, young millennials also like things with artistic value. Many artistic works can be found today, such as mural which is a wall painting and can be done both manually and digitally. By taking the advantage of millennial youth phenomenon's tastes, it is an opportunity for hotels to apply this theme concept in their interior designs, especially those that are having young people as their main target. This study aims to examine the application of contemporary artistic themes in one of the hotels that have young millennials as their target market, named Aloft Hotels, which is located in Central Jakarta.

Keywords: Aloft, artistic, contemporary, hotel, interior

1 INTRODUCTION

A hotel is a commercial accommodation that provides lodging as their main facility along with other supporting activities, such as restaurants, spas, fitness, etc. According to Ikhsan (2008:2), a hotel is an institution that provides guests to settle, where everyone can settle, eat, drink, and enjoy other facilities by making payment accordingly. Hotels have various types of target markets, ranging from tourists, business travelers, and families to young people. For now, there are still not too many hotels that provide facilities and themes that young millennials enjoy.

Millennial youths are people who were born between 1983 and 2001. Nowadays, millennial youths tend to like things in the form of unique exploration, decorative, more colorful, and don't want to look old-fashioned, commonly known as contemporary. In addition, millennial youths are now more sensitive and tend to like things involving artistic value, or what is commonly known as artistic in terms of art appreciation.

Aloft Hotels is one of the hotels that have millennial youth as the target market, supported with facilities that millennial youths tend to enjoy facilities such as a bar and live music, pet-friendly hotel, camping facilities, area for children, etc. This hotel adapted music and technology as their inspiration and also contemporary artistic as their interior concept to attract millennial youth visitors who are very fond of the enrichment.

2 RESEARCH METHODS

This research is using the qualitative method by using interviews, observations, and internet studies to collect data. Observations and interviews were conducted in person at one of the

Aloft Hotels branches, Aloft Jakarta Wahid Hasyim, that is located at Jalan Wahid Hasyim St No.92, Kebon Sirih, Menteng, Jakarta Pusat. Interviews were conducted with Aloft Jakarta Wahid Hasyim's Human Resource Development to obtain data related to the hotel's vision and mission, available facilities, as well as advantages and attractiveness of Aloft Hotels. Observations were made by documenting and observing the hotel's interior and the application of its themes and concepts. The literature review was conducted by doing an online search on the official Aloft Hotels website for additional information and images as research data.

3 RESULT AND DISCUSSION

3.1 *Contemporary meaning*

Contemporary interior has the meaning of room styling that always follows the trend. This style began to develop in the 1920s and was introduced by an architect at the Bauhaus School of Design, in Germany as an advancement of technology. This contemporary design tends to be more innovative, complex, varied, and flexible. According to Ardiansyah, S.T, M.T. (2017), usually, contemporary interiors have the characteristics of a more open room with harmonization between interior and exterior, and cultivation of geometric shapes with a clean impression. The interior elements usually use straight line details as well as the use of large windows and a neutral color palette interspersed with cheerful colors as the focal point.

3.2 *Artistic meaning*

Artistic has a meaning as an artistic value that can fulfill the satisfaction of human's beauty awareness. Artistic can be formed as two-dimensional art in the form of dimensions, such as mural art, painting, graphics, digital art, or can also be formed in three-dimensional forms such as sculptures, crafts, etc. One of the artistic works of art is mural which is a two-dimensional work of art using walls as the main medium. Mural art is made by painting directly on the wall or using the help of digital technology. Behind every mural lies the artistic meanings that the artist wants to convey. Artistic murals are also often found in today's interiors.

3.3 *Application of contemporary artistic in Aloft Jakarta Wahid Hasyim*

Aloft Hotel is a hotel that uses the theme of contemporary artistic concepts in its interior as their characteristic is also meant to attract young millennials.

3.3.1 *Lobby and receptionist*

A contemporary concept theme is being applied by having a round receptionist table formation, geometric interior elements on the upper wall, as well as using natural color

Figure 1. Lobby and receptionist.

tones such as gray, beige, and red on the receptionist table as the focal point of the room. As for the artistic aspect, it is applied to a typical digital Betawi mural that is applied to the upper wall of the room. This area uses gray ceramic material, beige wood patterned HPL walls, and decorative wall elements using hollow iron.

3.3.2 *W XYZ bar & live music*

In the bar and live music area, the hotel applies contemporary artistic concept themes as their selection of furniture shapes and interior elements are egg chairs, sofas, geometric partitions, and neon lights. As for the artistic part, it is applied to the walls decorated with hollow iron and digital mural images that contains Indonesia in it. As for interior materials, the floor is the same as the lobby area, which is gray ceramic with an increase in the floor level using a parquet. Walls are used as partitions, beige wood patterned HPL, cement walls, and exposed ceilings. The color tone in this room itself uses natural tones such as gray and cream, as well as several focal point colors such as red, yellow, and pink lights.

Figure 2. W XYZ Bar & live music.

3.3.3 *Re: Fuel*

Re: Fuel is a snack corner area in the form of a pantry that operates 24 hours with a self-service system. This area applies the theme of an artistic contemporary concept as its box-shaped dining area formation and the selection of beige color with pink as the focal point.

Figure 3. Re: Fuel.

From the artistic aspect, it is applied to digital painting as pictures on the walls. This area uses the same floor material as the lobby area, that is gray ceramic floors, walls, and ceilings that are blended using beige HPL.

3.3.4 *Meeting room*

The meeting room in this hotel has two forms that are quite contrasting. One of the rooms applies a contemporary concept with a U-shaped table, with beige and pink focal points in the presentation area. The ceiling in this room is designed in a linear pattern with carpeted floor materials. As for the second meeting room, the furniture does not reflect contemporary because it is more suitable for classic furniture with a round table layout that is spread out like a ballroom area; however, the interior elements are the same as the previous meeting room and the only differences is in the size of the room.

Figure 4.   Meeting room.

3.3.5 *Aloft guest room*

Aloft guest rooms consist of two types: twin and queen; however still have the same space formation. A contemporary artistic concept theme is applied to the room color selection of gray, beige, and focal points on the red-blue bench with contemporary shapes, also there is a digital mural with Betawi cultural values on the headboard of the mattress. For interior materials, this room uses carpet floors, beige HPL walls, gray paint, glass, and gypsum

Figure 5.   Aloft guest room.

ceilings. Moreover, the lighting of this room is completely indirect. Rooms in Aloft Jakarta Wahid Hasyim have no support facilities for guests to bring pets like other branches, but this hotel supports children's camps in rooms that can be requested at the reception area.

3.3.6 *Nook restaurant*

The restaurant at Aloft Hotels doesn't look too contemporary or artistic but looks more modern and elegant due to the simple and clean furniture that is supported by warm white lighting that creates an elegant impression. There is no artistic aspect to this hotel restaurant because there are no interior elements filled with murals or similar artistic images. This area has a hot neutral color tone supported by the gray and yellow color that dominates the room. For interior elements, it is being used floor carpets, gray walls and partitions, and white-colored down-ceilings. For this restaurant system, it uses a buffet dining system.

Figure 6. Nook restaurant.

3.4 *Analysis*

This hotel uses the application of contemporary artistic concept themes in every room. The contemporary aspect is very clear from the selection of varied furniture and exploration of geometric shapes to the use of cheerful colors as the focal point of a neutral room. The interior elements are also played with elements of geometric shapes to support a contemporary impression, some walls are also decorated with digital murals that have Betawi cultural values or the City of Jakarta. But some rooms still don't look artistically contemporary, such as one of the meeting room areas where furniture looks more classic, and a restaurant that looks more modern and elegant in terms of furniture or the play of colors and interior elements. Hence, it would be better if the meeting rooms and restaurants were designed in accordance with the theme of the contemporary artistic hotel concept.

4 CONCLUSION

Aloft Jakarta Wahid Hasyim is one of the hotel choices for millennials because it has facilities and contemporary artistic styling themes which are very popular because it follows the tastes of today's millennial youth. The supporting facilities at this hotel are the main attraction of the hotel, especially in the bar area and live music which looks very contemporary and artistic so that it can add to the atmosphere when guests are doing activities in the bar area.

REFERENCES

Akromullah, H. 2017. *Arti Dalam Nilai Seni.* 9(1), 15–27.

Budiati, I., Susianto, Y., Adi, W. P., Ayuni, S., Reagan, H. A., Larasaty, P., Setiyawati, N., Pratiwi, A. I., & Saputri, V. G. (2018). *Profil Generasi Milenial Indonesia.* 1–153. www.freepik.com

Hachim, S. I. (2021). Rimak International Journal of Humanities and Social Sciences. *RIMAK International Journal of Humanities and Social Sciences, 3*(1), 186–199. https://doi.org/10.47832/2717-8293.1-3.15

Istiawan, S. 2006. *Ruang Artistik Dengan Pencahayaan.* Jakarta: Griya Kreasi.

Putri, C. A., & Wahyudie, P. (2016). Studi Selera Generasi Muda Usia 18-35 Tahun terhadap Dekorasi Hotel di Jawa Timur. *Jurnal Desain Interior, 1*(1), 51. https://doi.org/10.12962/j12345678.v1i1.1467

Pratama, D., Tulistyantoro, L., & Mulyono, H. (2019). Perancangan Interior Restoran & Kafe Dengan Konsep Kontemporer Tradisional Khas Lombok di Lombok. *Jurnal INTRA, 7*(2), 781–792.

Obed Bima Wicandra dan Sophia Novita Angkadjaja. (2005). Efek Ekologi Visual Dan Sosio Kultural Melalui Graffiti Artistik Di Surabaya. *Nirmana, 7*(2), 99–108. http://puslit2.petra.ac.id/ejournal/index.php/dkv/article/view/16515

Studi, P., Arsitektur, T., Universitas, T., & Global, I. (n.d.). *Ardiansyah, s.t, m.t.*

Surakarta, K. D. I. (2018). *Desain Interior Art Space Dengan Konsep Seni.*

Susanto, D., Angelia, D. P., & Ningsih, T. A. (2018). Local Material as a Character of Contemporary Interior Design in Indonesia. *IOP Conference Series: Earth and Environmental Science, 99*(1). https://doi.org/10.1088/1755-1315/99/1/012021

The application of the philosophy and technical style of Batuan Bali painting to the modern narratives and process

D.K. Aditya* & I.D.A. Dwija Putra
Telkom University, Bandung, Indonesia

*ORCID ID: 0000-0002-2773-6814

ABSTRACT: The COVID-19 pandemic that we are currently experiencing has affected plenty of aspects of our lives, and also affected aspects related to art and culture. It not only affected the artist's life but also the contents and the work's narration. Apart from this phenomenon, this paper examines the creative process carried out in an artwork entitled *"Fierce Cattle Against Plague"*, which carries the theme of this pandemic-based issue. The painting itself adapted Bali's mythology that related to handling and dealing with disease outbreaks spiritually, as covered in the story of *Calon Arang*. The painting itself is prepared in digital enhancement, which is distinct from the common *Batuan* painting style. Through qualitative research by adhering to the theory of art criticism and adaptation theory, this study describes how the philosophy and creative process of the *Batuan* painting style can be used in the depiction of contemporary narratives such as the current COVID-19 issue and also still applied in the digital process.

Keywords: Batuan painting, Balinese art style, COVID-19

1 INTRODUCTION

The *Batuan* painting style is one of the most famous Balinese painting styles. The style is known since the 1930s. This style was born in the Batuan Region, Gianyar, Bali. The uniqueness of Batuan's visual style is its black and white style, along with depth and dimensions that form the impression of layers and the details of each component that are decoratively depicted in it. In contrast to other art styles, which are influenced by naturalist western aesthetics, perspective, and moment-taking, the rock style applies what is known as the late style. Professor Primadi Tabrani is a visual language that is flat in space and time (Tabrani 2005). There is no perspective in the *Batuan* painting style and also leaves no empty space. However, the beauty that is produced by painting with Batuan visual style does not only appear as a result of the artist's skill but also from the philosophy and meaning of each process carried out by the artist. In Balinese aesthetics, art is very dependent on the spirituality of the artist. The work is a reflection of contemplation and meditation that reaches the macro cosmos as well as the micro cosmos of the artist itself. The process is said to be bound by standard, rigid norms. Of course, this will be contrary to the dynamics of modern society, which changes rapidly and tries to simplify space and time, especially in design.

The problems of this study comprise that with the increase of digital art technology among young artists today, there are opinions that young and future artists are not interested in continuing the traditional arts. Meanwhile, the traditional elements are understood as ornaments only nowadays, without knowing why they were designed and illustrated in that way.

Some previous research and works had tried to do methods for preserving and introducing *Batuan* painting style to young generations, such as teaching and doing some workshops with exhibitions at many schools in Bali (Dewi 2020). However, they still use how to make paintings in Batuan style with traditional tools. Contrary to previous similar researches, this research develops more advance with digitalization in making *Batuan* style through digital ways without leaving its philosophy within the narrations and the process.

The purpose of this research is to find a synergy between the meaning and philosophy of the *Batuan* painting style's process with digital workmanship innovation. This will certainly provide new insights on the academic side, where the results achieved from this research can produce findings that can inspire innovation for young artists to work and continue to hold traditional cultural values by synergizing with technology.

2 RESEARCH METHODS

The method used in this research is the interpretative qualitative method in accordance with its essence, namely the content of the interpretation being the interpretation itself. Interpretation is to describe everything that is behind the existing data (Kutha Ratna 2010).

The preparations of the research are 1) Define approaches. These are to examine how Balinese *Batuan* paintings are made and to see the values and philosophies embodied in the workmanship of work through the *Batuan* painting style. After obtaining the philosophical value from each process, synergy is sought in the process of making the work in the process of digitizing, as well as in the process of making more contemporary design works, such as games and animation; 2) Determine Research Object. The object of this research is a painting in the style of *Batuan*; and 3) Data collection techniques and data types and sources. This research uses two kinds of data: a) Primary data sources: During a pandemic like COVID-19, going directly to an interview or coming to a location could be impossible. Therefore, the alternative to obtaining primary data is by asking people in the area of origin who are quite familiar with Balinese painting. Therefore, this study also involved lecturers and students from Bali in this study. b) Secondary data sources: Various literature studies in the form of reference books, articles, and theories about painting in the *Batuan* style. Data analysis is interpretive; the analysis is carried out simultaneously with the process of implementing data collection by describing everything that is behind the existing data.

3 RESULTS AND DISCUSSION

As previously stated, in general, *Batuan*-style paintings carry many themes of folklore, heroic stories, and also people's daily lives. Then, in subsequent developments, several artists also depict contemporary objects and raise the themes of contemporary life. However, painters like Ketut Sadia still use the visual style and techniques, typicaly the *Batuan* style. In general, *Batuan-style* painting is based on Hindu–Balinese philosophy, including belief in *Rwa Bhineda* and *Tri Hita Karana*. *Rwa Bhineda* is a dualism or two things that contradict each other but aim for world harmony. The philosophy of *Rwa Bhineda* describes the universe and its contents as composed or formed from two things that are different or contradictory, and complement each other for balance and the cycle of life (binary opposition). For example, male–female, sea-mountain, heaven-earth, joy and sorrow, good-bad, light-dark, black-and-white, top-down, *sekala-niskala*, left-right, and so on. In Chinese philosophy, *Rwa Bhineda* is known as "*Ying-Yang*", where on the white area there is a black dot and on the black area there is a white dot. *Rwa Bhineda's* philosophy is also seen in the use of black and white in *Batuan*-style painting (Granquist 2012) (Vickers 2012) (I Wayan Adnyana et al. 2017).

Meanwhile, according to Made Sujendra, the most common depiction that can be seen in Batuan's paintings is the image of the battle of *dharma* (goodness) with *adharma* (evil), in which dharma will surely win. Figures representing *dharma* are often represented by characters in wayang or also *tantra*, such as *Rama*, and *Pandavas*. Meanwhile, for the figures that represent *adharma*, they often use figures who are considered healthy, such as *Ravana* or

Kurawa. In addition, there are pictures of flora and fauna supporting decorations in the depiction of the battle (Adnyana *et al.* 2017).

Balinese aesthetics do not separate their daily or spiritual forms (Supir 2019). Balinese art treatises present this relationship as the unity between the cosmology of the 'great world' (*buwana agung*) as the macro cosmos and the 'small world' of mankind (*buwana alit*) as its micro cosmos. In the treatise, an artist meditates to unite the body and brush with the Almighty and his ancestors. Thus, painting is an act of meditation and artists are among those who reconcile the cosmic realm beyond the senses (*niskala*) with the everyday world of the senses (*sekala*) (Figure 1).

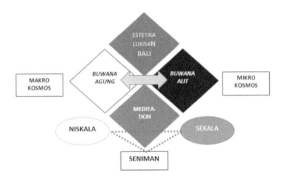

Figure 1. Schematic of the work process in balinese aesthetics (Source: Dimas K.A 2021).

The new narration's adaptation

The title of this work itself is *"Fierce Cattle Against Plague"* or the great war against the plague. The artist in this work attempts to adapt the narrative of the *Calon Arang* story into an illustration medium that combines the Batuan painting style with contemporary *manga* illustrations. The story of *Calon Arang* itself tells of a widow who mastered black magic, *Calon Arang* also had a beautiful daughter, Ratna Manggali. Unfortunately, no one wanted to marry Ratna Manggali because of her mother's evil nature. Enraged, *Calon Arang* finally made a witchcraft to spread disaster and disease that ravaged the peace of *Daha* kingdom. With the efforts of King *Airlangga* and his clerics and knights, *Calon Arang*'s black magic was finally defeated by Master Baradah's tactics.

In this work, the artist only records one chapter of the *Calon Arang* story. It takes place when *Patih Madri* led the people, civil servants, and martial arts experts to jointly attack the troops from *Calon Arang*. In several versions, it is narrated that *Patih Madri* died at the hand of *Calon Arang's* evil troops. The moment when *Patih Madri* took the lead was what the artist captured. To match the theme and actual phenomenon, each human object that appears is depicted wearing a mask and armed with soap bubbles, depicting a message that we don't forget to wash our hands if we want to keep away from the transmission of the COVID-19 virus. The election of the *Patih Madri* figure who died himself became a metaphor for hopes, plans, and achievements that had to be dropped or postponed or even canceled during this pandemic. The invitation to stay at home was seen by a mother who brought her child into the village to save from the dangers of the virus which was described as stealth or evil spirits.

Work process

Although using digital techniques to perform composition and visual styling using the visual style of manga, the artist uses the stages used in the Batuan painting technique, namely: *Ngorten*, *Nyawi Ngucak*, and *Nyenter*. Each step has its philosophy; so it can be said that

Table 1: The flow of the process of "Fierce Cattle Against Plague" creation.

Ngorten Sketching stage	*Nyawi* Stage of bolding of sketch lines	*Ngucak* Stage setup and layout	*Nyenter* Focus setting
At this stage, the artist sketches on Canson Marker paper first	Bold the outline of the sketch with ink. In general, this process uses Chinese ink. So to work on this, the artist uses Artline's Comic Pen	At this stage, the artist scans the finished images, then arranges the layer placement for each image sequence, its depth, and dimensions by using Layer Style on the Layer Menu in Adobe Photoshop CC software.	Setting the focus on the image object, this is also done using Adobe Photoshop CC
The philosophy of ngorten is that every life has a clear and precise direction.	The philosophy of nyawi is, when the planning and direction are clear, then to get maximum results, everyone needs to clarify the direction to be followed	Ngucek is the application of the philosophy of Rwa Bhineda, that life always contains dualism which is actually aimed at harmony	Nyenter also has the meaning of giving light and giving balancing to life

there are spiritual values in the process. If one of these processes is omitted, the Batuan painting will lose its essence (Adnyana *et al.* 2017). The sequence of stages and their explanations of the process and philosophies can be seen in Table 1.

The Final artwork can be seen in Figure 2.

Figure 2. The final artwork of *"Fierce Cattle Against Plague"*.
(Source: https://www.bumbungbudaya.id/en/events/pamerandaringspilc).

4 CONCLUSION

From the findings and implementation of the work, the researchers conclude that we can preserve traditional arts such as the visual style of *Batuan* painting, by updating the medium and its tools without having to replace or override the values contained in the *Batuan* painting style. The philosophy used in the digitalized work still refers to the philosophy of *Rwa Bhienada*, which implies that something is created in opposite ways, there is light and darkness, and there is evil and good, up and down (or binary opposition). This is also the reason why the colors used in the work are black and white, in order to create the illusion of dark and light, up and down, front and back coming alive. The use of digitization can be used, especially by the younger generations. Traditional art can synergize with contemporary narratives and can be maintained up-to-date, especially if the values contained in the creative process can go hand in hand with the processes of *Ngorten, Nyawi Ngucak,* and *Nyenter*, which are full of reflection in each process.

REFERENCES

Adnyana, D. I. W., M.Sn, D. I. M. B. Y., M.Sn, I. M. S., & S.Sos, W. S. (2017). *Seni Lukis Batuan*. http://repo.isi-dps.ac.id/2494/1/Seni_Lukis_Gaya_Batuan_Lengkap.pdf

Dewi, N. W. E. A. S. I. G. N. S. A. (2020). Alih Keterampilan Seni Lukis Gaya Batuan Oleh Komunitas Baturulangun Batuan. *Jurnal Pendidikan Seni Rupa Undiksha*, *10*(1). https://ejournal.undiksha.ac.id/index.php/JJPSP/article/download/28113/15917

Granquist, B. (2012). *Inventing Art, The Paintings of Batuan Bali*. Satumata Press. www.inventing-art.com

I Wayan Adnyana, K., I Made, B. Y., I Made, S., & Wayan, S. (2017). Seni Lukis Batuan. *Seni Lukis Batuan*. http://repo.isi-dps.ac.id/2494/1/Seni_Lukis_Gaya_Batuan_Lengkap.pdf

Kutha Ratna, N. (2010). *Metodologi Penelitian: Kajian Budaya dan Ilmu Sosial Humaniora Pada Umumnya*. Pustaka Pelajar.

Supir, I. K. (2019). Seni Lukis Realisme Sosial Batuan sebagai Seni Hibrid dan Gambaran Kehidupan Masrayakat Masa Kini. *Sandyakala: Prosiding Seminar Nasional Seni, Kriya, Dan Desain*, *1*, 301–309. https://eproceeding.isi-dps.ac.id/index.php/sandyakala/article/view/69

Tabrani, P. (2005). *Bahasa Rupa*. Kelir.

Vickers, A. (2012). *Balinese Art: Paintings and Drawings of Bali 1800–2010*. Tuttle Publishing.

La Kakao: Promoting small enterprises through visual storytelling based on the interpretation of local wisdom

S.M.B. Haswati & K.S. Ahada
Telkom University, Bandung, Indonesia

M. Buana
Leiden University, Leiden, The Netherlands

R.R.F. Ramli
Hasanuddin University, Makassar, Indonesia

ORCID ID: 0000-0003-1606-1820

ABSTRACT: As public awareness for the growth of local industry 4.0 flourishes in Indonesia, business owners learn to build a deeper connection with the consumers through several approaches. One of the most prominent communication strategies is visual storytelling. This paper discusses a case where designing visual storytelling is formulated to enhance local product promotion strategy. *La Kakao* is a set of fictional characters designed to re-introduce ChoKoo as a brand of chocolate drink and simultaneously attract the attention of the targeted audience by interpreting the richness of cosmopolitan cultural heritage in South Sulawesi. Within qualitative methodology, the research instruments include observation, content analysis, interviews, and questionnaires. The result is the illustration of characters, packaging labels, and strip comics to nourish the story of La Kakao. Those visual assets are ready to be applied by ChoKoo to relaunch the brand after the pandemic.

Keywords: Brand Communication, Character Design, Local Brand, Visual Storytelling

1 INTRODUCTION

In Indonesia, local brands from small enterprises have multiplied since 2010. Several factors trigger this development, 1) the widespread use of ICT (Information and Communication Technology), 2) accessibility of venture capital, and 3) the reduction of the income tax for business owners (Lathifa 2019). When the Covid-19 pandemic hit the global economy in 2020, it also caused a negative impact on the industry. During this time – 56.8% of small enterprises in the Jabodetabek (Jakarta-Bogor-Depok-Tangerang-Bekasi) area experienced unfavourable effects on their business (Katadata 2020). These problems include the scarcity of raw materials for production, the collapse of purchase power, and delays in production and distribution (Catriana 2021).

Chokoo (formerly Kokoronotomo) is a local brand of chocolate drink that decided to close its new stall only a few months after it was freshly launched in 2020 due to the Covid-19 outbreak. The owner had no choice but to return to serving drinks from the home kitchen as it started in 2018. Before installing the stall, Chokoo had already settled by receiving the consumer pre-orders. The production can prepare up to 100 bottles of chocolate drinks daily for the resellers. Despite the continuous efforts by the management, opening a new stall throughout a pandemic was not always the best consideration for a young brand. Ultimately, Chokoo paused the business in mid-2021, although its customers still asked for the products.

Figure 1. Chokoo labels and social media account 2020–2021 with 315 followers.

In 2021, research at Telkom University was conducted to help Chokoo owners remodel the business. The intention was simple, to collaborate and design creative strategies to boost the enterprise starting from packaging design. Yet, as the research progressed – the problem identification developed into a redesign of the key visual and visual assets for promotion media through implementing the visual storytelling concept. Consequently, the segmentation of Chokoo's previous target audience must be expanded from millennials only (1981–1996) to four generations: gen X, Y, Z, and Alpha. It is expected that Chokoo could be purchased to share with friends and family or bought as an occasional food gift.

In recent years, several studies have discussed how constructing storytelling based on cultural narration through illustration on product packaging could bring consumers more positive impact and a sense of cultural identity. It is clearly stated in the study case of local chocolate packaging from Yogyakarta by Yuwono and Wibawa (2022). While Long (2019) explained one of the advantages of illustrated tea packaging based on Chinese culture is to build a connection or consumer experience between the past and modern times. Also, Erlyana and Betsymorla (2020) analyzed the illustration contest to celebrate 75th Indonesia Independence Day by *Teh Botol Sosro*. It is perceived that the eight winning illustrations successfully represent the theme "*Keragaman Nusantara*" for each of the eight main islands in Indonesia.

Need to be highlighted that building a sense of cultural identity in an evolving city like Makassar is not only an attempt to achieve sustainable cities and communities as it is listed in Sustainable Development Goals. As long as public awareness about preserving heritage is settled, it also fuels the local industry by providing unlimited content from unlimited resources. For instance, Japanese packaging that depicts their cultural or social characteristic is quite stand out, aesthetically pleasing and unique for tourists, especially when compared to non-Japanese (Saito 2007). This particular design approach is inclusive and teachable, including for the SMEs in Indonesia.

2 RESEARCH METHODS

Qualitative methodology in the forms of interviews, content analysis, and observation is conducted in this research to process raw data into creative outputs. Since the research concept is set following the Stages of Design Thinking (Ambrose and Harris 2009), the graphic below explains the framework of research from collecting to data analysis, then finished by making a prototype for the design. La Kakao is meant as an innovation for a chocolate brand that merges promotion strategy with preserving Indonesian heritage and cosmopolitanism through visual storytelling.

2.1 Data collecting and analysis

The required information for this research is collected through observation and interview at the stall of Chokoo in Jl. Sadakeling on March 2021. Data is also collected by asking Dyo Pamungkas as the owner, dozens of consumers who have tasted Chokoo, and some who ordered Chokoo through food delivery apps themself. The interviews were running again during the research to gain insight into the business pattern and consumer behaviour. Finally, when the prototype of Chokoo drinks and the new design were finished, other feedback was gathered through product sampling. As an observational method of media representing particular issues (Bell 2001), content analysis was performed as the fundamental base for the character design of La Kakao: Some parts of ancient Mayan graphics from the Dresden Codex (Gates 1932), the illustration styles from local Mexican illustrator (Amatlapalli 2021), several studies about the history of cacao (McNeil 2009) (Ardren 2020) and heritage storytelling method (Diego 2020) to build up the narration.

3 RESULT AND DISCUSSION

The figure below illustrates the analysis of the creative process of designing package labels for Chokoo Drinks, formulated on the seven stages of the design method (define, research, ideate, prototype, select, implement, learn) where each step has been tested and evaluated. This primary method could be followed, modified or even developed by other brands from SMEs regarding performing a similar model.

3.1 *La Kakao*

Many ancient and traditional cultures have cultivated folklore that resembles each other. In South Sulawesi, people used to believe that the seasonal harvest depended on the favour of *Sangiang Serri* (A Goddess of agriculture and fertility in Bugis culture, similar to *Shri Laksmi* from India, *Dewi Sri* from Javanese culture, or *Demeter* from the ancient Greek pantheon). *Sangiang Serri* was highly respected and figured in nurturing the fields from seeds to crop yield (Rahayu & Devi 2021) for all types of plantations. It was crucial for Buginese people at that time since agriculture also portrayed the prosperity level of their community. Furthermore, Buana (2020) mentioned that even during the Dutch attack on South Sulawesi in the 17th century, horror for the rage of *Sangiang Serri* convinced the Buginese people to protect the paddy fields from their strike.

On the other side, the Spanish brought cocoa seeds to Minahasa in Northern Celebes for the first time around the 16th century. Later, Dutch Government established cocoa agroforestry as crucial mass production in the 19th century. Through decades, cocoa has grown as one of Indonesia's most profitable export commodities—especially in Sulawesi. It makes Indonesia the biggest cocoa producer in Asia and the third in the global cocoa market after Ivory Coast and Ghana (Kurniawan 2016).

Regardless of the multiple versions about the first arrival of cocoa in Indonesia, the main idea of La Kakao itself is focused on character-based visual storytelling. Created by Lontara Project, the story narrates the birth of the cocoa spirit named *La Kakao* for the first time in Celebes after a long journey from Mesoamerica. His delivery is witnessed by *Sangngiang Serri* joyfully. She is accompanied by spirits such as *Meong Mpaloé* and *Miko-miko* (the domestic cats found in La Galigo script) and *La Barellé (*the spirit of the Corn plant). In the other scenes, *La Kakao's* two companions, inspired by ancient Mayan paintings: *Chuchu* (Spider Monkey) and *Batso* (Howler Monkey), also appear.

Figure 2. The main characters of La Kakao are illustrated by Kania Sabitha from left to right side: *La Kakao, La Barellé, Sangngiang Serri, Méong Mpaloe, Miko-miko, Chuchu,* and *Batso.*

3.2 Visual storytelling as promotional content

The first product of La Kakao is an unpublished digital illustration with the same title, "The Birth of La Galigo" (Bugis: *"Iyanaé ajajianna La Kakao ri Tana Sulawesi"*), by Maharani Budi in 2021. The image depicts three major scenes: 1) A Mayan priest is sitting down while worshipping a Cocoa tree; 2) A Spanish ship containing wooden boxes to bring cocoa is sailing toward the land of Celebes; 3) *La Kakao* comes out from an opened giant cocoa pod surrounded by other characters. To bring the characters of La Kakao to life, two university students—Kania Sabitha (Telkom University) and Rizka Ramli (Hasanuddin University), drew four comic-strip episodes based on the story by Louie Buana from Lontara Project.

Figure 3. The scene from comic strips illustrated by Rizka Ramli (Left) & Kania Sabitha (Right).

Unarguably, designed characters and comics can be managed as a powerful asset in visual storytelling. They could be generated as limitless creative products like movies, music videos, or games. As the art of persuasion, it has persuaded advertising practitioners to use the same strategy to communicate the brand's creative values, attract more public attention, and expand the customer experience (Zatwarnicka-Madura & Nowacki 2018).

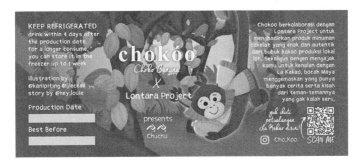

Figure 4. The label of Choco Banana with Chuchu, a spider monkey, was inspired by Mayan folklore of Hun Chouén.

Chokoo practises the same method by improvising the story's origin with further steps. The business expands commercial and heritage preservation benefits through collaboration with Lontara Project (a youth community that stands for the creative conservation of cultural heritage). The new design of Chokoo's packaging label introduces the characters of La Kakao and arranges them by flavours. The 500ml bottle packaging displays five different labels for each variant: Authentic Chokoo with *La Kakao*, Choco Banana with *Chuchu*, Salted Caramel with *Sangiang Serri*, Hazelnut by *Meong Mpaloe*, and the other flavour (for example, Matcha *Chizu* or *Avocado Chizu* only made by demand) with *La Barellé*. For 250ml bottle packaging, a blank space would be available on the label so the seller could fill the flavour manually with a marker.

Referring to Figure 5 above, customers may use their smartphone to scan the QR Code linked to lontaraproject.com which contains visual materials about La Kakao, La Galigo, and other heritage articles. The Instagram account is displayed on the label, directing the customer to visit the official account of Chokoo and follow it. Instagram account for business works like a digital catalogue that enables a brand to display news and connect to its customer effortlessly.

Figure 5. New label packaging of Chokoo x Lontara project for La Kakao edition.

Chokoo's promotional media integrates both traditional and digital platforms. The consumer will find the basic information on the packaging label printed-out media featuring the La Kakao characters and the other key visuals, as shown in Figure 6. Besides product photography, the different plans for Chokoo media promotion in the short term are to print out the comics, product catalogues as flyers, and Instagram content. It is also predicted that comics can be converted to digital format to broaden coverage, for example, on the website. Hence, all designs could be applied to merchandise and exhibition for a longer term.

4 CONCLUSION

Despite the tremendous challenges of rebuilding small enterprises and young brand businesses post-pandemic, Chokoo had taken a significant decision by collaborating with the heritage community through Lontara Project. A similar concern from both sides has brought the business to expand its value. Not only about Chokoo making a tasteful chocolate drink for consumers, but it also invites them to explore the visual story of La Kakao based on cultural heritage in South Sulawesi. The most significant limitation of this research comes from time. Still, in the future, it is expected that the following articles could share the other findings related to new branding and the advertising strategy for Chokoo.

REFERENCES

Amatlapalli, C.I. [@micorazonmexica]. (n.d.) *Posts* [Instagram profile]. Instagram. Retrieved August 25, 2021, from https://www.instagram.com/micorazonmexica/
Ambrose, G. & Harris, P. 2009. *Basics Design 08: Design Thinking*, Bloomsbury Publishing.
Ardren, T. 2020. *Her Cup for Sweet Cacao: Food in Ancient Maya Society*, University of Texas Press.
Bell, P. 2001. Content Analysis of Visual Images. *Handbook of Visual Analysis*, 13.
Buana, M. L. 2020. Die Verehrung Der Reisgöttin Dewi Sri. *südostasien–Zeitschrift für Politik• Kultur• Dialog*, 36.
Catriana, E. 2021. Kaleidoskop 2021: Tahun Penuh Harapan Bagi Pelaku Umkm. *Kompas.Com*, 29 December.
Diego, R. 2020. Heritage Storytelling, Community Empowerment and Sustainable Development. *Правоведение*, 64, 57–67.

Ehrenreich, B. 2019. 'Humans Were Not Centre Stage': How Ancient Cave Art Puts Us in Our Place. *Theguardian.com*, 12 December.

Erlyana, Y. & Betsymorla, B. 2020. Tinjauan Visual Kemasan Teh Botol Sosro Edisi Khusus Hut Ri Ke-75. *DESKOMVIS: Jurnal Ilmiah Desain Komunikasi Visual, Seni Rupa Dan Media*, 1, 173–183.

Gates, W. 1932. *The Dresden Codex*, Maya Society at Johns Hopkins University.

Katadata 2020. Digitalisasi Umkm Di Tengah Pandemi Covid-19. *Jabodetabek: Katadata Insight Center*.

Kurniawan, A. 2016. *Cacao Catalogue. In:* Foundation, B. (ed.). Menteng, Jakarta: Yayasan Belantara.

Lathifa, D. 2019. Meninjau Perkembangan Umkm Di Indonesia, Bagaimana Kondisinya? Available: https://www.online-pajak.com/tentang-pph-final/perkembangan-umkm-di-indonesia [Accessed 22 June 2022].

McNeil, C. L. 2009. *Chocolate in Mesoamerica: A Cultural History of Cacao*, University Press of Florida.

Rahayu, N. W. S. & Devi, N. K. T. S. 2021. Pemujaan Sangiang Serri Di Tanah Bugis. *Widya Katambung*, 12, 63–69.

Saito, Y. 2007. The Moral Dimension of Japanese Aesthetics. *The Journal of Aesthetics and Art Criticism*, 65, 85–97.

Yuwono, E. C. & Wibawa, I. S. 2022. Representation of Yogyakarta's Identity through Graphic Elements of Local Chocolate Packaging. *KnE Social Sciences*, 358–367.

Zatwarnicka-Madura, B. & Nowacki, R. *Storytelling and Its Impact on Effectiveness of Advertising*. ICoM 2018 8th International Conference on Management, 2018. 694.

The role of compact furniture design in the efficiency of floor area used of micro residence. Case study: Loft-style apartments in Surabaya

Mahendra Nur Hadiansyah
Telkom University, Bandung, Indonesia

ORCID ID: 0000-0003-1553-0904

ABSTRACT: Surabaya is one of the largest and most populous cities in Indonesia after Jakarta. The lack of land availability in urban areas directs the design to residential vertically and micro. One of the micro residences is a loft-type apartment. Limited space in loft-type apartments presents a challenge for interior designers to create compact furniture but does not reduce the fulfillment of residents' needs. This study aims to find out how the system is used in compact furniture and the utilization of the floor area used to streamline the use of limited space. The method used is observation and analysis through scalable digital simulations to measure the efficiency of space use. To be more focused on research using a case study approach, namely one of the loft-type apartments in Surabaya. The results showed that there are four types of compact systems applied to furniture, namely multi-function, utilization of dead space, maximizing the volume of cavities in furniture, and utilizing the height of the space. In conclusion, the compact system that is applied can significantly reduce the use of floor area for furniture. The compact system of the four types in total can save 52.75% of the space used to arrange furniture and add to the comfort of circulation of activities in the loft-type apartment.

Keywords: Micro-house, loft-type apartment, limited space, compact furniture, efficiency of space

1 INTRODUCTION

Limited land has become a major issue in large cities in Indonesia, including Surabaya. Urban communities must indirectly be flexible and adapt to these conditions. In the end, the development of residential design adapts to these conditions more and more residential of limited space and vertically such as flats, apartments, and pavilions. Housing with limited space is commonly called micro-housing. The micro-house is a residence with a small space due to limited land in the city, which is specially designed so that it can flexibly accommodate all residents' activities in space (Zang & Yang 2016). One type of micro-house is a loft-type apartment, where this apartment has two floors in one room. On the second floor, it resembles a mezzanine. The loft is the result of utilizing the optimization of the volume of the building under the roof of the house initially for storage space. The development of design and architecture is increasingly leading to the optimization of volume in buildings by presenting new functions in loft-style structures (Hizli & Mizrak 2015). One of these loft-type apartments is the Royal City loft located in Surabaya, which is a case study in this research.

The limited space ultimately presents challenges for interior designers to design the right furniture so that circulation and activities in the space can still be accommodated properly without compromising aesthetics and a comfortable environment. Compact furniture is considered the most appropriate and relevant to be applied to micro residences such as loft-style apartments. This is in line with the results of previous studies showing that the use of compact, multifunctional, foldable, and convertible furniture in limited spaces such as studio-type apartments is needed in saving space. For example, such as multifunctional furniture, several functions that can present large dimensions in space can be summarized into one design that will not take up much space (Natalia *et al.* 2021). This multifunctional system can summarize the needs of residents in one piece of furniture, leaving space for other activities. (Husein 2021). One type of compact furniture is fold, the folding system on furniture in the interior will be able to maximize space utilization without taking up a lot of space but not reducing the function of the furniture itself when used or not (Li & Wu 2021). This study aims to find out how the systems and methods used in compact furniture design are applied to case studies so that they are able to streamline the limited space and how much is the percentage of saving in the floor area used in the placement and arrangement of compact furniture.

2 RESEARCH METHODS

This study uses two methods, namely a blend of qualitative and quantitative methods. Qualitative methods are applied in data mining through the selection of case studies, namely the Royal city loft apartment located in Surabaya. Processing and data analysis is carried out through scalable digital simulations using software namely SketchUp. Simulation is the engineering of actual circumstances and conditions either using supporting media or demonstrated by similar substitutes (Indriasari 2016). While the quantitative method through simulation using SketchUp serves to find out in detail and measurably the comparison of the use of floor area when furniture do not use a compact system and calculates the floor area that can be saved in terms of furniture placement and arrangement.

2.1 *Methods of data collecting*

The data were collected through direct observation in case studies measuring all dimensions of compact furniture, observing the system applied to each piece of furniture, and documenting by taking photos of every detail of the observations made for the next stage of analysis.

2.2 *Methods of data analysis*

The analysis data was carried out by describing the results of observation, namely the type of compact system applied to each furniture in the space of the case study. This was followed by a scalable digital simulation using SketchUp to calculate the comparison of the use of compact furniture on the floor area with those of the standard ones. Additionally, the percentage of floor area savings as a whole was known.

3 RESULT AND DISCUSSION

The Royal City Loft Apartment has a main accessibility character through the entrance of each residence which is directly connected to the building corridor on each floor. One of the interior designs of the apartment that is a case study has two floors consisting of four rooms containing compact furniture, namely the dining room, living room, master bedroom, and kid's bedroom.

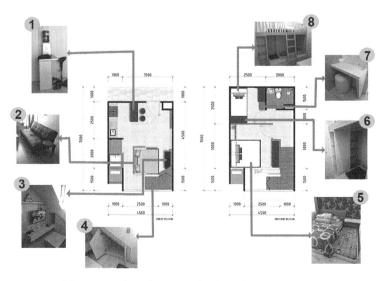

Figure 1. Lay out and documentation of compact furniture in royal city loft apartment.

When you first enter through the main entrance on the right side, there is a kitchen area in front of which there is furniture such as a mini-bar (1). Its real function is not as a bar but as a dining table. This furniture has multifunctional properties and have three functions, namely the dining table, serving table, and storage of cutlery. The dimensions of the floor area when furniture is not in use are 60 x 120 cm. On the side, there is a knob that can be pulled to extend the table 80 cm long by 50 cm wide which is used to serve food or additional users for eating activities. Therefore, the multifunction system has saved space by 0.4 m^2, which is 35.71% of the total floor use area if this furniture is not compact.

After passing through the dining area, there is a living room with three kinds of compact furniture. First is the sofa bed (2). This sofa is multifunction and aimed at saving space when not functioning as a bed. When it becomes a bed, the floor area used is 200 x 120 cm. When the backrest is enforced to become a sofa, the use of the floor area is reduced, which is only 200 x 80 cm. Thus, this sofa-bed can save space by 0.8 m^2, which is 33.33% of the total floor area usage when you don't use compact furniture.

The other two compact pieces of furniture in the living room are the Television backdrop (3) and the built-in cabinet under the stairs (4). The compact system used is to utilize dead space. The dead space in the interior is never reached by residents because it is very not ergonomic for activities. Meanwhile, in this apartment, the dead room under the stairs is used for the placement of furniture whose activities and range are close to the living room area. Thus, the area under the stairs is utilized 100% of the total floor area of 4 m^2 which is used as a staircase area so that it does not use the floor area of the active space.

Another compact furniture is located in the main bedroom, on the second floor, which is a bed for two people (5). In this furniture, the system used is the utilization of the cavity under the bed which is used to store additional beds for one person. It is used by pulling the bottom of the bed towards the side. Thus, the utilization of bed storage for one person covering an area of 90 x 200 cm and the use of the floor under the bed for two people covering an area of 180 x 200 cm has saved 33.33% of the total floor area of the bed usage, which is 1.8 m^2 when not in use.

Still, on the second floor, there is a kids' bedroom containing two compact pieces of furniture. The first is a bed for two children (8). Due to the limited area, this bed design utilizes the volume of space height so that the bed is made stacked. The system makes the two children not sleep next to each other but on a bed with the same floor area of 230 x

90 cm, only at the bottom and at the top. Thus, the bed floor area saves 50% of the total bed area that should be for two children, which is enough being 2.07 m^2.

The bed in the kid's bedroom also maximizes the remaining dead space by making a display as well as a toys cabinet (6) on the side of the bed at the bottom. The dead space is above the floor measuring 90 x 30 cm, and thus it can save 6.52% of the total floor area used by the combination of furniture bed and toy cabinet if it is not compact, which is 4.41 m^2.

The last piece of furniture is a table with dimensions of 120 x 50 cm that can store stools underneath (7). The system on this furniture utilizes the volume of the bottom cavity of the table for storage of stools when not in use, so that the floor area can be saved by 0.25 m^2 or 29.41% of the total floor area of table placement and stool. The eight compact furniture can be analyzed through a recapitulation with Table 1.

Table 1. Recapitulation of the use of floor area in furniture.

No	Furniture	Total floor area if not compact (m^2)	Total floor area if compact (m^2)	Total floor area saved (m^2)	Floor area saved percentage (%)	Floor area saved a percentage of all (%)
1	Mini-bar	1,12	0,72	0,4	35,71	2,20
2	Sofa Bed	2,4	1,6	0,8	33,33	4,40
3	Backdrop TV (on dead space)	3	0	3	100,00	16,50
4	Built-in Cabinet under the stairs (on dead space)	1	0	1	100,00	5,50
5	Bed of the main bedroom	5,4	3,6	1,8	33,33	9,90
6	Toy Cabinet and Display (on dead space)	0,27	0	0,27	100,00	1,49
7	Table	0,85	0,6	0,25	29,41	1,38
8	Bed of Kids Bedroom	4,14	2,07	2,07	50,00	11,39
	Total	18,18	8,59	9,59	Total	52,75

Based on the results of the recapitulation in the table, it shows that the total floor area used, if not using a compact system, is 18.18 m^2. The total floor area used by all existing furniture with a compact system is 8.59 m^2. The floor area that can be saved is 9.59 m^2 with a percentage of 52.75% of the total floor area used by furniture if it is not compact.

4 CONCLUSION

There are four types of systems applied to compact furniture that fill Royal City Loft Apartments which are the topic of our case studies; these types are multifunctionality, good utilization of dead space, maximizing the volume of cavities in furniture, and utilization of space height. The four types of systems of compact furniture that are applied show a very significant efficiency in using floor area, which is 9.59 m^2 of the total floor area if using a non-compact furniture design, which is 18.18 m^2. In percentage terms, compact furniture can reduce the use of floor area for placement and arranging furniture by as much as 52.75% in the interior of the royal city loft apartment located in the city of Surabaya. Floor area savings can eventually be maximized for circulation which affects the comfort of occupant activities in the space.

REFERENCES

Hizli, N. & Mizrak, B. 2015. *Re-Thinking Loft Buildings in the Scope of Housing Production in Turkey*. Megaron 2015;10(4):479–493. Department of Architecture, Yildiz Technical University, İstanbul, Turkey.

Husein, H. A. (2021). Multifunctional Furniture as a Smart Solution for Small Spaces for the Case of Zaniary Towers Apartments in Erbil City, Iraq. *International Transaction Journal of Engineering, Management, & Applied Sciences & Technologies*, *12*(1), 12A1H, 1–11. http://TUENGR.COM/V12/12A1H.pdf DOI: 10.14456/ITJEMAST.2021.8.

Indriasari, F.N. 2016. *Pengaruh Pemberian Metode Simulasi Siaga Bencana Gempa Bumi Terhadap Kesiapsiagaan Anak di Yogyakarta. Jurnal Keperawatan Soedirman, The Soedirman Journal of Nursing*, Volume 11, No. 3, November 2016.

Li, Z. & Wu, J. 2021. Research on the Design of Small Interior Space. *E3S Web of Conferences 308, 01002 (2021) MSETEE 2021*. https://doi.org/10.1051/e3sconf/202130801002

Natalia, V., Siwi, S.H., Lianto, F., Susetyarto, M.B. 2021. *The Utilization of Compact-Convertible Furniture Module in Studio-Type Apartment in M-Town Residence at Summarecon Serpong* Advances in Social Science, Education and Humanities Research, volume 570. Proceedings of the International Conference on Economics, Business, Social, and Humanities (ICEBSH 2021). Atlantis Press.

Zang, W. & Yang, J. 2016. International Forum on Management, Education and Information Technology Application (IFMEITA 2016), *Development Strategy of Micro House*. Atlantis Press.

Investigating commercial interior space presented in tourist Instagram posts

A. Kusumowidagdo* & M. Rahadiyanti
Universitas Ciputra, Surabaya, Indonesia

*ORCID ID: 0000-0001-5137-6815

ABSTRACT: This research's purpose is to study tourists' Instagram posts which represent commercial interior areas in the residential area of Sade village, Lombok. There are over 200 Instagram posts in this research which were posted by domestic tourists of Sade village. The research parameter discovery on the Instagram post was the physical interior aspect as well as social. Data analysis relied on content analysis and the result informed that for physical category discovery, interior space, the most in a row are on interior scope (layout, ambiance, natural materials, color, and textures), product display, and product placement. On the other hand, in social aspects, visually, ethnic as well as tourists and merchants' lifestyles were the major discoveries. From the texts, there were many similar visual discoveries disclosed.

Keywords: Instagram post, commercial interior space, indigenous village

1 INTRODUCTION

Instagram is a social media with lots of users. Its existence can bring additional value to tourist attraction areas. Visitors' posts on their Instagram accounts work as a promotional platform to increase the number of visitors. As one of the tourist villages, Sade Village Lombok, with its indigenous residential cluster, has its own attractive value. The interiors of the private rooms are then publicly displayed. This area became an interesting area as the research's object. Sade Village's visitors posted a lot of pictures of the area and presented them according to their own personal preferences.

A lot of research surrounding Instagram had been done, especially with the connection to spatial experiences (Giorgi *et al.* 2020; Jorgensen & Stedman 2011; Puren et al. 2018), tourists attraction representation (Gon 2021; Iglesias-Sánchez *et al.* 2020; Parsons 2017) as well as commercial areas (Kwon & Lee 2021). However, the focus on interior spaces of indigenous houses, such as in this Sade Village research, barely exists. This writing aims to fill the research gap. The research novelty lies in the study of commercial interior space in an indigenous village and then linked to social media which is Instagram.

Going further into the investigation of this commercial space interior, it will be reviewed for physical and social factors be it for Instagram posts whose characteristics are both visual and text. Physical and social factors are two things that built the place's uniqueness (physical factors of commercial areas that can be observed are, for example, zoning and interior space, layout display, street width, material, and ambiance (Kusumowidagdo *et al.* 2022). More studies (Rahadiyanti *et al.* 2019) stated that interior, such as natural material, display merchandise, and product's attributes, is one of the most crucial factors in building sense of place in Sade Village. Whereas social factors of the parameter area are the local and comer merchants which are shown in clothing, languages, and accents (Rahadiyanti *et al.* 2019).

2 RESEARCH METHODS

This research used a single object, Sade Village, which is located in a super-prioritized tourist attraction in Indonesia. It is a quantitative content analysis by studying commercial interior research subjects, which are shared on Instagram 200 times, of Sade Village. Previous theories (Kusumowidagdo *et al.* 2022; Rahadiyanti *et al.* 2019) completed with discoveries from direct observation were the analysis tools.

Data were collected from the visitors' Instagram posts having public accounts. The pictures/photos and names from this research had been blurred to maintain privacies. For validating this research, a triangulated source was applied and several sources were used to maintain the research's validity. Sources were not just from Instagram posts but also interviews with the Village's Head (Mr. Kurdap Selake), merchants, and visitors. This research focuses on Instagram posts, namely visual posts/pictures and text/narratives. In the content analysis, images and narration posts were reviewed, to analyze the physical and social elements of the sense of place. The physical factors appear in images and text, including interior scope, product placement, product display, natural materials, interior ambiance, and products' colors and textures. Meanwhile, the social elements in the images and narration show ethnic, merchants' lifestyle, tourists' lifestyle, interaction between merchants and tourists, attributes, and other elements. This categorization was obtained through the coding process of the 200 posts.

3 RESULT AND DISCUSSION

3.1 *Physical analysis: Interior space*

In the visual and text analysis, some visual categories are discovered such as interior scope, product placement, product display, natural material, interior ambiance, and products' colors and textures. The discoveries mentioned above supported the discoveries by Kusumowidagdo *et al.* (2022).

Figure 1 shows some illustrations of Instagram posts for both text and visual.

Figure 1. Open displays rooms (right), products' attractive colors and textures (left). (Anrgraphy 2021 & Nita_Agam 2021).

3.1.1 *Visual posts*
From previous social media research related to the visual category of Sade Market interior, the following results are received, as shown in Table 1.

Table 1. Physical factor of Sade Market "visual interior categories".

Visual Interior Categories	Account	Percentage
1. Interior Scope (layout, ambiance, color, and textures)	144	72%
2. Product placement	11	5.50%
3. Product display	45	22.50%
Total	200	100.00%

According to Table 1, it can be seen that around 72% of social media accounts discussed interior scope in Sade Market, around 22.50% discussed product display, and around 5.50% discussed product placements.

3.1.2 *Narrative posts*

From previous social media research related to the narrative category of Sade Market interior, the following results are received, as shown in Table 2.

Table 2. Sade Market's Physical factors "narrative interior categories".

Narrative Interior Categories	Account	Percentage
1. Interior scope (layout, material, ambiance, color, and textures)	196	98.00%
2. Product placement	1	0.50%
3. Product display	3	1.50%
Total	200	100.00%

From the last observation (Table 2), it is shown that around 98% of Instagram accounts discussed the interior scope, around 1.50% discussed product display, and around 0.5% discussed Sade Market Lombok's product placement.

3.2 *Social analysis: Interactions between merchants and tourists*

In the social analysis for both visual and text regarding interactions between merchants and tourists, some visual categories are found, such as ethnic, merchants' and tourists' lifestyles andinteractions, and other attributes and contexts. These findings complete the previous finding from Rahadiyanti *et al.* (2019).

Figure 2 shows some illustrations of Instagram posts for both visual and text.

Figure 2. Visitors' lifestyles (right) and merchants' lifestyles (left). (Deby_Skaisti 2022 & Muhajirefendysecond 2022).

3.2.1 *Visual posts*

From previous social media research related to visual categories on the interactions between merchants and visitors of Sade Market, Lombok, the following results were obtained, as shown in Table 3.

Table 3. Sade market social factors "visual categories of interactions between merchants and tourists".

Visual Categories of Interactions Between Merchants and Tourists	Account	Percentage
1. Ethnic	121	60.50%
2. Merchants' lifestyle	19	9.50%
3. Tourists' lifestyle	40	20.00%
4. Interactions between merchants and tourists	14	7.00%
5. Attributes	3	1.50%
6. Others	3	1.50%
Total	200	100.00%

From previous research on social media accounts shown in Table 3, it can be seen that around 60.50% discussed ethnic, around 20% discussed tourists' lifestyles, around 9.50% discussed merchants' lifestyles, around 7% discussed interactions between merchants and tourists, and around 1.5% discussed the attributes and other contexts.

3.2.2 *Narrative posts*

From previous research through social media related to a narrative category on interactions between merchants and tourists in Sade Market Lombok, the following results were obtained, as shown in Table 4.

Table 4. Sade market social factors "narrative categories of interactions between merchants and tourists".

Interactions Between Merchants and Tourists	Account	Percentage
1. Ethnic	171	85.50%
2. Merchants' lifestyle	12	6.00%
3. Tourists' lifestyle	10	5.00%
4. Interactions between merchants and tourists	5	2.50%
5. Attributes	0	0.00%
6. Others	2	1.00%
Total	200	100.00%

According to Table 4, it can be seen that around 85.50% discussed ethnic, around 6% discussed merchant's lifestyle, around 5% discussed tourist's lifestyle, around 2.50% discussed interactions between merchants and tourists, and around 1% discussed other contexts. No display of attributes found in Instagram text posts.

4 CONCLUSION

During this investigation, some facts are discovered such as: (1) physical and social discoveries have different percentage compositions because not all visual pictures are related to texts. Visitors' preferences factor is their key factor; (2) the most physical visual discoveries for interior rooms are interior display scope, products' colors, and textures, while in text form, the interior scope was mostly found; (3) the most physical visual discoveries for social aspects are ethnic, merchants' and tourists' lifestyles, as well as in the text discoveries. These discoveries are key factors that need to be maintained and improved by the policymakers of Sade Village to maintain its cultural sustainability. In the future, this research can be widened to new objects in different areas, different methods, as well as different presentation technologies.

REFERENCES

Giorgi, E., Bugatti, A., & Bosio, A. (2020). Social and Spatial Experiences in the Cities of Tomorrow. *Societies*, *10*(1), 9.

Gon, M. (2021). Local Experiences on Instagram: Social Media Data as a Source of Evidence for Experience Design. *Journal of Destination Marketing & Management*, *19*, 100435.

Iglesias-Sánchez, P. P., Correia, M. B., Jambrino-Maldonado, C., & de las Heras-Pedrosa, C. (2020). Instagram as a Co-creation Space for Tourist Destination Image-building: Algarve and Costa Del Sol Case Studies. *Sustainability (Switzerland)*, *12*(7), 1–26. https://doi.org/10.3390/su12072793

Jorgensen, B. S., & Stedman, R. C. (2011). Measuring the Spatial Component of Sense of Place: A Methodology for Research on the Spatial Dynamics of Psychological Experiences of Places. *Environment and Planning B: Planning and Design*, *38*(5), 795–813. https://doi.org/10.1068/b37054

Kusumowidagdo, A., Ujang, N., Rahadiyanti, M., & Ramli, N. A. (2022). Exploring the Sense of Place of Traditional Shopping Streets Through Instagram's Visual Images and Narratives. *Open House International*. https://doi.org/10.1108/OHI-01-2022-0009

Kwon, K., & Lee, J. (2021). Corporate Social Responsibility Advertising in Social Media: A Content Analysis of the Fashion Industry's CSR Advertising on Instagram. *Corporate Communications: An International Journal*, *26*(4), 700–715. https://doi.org/10.1108/CCIJ-01-2021-0016

Parsons, H. (2017). *Does Social Media Influence an Individual's Decision to Visit Tourist Destinations? Using a Case Study of Instagram*. Cardiff Metropolitan University.

Puren, K., Roos, V., & Coetzee, H. (2018). Sense of Place: Using People's Experiences in Relation to a Rural Landscape to Inform Spatial Planning Guidelines. *International Planning Studies*, *23*(1), 16–36.

Rahadiyanti, M., Kusumowidagdo, A., Wardhani, D. K., Kaihatu, T. S., & Swari, I. A. I. (2019). The Sense of Place: Sade Shopping Corridor. *The European Proceedings of Multidisciplinary Sciences*, 740–749. Future Academy www.FutureAcademy.org.

Image sources:
Anrgraphy, 2021. "#lombok #lombokisland" [Instagram], December 31, available at https://www.instagram.com/p/CXBc3uLhYIh/ (accessed 1 July 2022)

Deby_Skaisti, 2022. [Instagram], available at https://www.instagram.com/deby_skaisti/ (accessed 1 July 2022)

Muhajirefendysecond, 2022. "#sade #desasade" [Instagram], May 27, available at https://www.instagram.com/p/CeC3I7JBn5Z/ (accessed 1 July 2022)

Nita_Agam, 2021. "#travelphotography #travel" [Instagram], November 22, available at https://www.instagram.com/p/CWkzbvfJlps/ (accessed 1 July 2022)

IRES marketing public relations strategy to raise children's environmental awareness through *KOMIK SOTA*

P.N. Larasaty*, M. Lemona, N.P. Pangaribuan, D. Widowati & M.B. Simanjuntak
LSPR Institute of Communication and Business, Jakarta, Indonesia

*ORCID ID: 0000-0002-3734-1760

ABSTRACT: This paper is aimed to analyze what IRES, a Jakarta-based NGO that concerns about the environment, climate change, and disaster mitigation, does to educate 5–12 years old kids about disaster and environment protection through Sota Comic Characters (a honey fox that is very resilient). The main theory used in this research is Whalen's seven Strategic Planning steps: situation analysis, objective, strategy, targeting, message, tactic, and evaluation. Interview was done as the data collection to find out the seven steps done by IRES to deliver the message through their comic, and analyzed using a qualitative approach. The result has shown that IRES is taking a systematic method to introduce the comic and reach the audience. The comic also highlighted the use of visual communication to deliver the message to the target audience. The result also revealed that an evaluation needs to be done by IRES to measure the effectiveness of their strategy.

Keywords: marketing communication, visual communication, comic character, environment

1 INTRODUCTION

Nairat *et al.* (2020) stated that along with the massive use of technology as a substitute for arts and traditional design and drawing tools, comic artists have their own challenge in creating attractive content and design. Samara (2007) similarly added that comic is a unique synthesis of words and images, making them a distinctive and powerful visual form of arts and creativity. Unlike television and films, comics combine words and still images that give another experience to the reader to explore and imagine the story and message freely to each individual who reads it. Understanding this uniqueness of comics as mass media, many content creators and illustrators are taking the opportunity to reach potential readers by disseminating particular messages. Comics have their own positioning to attract and persuade readers through their visualization and story development (Escalas 2004). What is also important in a comic is how the characters are put together to create a situation and context that take readers to be part of the story and even emotionally involved in the story developments.

As an NGO whose works are related to environmental issues, climate change, and disaster mitigation, IRES is aware of the benefits of how comics can be an alternative campaign tool for them when communicating with children. Titled as "Komik Sota", IRES framed the series into an educational series about disaster and environmental protection for kids aged 7 to 12 years old, especially for elementary level audiences. This research is focusing on how IRES tries to communicate its message through comics and its visual content to children. The objective of this research is to examine what steps or tactics IRES uses to reach its audience and the measurable results that can be evaluated.

2 LITERATURE STUDY

2.1 *Marketing public relations*

According to Kotler and Keller (2012), marketing public relations is an activity that is seen to obtain editorial space, paid in print and broadcast media to promote a product or service. According to Kotler and Keller, the MPR dimensions include publications, events, news, sponsorship, speeches, public service activities, and identity media. PR marketing was originally better known as publicity, but now MPR has more functions than that including (a) supporting product launches, (b) assisting in product repositioning, (c) popularizing certain product categories, (d) influencing certain target groups, (e) defending a product that is in trouble, and (f) building a corporate image so that it can improve the image of its products.

To maintain a good relationship with the public, Marketing Public Relations can be done by organizations through three strategies: push, pull, and pass (Keller & Kotler 2012). Push strategy is aimed to promote the product or brands to distributors and retailers. For instance, by putting the product in point of sales to increase visibility and exchange form in goods or money. Pull strategy, on the other hand, is aimed to attract customers through various media and target end users of the specific product. This strategy is usually done by conducting events, product publication and advertising, sampling, survey, newsletter, websites, and many more. Lastly, pass strategy is aimed to create a favorable public opinion. It also aims to government, NGOs, policymakers, and key opinion leaders to create image and understanding. It's very common to pass strategy by collaborating with society through social activities and social events that involve public figures and opinion leaders.

According to Grunig (2013), three PR publicity types have marketing orientation, such as (a) Product release, as an effort to announce the product and spread product knowledge, (b) Executive Statement Release, as the official words from CEO and other company executives to reflect significant credibility to the product, and (c) Feature Article, a detailed explanation about a product of a program that can be released as a news story by media, both in print and online platforms. According to these PR Publicity types, SOTA Comic can be categorized as Feature articles.

To build brand awareness, some steps can be useful to see the brand's value in society. The Whalen Seven Steps Strategic according to Harris & Whalen, (2006) is a marketing public relations concept that includes seven strategic planning that can boost and promote the target achievement. The seven steps are situation analysis, setting objectives, defining strategy, setting the target, creating the message, planning the tactic, and conducting the evaluation.

2.2 *Storytelling marketing*

Storytelling marketing is an outgrowth of the experiential marketing trend that is used to make consumers' feelings, thoughts, senses, and actions work (ZA *et al.* 2021). A story is a journey that will affect the listener and put them into an imaginary motion. When the listener decides to go on that journey, they will feel something different and the result is persuasion or even action as they perceive the story emotionally (Aaker *et al.* 2012).

McLellan (ZA *et al.* 2021) state that storytelling is a means of describing a message to be understandable, containing a meaning, and can create certain memory in the head of the readers. Therefore, strong brands are built on emotional connections between the consumers and the brands' clear values, related and poured into stories. That being said, storytelling and consistent brand images along with emotional bonds can be enhanced in a creative way (ZA *et al.* 2021). The elements that stories are character, setting, plot, theme, and tone. Samudra (2019) researched that a story must have characters with specific roles; a plot and a timeline of the story from the beginning until the end, as well as a conflict to promote the story's interesting point to the audience.

2.3 *SDG aspect*

The agenda of the UN member states have stated that the commitment is to protect the earth from further destruction, especially in the environmental aspect and to prevent the worse impact of climate change and global warming (UN 2022).

3 RESEARCH METHODS

Using the qualitative approach, data in this research are gathered through interviews with informants: Hari Apriawan Akbar (Director of IRES) and Farikha Khotimah (IRES SOTA Comic illustrator). An in-depth interview was conducted to gather information regarding what steps are taken by IRES to introduce the comic based on their target audience. Secondary data are also collected from the SOTA Comic, IRES documentation, and other literature studies. Using Miles *et al.* (2018), data analysis is done by data collection then condensed, analyzed, displayed, and finally, data is displayed.

4 FINDINGS AND DISCUSSION

4.1 *Visual elements as communication tools*

Character is one of the most important elements in the dramatic work of Egri 1960 (Nairat *et al.* 2020). It is the main motive and force that moves the storyline ahead, and the vital part of any memorable story. Its development is the key element in story creation. The more relevant the readers can identify the elements, the more understandable and enjoyable the story is. Characters often remain in our minds long after the setting and theme of the plot are forgotten. The concept of character refers to a textual representation of a human being or other creature (Nairat *et al.* 2020).

Since the target market is 7–12 years old kids affected by natural disasters, SOTA Comic's main characters were created by using three friendly and sympathetic figures. They are one curious honey fox (Sota), one diligent young man (Bumi), and one considerate young girl (Raya).

According to Samara (2007), there are few visual stimuli as powerful as color; it is a profoundly useful communication tool. Using relatively light colors, thick-lined illustrations, and handwritten-san serif typefaces, the overall visuals in SOTA Comic are easy to read and convenient for their young readers. Dominated with green, yellow, and brown as the color palette, IRES aims to give consolation to its SOTA Comic reader about nature and the environment. Analogous color schemes—those close to each other in the color wheel—create a feeling of harmony: for instance, yellow and green have softer visual associations; green color helps people to keep calm and is associated with spring, youth, and the environment (Dabner *et al.* 2017).

4.2 *Whalen's seven steps of strategic planning*

Based on the interviews done with informants, IRES strategy in introducing Komik Sota can be mapped out into Whalen's 7 steps of strategic planning as follows.

4.2.1 *Situation analysis*
IRES collected the database of disaster victims across Indonesian disasters from 2018 to 2022 and figured out that children are one of the most vulnerable parts of society affected by disasters. Children become the main targets as they barely provided with enough knowledge by any other stakeholders about natural disasters and their prevention. What children have received so far in school and other informal educational activities is disaster simulation and

basic knowledge on how to survive during a disaster. This, however, is still not felt not comprehensive and situation-based training for the young generation. Farikha, the illustrator of Komik Sota, stated that "By creating a comic book, children see this education media in an entertaining setting and material. Hopefully they receive the lesson taught by the characters in this story". That being said, IRES acknowledges that children tend to receive things easier when the message comes with entertaining and exciting packaging.

4.2.2 Setting objective

The objective of this comic is to create awareness among children in elementary schools and build an understanding of environmental issues. The stories included also tell simple actions that are possible to be done daily to prevent the further impact of environmental damage. Children are introduced to general knowledge and analogy of climate change phenomenon and how they can participate in preventing worse effects. Given this understanding of children, it is expected that from a very young age, children are aware of their environment and surroundings to see the potential of environmental damage. Through visual elements in the comic, children are introduced to imaginative, but relevant, characters that can help them imagine the situation better in their reality.

4.2.3 Defining strategy

To reach the massive but segmented audience, covering children aged 7–12 years old, IRES has decided that they are using a pass strategy. The pass strategy used by IRES means planning and conducting the Road to Launching (RTL) campaign that includes school visits, library visits, and a launching event. In the event, IRES is planning to include interactive games and quizzes during the events for the participants who attend the event. This strategy is designed to deliver the main objective of the comic launching and the main message regarding environmental awareness in children's level of understanding.

4.2.4 Setting the target

The target for this initial program through comics is to raise awareness for children about environmental issues and natural disaster prevention in the simplest way through visual communication in comics. Within 3 to 6 months, Komik Sota is targeting to reach at least public and private libraries, as well as interesting delivery methods such as interactive games and storytelling.

4.2.5 Creating message

The main message "Menjadi Tangguh untuk Lindungi Lingkungan" (become resilient to protect the environment) is explicitly stated as the title of the comic. Along with the symbolization of Sota character, a honey fox that is known to be very resilient in many environments, IRES creates a simple and relatable message through the story and characterization of Raya and Kak Bumi. Through the characters of Sota, Raya, and Kak Bumi, children can process the message and relate it to their daily life in order to influence their way of thinking. Environment-protecting actions such as throwing garbage in the trash bin, saving energy, and planting as many greens as possible at home, are promoted through the characters in the comic.

4.2.6 Planning the tactic

By conducting RTL events from June to July 2022, IRES set the tactic to reach as many children and educators in Jakarta, Bogor, Depok, Tangerang, and Bekasi areas as their short-term objective. IRES involves many stakeholders such as environmentalists, government bodies, school teachers, and other educational institutions to be part of the RTL journey. These stakeholders are playing an important role as the commentators, presenters, and promoters of the comic, especially to deliver the objective of the message which is to preserve and protect the environment.

4.2.7 Conducting evaluation

Followed by the RTL, IRES will arrive at the summit of the marketing public relations event which is the grand launching on 29 July 2022. This grand launching event will be the moment to collect the survey and monitoring efforts throughout the two promotional months. This is to analyze the trend, reception, and public expectation to the public.

IRES will also consider monitoring the media publication to check if there is any news coverage that helps the dissemination of the message or the program from this comic launch. As this is not the first time IRES receives a publication after previously their campaign was exposed during the pandemic era (Wahyudi 2021). This is what Aaker *et al.* (2012) claimed about how people acknowledge and be aware of the existing brand or product.

4.3 SDG goal 13 climate action

Realizing the important aspect of SDG conducted by every organization across the globe, IRES is seemed to aim for reaching SDG No. 13 Climate Action by creating this comic and aiming for the young generation that will feel the impact upon their growing up period. This comic contains stories and real-life day-to-day illustrations of how the young generation especially kids take part in creating a better environment and become more adaptable to climate change. The interview also excerpts that IRES is taking the step to contributing to SDG goal 13 by educating and influencing the children through artsy comic media.

5 CONCLUSION

By utilizing comics as the medium to communicate with children, IRES is attempting to create more awareness and understanding in the public about environmental issues. Their first comic, *Serial Belajar Bencana Bersama Sota: Menjadi Tangguh untuk Lindungi Lingkungan*, becomes the initial steps to educate children about the importance of resilience during and after the disaster, as well as promoting the simple idea of disaster prevention. To achieve this objective, IRES worked out the steps of Whalen's seven Strategic Planning steps: situation analysis, objective, strategy, targeting, message, tactic, and evaluation. The only part that has not yet been done is conducting the survey regarding their RTL event to gain feedback and know areas to be improved in the future for Sota Comic. Furthermore, recommendations can be made such as measuring the effectiveness of this campaign by conducting quantitative research from the perspective of the receiver of the comic or the relevant stakeholders.

REFERENCES

Aaker, J. L., Garbinsky, E. N., & Vohs, K. D. (2012). Cultivating Admiration in Brands: Warmth, Competence, and Landing in the "Golden Quadrant." *Journal of Consumer Psychology*, 22(2), 191–194. https://doi.org/10.1016/j.jcps.2011.11.012

Dabner, D., Stewart, S., & Vickress, A. (2017). *Graphic Design School: The Principles and Practices of Graphic Design* (D. Dabner, Ed.). John Wiley & Sons.

Escalas, J. E. (2004). Narrative Processing: Building Consumer Connections to Brands. *Journal of Consumer Psychology*, 14(1–2), 168–180.

Grunig, J. E. (2013). *Excellence in Public Relations and Communication Management*. Taylor & Francis.

Harris, T. L., & Whalen, P. T. (2006). *A Marketer's Guide to Public Relations in the 21st Century*. Texere.

Keller, K. L., & Kotler, P. (2012). *Marketing Management*. Pearson Education.

Miles, M. B., Huberman, A. M., & Saldaña, J. (2018). *Qualitative Data Analysis: A Methods Sourcebook*. Sage Publications.

Nairat, M., Nordahl, M., & Dahlstedt, P. (2020). Generative Comics: a Character Evolution Approach for Creating Fictional Comics. *Digital Creativity*, *31*(4), 284–301. https://doi.org/10.1080/14626268.2020.1818584

Samara, T. (2007). *Design Elements: A Graphic Style Manual*. Rockport Publishers.

Samudra, I. B. (2019). Media Integrated Marketing Communication Pada Komik Anti Golput Bergenre Slice of Life Sebagai Kampanye Bagi Pemilih Pemula Berusia Muda di Kota Surakarta. *Doctoral Dissertation*.

UN. (2022). *Climate Change*. United Nations.

Wahyudi, A. (2021). *Indonesia Resilience Membuka Layanan Dokter Virtual Gratis Bagi Pasien Isoman*. Barisan.Co.

ZA, S., Tricahyadinata, I., Robiansyah, R., Darma, D. C., & Achmad, G. N. (2021). Storytelling Marketing, Content Marketing, and Social Media Marketing on the Purchasing Decision. *Budapest International Research and Critics Institute (BIRCI-Journal): Humanities and Social Sciences*, *4*(3), 3836–3842.

@kiossahabat.id as social media marketing by Asperger individuals

S.U. Suskarwati, E.A. Puruhito, A.H. Simanjuntak, A.J.C. Sagala & B.R. Satria
Institut Komunikasi dan Bisnis LSPR, Jakarta, Indonesia

ORCID ID: 0000-0002-7932-7334

ABSTRACT: The use of social media, which was originally a network of friends, has now been used as a marketing communication model. One of them is used by the London School Beyond Academy (LSBA) which trains students and graduates of LSBA named @kiossahabat.id. This is an Instagram account that is managed as Social Media Marketing (SMM). LSBA accepts adolescents with special needs especially those categorized as Asperger. There are four steps in SMM: Listen, Communicate, Engage, and Collaborate. This study aims to understand how the implementation of these steps is carried out by Asperger individuals. A case study of a descriptive qualitative research methodology on @kiossahabat.id stated that the ability of Asperger individuals can be directed to carry out marketing activities mentored by their supervisor. In the communication concept, it was found that this concept is not only focused on the content material produced but responses to content itself are a form of communication.

Keywords: social media marketing, Asperger individuals, digital era

1 INTRODUCTION

One of the most popular applications in the new media era is what people called Social Networking sites, which have a significant role as one of interpersonal communication media. This particular communication channel connects one person to others and enables them to share pictures, and distribute links and videos (Lister *et al.* 2009). Internet has become an intermediary for the emergence of new media. The Internet is characterized as a computer-based technology, hybrid in character, not dedicated, flexible, and potentially interactive. Other characteristics are that it can function publicly and privately, regulations are not strict, are interconnected, do not depend on location, can be accessed by individuals, and become a medium of mass and private communication (McQuail 2011).

These characteristics are currently used in various aspects of communication, both for personal communication purposes and communication for business purposes. One of them is used for marketing purposes known as Social Media Marketing (SMM). It can be used to maintain or expand a brand that is tailored to marketing objectives (Chaffey & Smith 2017). SMM is the business use of selected social media to understand customers and to engage them in communication and collaborations in ways that lead to the achievement of ultimate marketing and business goals (Roberts & Zahay 2012).

Every marketing person has used this media, some of which are the Toyota, Daihatsu, and Honda Car Dealers in Bandung who apply social media marketing to increase sales (Akbar & Helmiawan 2018). It is also stated that sharing information with visitors or followers is not the only advantage of using social media for a business. The main thing is that a product sells in the market and is in great demand by customers (Valentika *et al.* 2020). Meanwhile, according to Strauss & Frost (2011), social media focused on providing information, photo, video, or any other content to the audience who wants to learn something new or need a source of entertainment. Through social media, audiences can be more interactive with one another and will be able to build social and business relationships based on existing interactions through social media.

SMM has also been utilized by beyond creativity as a business unit under the London School Beyond Academy (LSBA). Students used online media in carrying out marketing activities and customer service. LSBA is a skills training institution established by The London School of Public Relations (LSPR) Institute. LSBA accepts adolescents with special needs, especially people categorized as an Asperger, Attention Deficit Hyperactivity Disorder (ADHD), and Down syndrome to be trained in order to have working skills (Wempi & Chrisdina 2020).

Currently, LSBA is building a new communication medium used by students to learn to run a business. LSBA created an account on Instagram, namely @kiossahabat.id (KS) which is the first digital kiosk for friends of individuals with special needs (*@kiossahabat.Id • Instagram Photos and Videos* 2021). Understanding how LSBA students use this platform as SMM is certainly an interesting thing. Marketing through this new media is common and learned by non-disabled people, but this is something new for people with disabilities (Andina *et al.* 2020). Sucahyo stated that gadgets and technological developments are one of the factors that can break the chain of impoverishment for disabled groups to access information and communication and can open up space for empowering their groups to be independent in their livelihoods (Christiani *et al.* 2021).

According to Roberts & Zahay (2012), there are four steps of SMM Strategy: Listen, Communicate, Engage, and Collaborate. First step is Listen, meaning that it is mandatory for marketers to identify the social habitats of their defined target market as well as to pay attention to what people do when they are visiting social sites. Second is Communicate, which means that companies need to position themselves as creators and communicators of content. Customers are willing to listen to the meaning of the products, the benefits, and the experience as opposed to just communicating about the product itself. The third step is Engage, which is defined by actions done by the customers as desired by the marketers. The action of the customer is mainly the result of attitude-possibly brand loyalty and a positive attitude toward the brand. The last step is Collaborating, in which the content provides importance not just for the customers but also for the social media marketers. These steps will guide the author in understanding how to implement KS as an SMM managed by special needs students of LSBA.

2 RESEARCH METHODS

The research methodology in this research is the qualitative method. According to Bogdan and Taylor on Moleong (2011), qualitative methodology is a research procedure that will result in descriptive data in a form of written words or verbal from individuals and the behavior that is being observed. Data collecting for this research is done through an interview, which is the process of obtaining information for research purposes by doing face-to-face questions and answers between the interviewer with respondents/people interviewed, with or without using interview guidelines. The interviews are usually carried out either individually or in a group form so that the researcher can get oriented informatics data (Yusuf 2016). Secondary data sources can be obtained through observation and documentation.

The methodology used in this research is a case study. Meanwhile, Creswell (2010) describes a case study as a research strategy in which the author carefully investigates a program, event, activity, process, or group of individuals. The cases were limited by time and activity, and the authors collected complete information using various data collection procedures based on the allotted time.

3 RESULT AND DISCUSSION

The development of communication technology is currently utilized by various elements in the industry and provides opportunities for users to reach the market. Based on an interview with Chrisdina, the need for alternative employment opportunities for Asperger individuals, the digitalization era opened up opportunities to form a digital kiosk which was later named @kiossahabat.id (KS). Sembako (staple food) is a product offered by KS because it has a wide market and a long product shelf life. KS is unique because there are no product sales activities through digital platforms that involve Asperger individuals. This KS provides a forum for teamwork training for students and graduates of LSBA. One of the biggest difficulties for these Asperger individuals is working as a team. But with this KS, they become productive. The whole process consistently involves them starting from receiving orders, making product purchases, receiving orders, calculating how many items are ordered even packing and delivering orders. KS, which was formed in 2021, is a communication medium for marketing activities that trains them to work using digital platforms. Moreover, to introduce to the public, in this digital era, individuals with special needs can also work using digital applications (Chrisdina 2022).

The first step of the SMM strategy is to analyze the needs of the target market. The pandemic situation makes the market focus on basic daily needs. In addition, job opportunities for some people have also decreased, especially for Asperger individuals and their parents. Based on the interview with Hersinta, KS then became a source of good news because she needed information about products that had become her basic daily needs and was also very concerned with the activities of Asperger individuals. When she found about Instagram @kiossahabat.id, she immediately became a follower at that time (Hersinta 2022). This is relevant to the first step in SMM, whether they are your customers or people you want to attract as customers (prospects), the marketer must listen before speaking. Hermawan (2022) stated that the target market for @kiossahabat.id is the general public because they want to reach a wider market. Along with the parents of Asperger individuals, this platform is one of the implementations of learning activities that students get from LSBA. The student supervisors inform the intended target market and discuss with a student who involves to manage @kiossahabat.id.

Over Instagram id of KS, we could see that the marketers understand the target audience which are mostly people with special needs. The design and the user interface of each post have a very simple design, with easy-to-read information as well as color choices and fun templates. This is to cater to the awareness of the customer and to attract their attention to the KS content. Further, if the customer with special needs has questions, they could ask for assistance from their friends or parents who could also understand the content properly because it was designed in a simple matter.

The logical consequence of understanding what is needed by the market on the concept of listening in SMM is the basis for building the form and content of communication which will then be poured into KS. Product promos, new products, and testimonials are some of the types of content that appear on the features on Instagram KS, which resembles quite simple content. Here, it can be seen that KS Instagram content does not only mean the distribution of content material but how many people access and react to the content material is content. As quoted in the second step in SMM, it is no longer a one-marketer-to-many-audience members type of communication; it is many-to-many. Marketers no longer establish the rules of communication, which is more difficult, however, audience members do. The feeds in the KS Instagram facilitate communication in both ways. The customer could browse the posts on the feeds as well as the stories that are saved in Instagram highlights. When customers need further information, they could easily click the linktree that is stated on the profile section and it will be directed to several communication channels such as Whatsapp, E-Katalog for product offerings, and Instagram itself.

Figure 1. User interface of @kiossahabat.id.

Hermawan, one of the supervisors and marketing LSBA in producing content material on @kiossahabat.id, provides direction to students. These directions are tailored to the skills they have, such as creating content designs through Canva or creating captions. While providing direction, no significant obstacles were found because the direction had been adjusted to the abilities of the students. Previously, these students also received learning sessions in class, so that the supervisors could assess their skills.

Reactions and comments on content that are informed through KS's Instagram are a form of engagement. Likewise, the testimonials given by customers can build engagement. This is, of course, the meaning of the engagement concept described in the third step of SMM, where Instagram followers can take advantage of the Instagram feature to respond as a form of reciprocity from the content submitted. Based on the KS Instagram feeds, out of 33 posts, there was only one post was commented. The post was about the amount of Paket Ramadan the LSBA students produced and it showed how proud they are while working in a solid team. The post has 5 encouraging comments with a positive tone.

After consumers take advantage of KS services to meet their needs, consumers do not hesitate to provide testimonials. Testimonials given by KS consumers are a form of collaboration. This means that KS consumers have also carried out marketing activities. KS Instagram also encourages collaboration between the customers as well as the marketers. It was shown on the testimony highlighted how the customer had received the order safely and was satisfied with the products received. The testimony was reposted by the admin of KS Instagram. For every activity at @kiossahabat.id, the supervisors always involve students, including in reposting testimonials. Overall, the activities carried out in managing @kiossahabat.id were from beginning to end. The goal is to provide them with learning to understand the use of digital platforms in marketing activities. Therefore, after they graduate from LSBA, they can carry out this activity independently.

4 CONCLUSION

Managing @kiossahabat.id as SMM by Asperger individuals is not a simple thing. There are four steps in developing SMM through the KS digital platform that requires assistance. The features attached to digital communication media such as KS have indeed helped marketing activities in selling product (Sembako) that trains them to work by using digital platforms. In carrying out all these marketing activities, Asperger's individuals are guided and directed according to their abilities.

Meeting the needs of the intended target market is likely to be successful if the steps of SMM are consistently executed correctly. The novelty offered by this research is that the communication concept in the second step of SMM provides an understanding that when distributing content, we do not only interpret the content itself, but also the reactions and comments it elicits. It sharpens knowledge about existing forms of concept communication in SMM. The potential research is to analyze the use of quantitative methods to discover the impact of Instagram posts in increasing the awareness and intention to buy KS.

REFERENCES

@kiossahabat.id • *Instagram Photos and Videos*. (2021). https://www.instagram.com/kiossahabat.id/

Akbar, Y. H., & Helmiawan, M. A. (2018). Penerapan Strategi Social Media Marketing Untuk Meningkatkan Penjualan Pada Dealer Mobil Toyota, Daihatsu dan Honda Bandung. *Infoman's: Jurnal Ilmu-Ilmu Manajemen Dan Informatika*, *12*(2), 115–124.

Andina, A. N., Zahra, E. F., & Munirruddin, M. (2020). Pemanfaatan Sosial Media untuk Digital Marketing Bagi Anak Berkebutuhan Khusus di SLB B Yakut Purwokerto. *Jurnal Pemberdayaan: Publikasi Hasil Pengabdian Kepada Masyarakat*, *4*(3), 245–250.

Chaffey, D., & Smith, P. R. (2017). *Digital Marketing Excellence: Planning, Optimizing and Integrating Online Marketing*. Routledge.

Chrisdina. (2022). *Personal Interview*.

Christiani, L. C., Ikasari, P. N., & Nisa, F. K. (2021). Pengembangan Kemandirian Kelompok Difabel Melalui Pemanfaatan Pemasaran Digital di Kota Magelang. *Dinamisia: Jurnal Pengabdian Kepada Masyarakat*, *5*(2), 276–286.

Creswell, J. W. (2010). *Research Design Pendekatan Kualitatif, Kuantitatif, dan Mixed*. Yogyakarta: Pustaka Pelajar.

Hermawan, I. (2022). *Personal Interview*.

Hersinta. (2022). *Personal Interview*.

Lister, M., Dovey, J., Giddings, S., Grant, I., & Kelly, K. (2009). *New Media: A Critical Introduction 2nd Edition*.

McQuail, D. (2011). *Teori Komunikasi Massa*. Salemba Humanika.

Moleong, L. J. (2011). *Metodologi Penelitian Kualitatif, Cetakan XXIX*. Bandung: PT. Remaja, Rosdakarya.

Roberts, M. Lou, & Zahay, D. (2012). *Internet Marketing: Integrating Online and Offline Strategies*. Cengage Learning.

Strauss, J., & Frost, R. (2011). *E-Marketing*. Prentice Hall Press.

Valentika, N., Zenabia, T., Muslim, M., Rosini, N. I., & Nining, N. (2020). Implementasi Sosial Media Marketing Dalam Meningkatkan Jaringan Pasar. *Jurnal Pengabdian Kepada Masyarakat (JPKM)- Aphelion*, *1*(01), 68–74.

Wempi, J. A., & Chrisdina. (2020). Human Communication Online Social Entrepreneur By Special Needs Students. *Searching for the Next Level of Human Communication: Human, Social, And Neuro (Society 5.0)*, 320–330. http://ic.aspikom.org/wp/accepted/

Yusuf, A. M. (2016). *Metode Penelitian Kuantitatif, Kualitatif & Penelitian Gabungan*. Prenada Media.

Forging ASEAN's cultural identity through digital museum diplomacy

B. Riyanto, A.H. Assegaf, Y.W. Kurniawan, C. Mawuntu & G. Aulia*
Institut Komunikasi dan Bisnis, LSPR, Jakarta, Indonesia

*ORCID ID: 0000-0001-6330-0308

ABSTRACT: Regional identity increasingly becomes an important concept to construct ASEAN Community. Cultural diversity is a challenge in itself in realizing a single ASEAN identity based on the interests of the people. Through the concept of museum diplomacy, this paper seeks to dig deeper into the role of the ASEAN Cultural Heritage Digital Archive (ACHDA) in preserving cultural heritage, as well as bridging intercultural understanding, promoting regional identity, and encouraging collaboration. Using the descriptive qualitative method, this study explores the construction of the concept of ASEAN identity in documents and policies agreed by ASEAN leaders, as well as the strategic position of ACHDA in opening wider access to cultural contacts and cooperation between communities through digitalization. Although still in its early stage since the inception, ACHDA's role in promoting ASEAN Cultural Identity is promising in two ways: increasing cultural interaction based on contacts between ASEAN peoples, and encouraging a greater international role for the museum through exhibitions and cultural research.

Keywords: ASEAN Cultural Identity, Museum Diplomacy, Cultural Heritage, Digitalization

1 INTRODUCTION

Since its establishment in 1967, ASEAN has been trying to find a shared identity as the glue that drives the region. Although the formation of ASEAN was dominated by elite interests, and great concern was the issue of regional security amid the Cold War. The need to build a regional identity that transcends geographical proximity is needed to further explore the role of wider cooperation. The need for a common identity is increasingly pressing, especially with the end of the cold war which previously defined common interests at the beginning, coupled with the joining of countries such as Cambodia, Laos, Myanmar, and Vietnam (CLMV) in the 1990s which were almost without standardization in their membership. This change is a challenge for ASEAN in finding common ground in the next cooperation scheme (Severino 2008).

Despite the diversity of historical backgrounds, cultures, economic affiliations, and political systems, ASEAN has evolved into a regional organization capable of creating a conducive climate for regional cooperation, stability, and peace (Mahbubani & Sng 2017). Recognizing the development of increasingly complex issues and challenges, the need to build a more solid shared identity is explicitly stated in the ASEAN motto agreed upon by ASEAN leaders in 2005: "One Vision, One Identity, One Community." The motto departs from the need to build a regional organization that is more inclusive and reflects the interests of the people of ASEAN in a true ASEAN Community.

To create a narrative of shared identity at the community level, steps are needed to bring the community closer, especially in the cultural realm. In contrast to political, security, and economic issues which are largely dominated by the states, cultural diplomacy places more emphasis on public participation in relations in the region and therefore has more potential to create an ASEAN identity centered on the interests of the ASEAN public. ASEAN's efforts to bring cultural heritage closer together by establishing The ASEAN Cultural Heritage Digital Archive (ACHDA) is an interesting strategic step to realize the ASEAN identity.

This research seeks to dig deeper into the potential of digital museum diplomacy adopted by the ACHDA in bringing cultural diversity closer through digitizing cultural heritage artifacts presented from various museums, galleries, and national libraries in the ASEAN region. Since ASEAN launched the ACHDA website on 27 February 2020 at the ASEAN Secretariat, the Deputy Secretary-General for ASEAN Socio-Cultural Community, Kung Phoak said that digitalization is part of an important milestone for preserving cultural heritage and instilling a greater regional sense of belonging toward fostering ASEAN identity (NTT DATA Corporation 2020).

This paper will be divided into three parts. First, discussing the concept of ASEAN's identity in facing its strategic challenges while also looking at the narrative point of view of identity both internally and externally. Second, looking at ACHDA's strategic role in preserving cultural heritage through digitizing museums, galleries, and libraries. Third, linking the promise and pitfall of the digital museum diplomacy as an effort to realize ASEAN identity.

2 RESEARCH METHODS

This study uses descriptive qualitative methods to understand more deeply about the concept of ASEAN identity, as well as an overview of the role of museum digitization at the regional level to create an ASEAN Community. The document study was chosen to look at the historical stages in the construction of the ASEAN identity concept, narratives about the importance of ASEAN identity to realize a people-centred community, and the role of ACHDA in encouraging a sense of belonging to regional communities with people-to-people contact. Several official documents released by ASEAN such as ASEAN 2025: Forging ASEAN Ahead, The Narrative of ASEAN Identity, as well as a number of official statements from officials and heads of state, related to ASEAN Identity and ACHDA, both through releases and media statements are also the focus of this research. Due to limited time and funding, this research will also use a document from previous studies released from credible institutions to understand people's perceptions of ASEAN Identity, ASEAN Community, people-to-people contact, and cultural understanding.

3 RESULT AND DISCUSSION

3.1 *The construction of ASEAN identity*

ASEAN agreed on a strategic step through the document, The Narrative of ASEAN Identity (NAI), which was adopted at the 37th ASEAN Summit, on 12 November 2020. Regarding the commitment to deepen regional integration through the three pillars of the ASEAN Community, namely political and security, economy, and socio-cultural communities, the document states: "ASEAN Identity shall strengthen the ASEAN Community. ASEAN Identity will enhance common values with a higher degree of we-feeling and sense of belonging and sharing in all the benefits of regional integration" (ASEAN Secretariat 2020).

The NAI was first initiated by Indonesia when Indonesia was the Chair of Indonesia at the ASEAN Senior Officials Meeting for Culture and Arts (SOMCA). The Indonesian Minister of Foreign Affairs, Retno Marsudi, also conveyed three important points in the opening of the "ASEAN Virtual Cross Pillars Consultation on The Narrative of ASEAN Identity." In this activity, she raised three points including (1) ASEAN identity must be the unifying soul of ASEAN, (2) NAI is ASEAN's source to convey better information about ASEAN to the general public and serve as the basis for ASEAN relevance, (3) ASEAN identity should also be promoted outside the ASEAN region (Ministry of Foreign Affairs of The Republic of Indonesia 2020).

The concept of ASEAN identity could be traced back to the 10th General Principle of the Bali Concord II, adopted in 2003, which states that "ASEAN will continue to develop caring communities and promote a shared regional identity." Among the objectives enshrined in the ASEAN Charter adopted in 2008: "To promote ASEAN identity through the development of greater awareness of the region's cultural diversity and heritage" (ASEAN Secretariat 2008). Since then, ASEAN has consistently emphasized the slogan "One Vision, One Identity, One Community" in many of its official statements and documents (ASEAN Secretariat 2015).

According to Acharya (2018), there are two critical attributes in a community. The first is trust, friendship, complementarity, and responsiveness, and the second is people with the same cultural and physical characteristics, showing mutual responsiveness, self-confidence, and self-esteem, which can identify them as a unit. Identity can not only be formed through social construction using socialization and habit-forming instruments but also through cultural and historical ties. Therefore, it is necessary to have a shared sense of identity to build a community so that what is done or done has legitimacy and strong reasons to be carried out or applied collectively (Acharya 2018).

Acharya's statement is in line with NAI which states that the regional identity built to support the realization of the ASEAN community is based on constructed values and inherited values. Acharya stated that the ASEAN identity is more a part of the form of constructed values that are constantly changing through elite socialization, and thus more sensitive to social and political manipulation. This is different from the Southeast Asian identity which comes from a long historical interaction in response to the strategic environment, and is, therefore, more permanent. However, the two complement each other in fostering a sense of community (Acharya 2018).

3.2 *Digitalization Southeast Asian cultural heritage*

As a region consisting of countries that have cultural similarities, the ASEAN created a digital museum containing historical objects from its member countries called ASEAN Cultural Heritage Digital Archive (ACHDA) which can be accessed at heritage.asean.org. ACHDA was launched in February 2018 under the auspices of the ASEAN Senior Officials for Culture and Arts (SOMCA) and is funded by the Japanese government through the Japan-ASEAN Integration Fund (JAIF).

ACHDA has currently digitized 279 items from Indonesia, Malaysia, Cambodia, Myanmar, and Thailand. ACHDA will expand its digitalization efforts in the future to include cultural artifacts from other ASEAN countries. Its existence honors ASEAN's rich and diverse cultural heritage by displaying a veritable treasure trove of outstanding artifacts and history.

ACHDA hopes to raise greater awareness and appreciation of ASEAN's shared cultural heritage by sharing ASEAN's beloved cultural treasures online and also ASEAN identity as the main goal. Recently ACHDA collects artifacts from various museums, galleries, and libraries from Thailand, Malaysia, Indonesia, Cambodia, and Myanmar and still continuing. The ASEAN Cultural Heritage Digital Archive (ACHDA) trying to showcase many different models. Until now, there are 194 images, 84 3D visuals, and 1 audio/visual. If divided by form there are 134 objects, 126 artworks, 18 books and documents, and 1 building. During the pandemic, ACHDA launches its first e-exhibition, with Forging History: Metals in the Crucible of ASEAN's Transformation as the main theme.

Globalization and the development of ICTs encourage the expansion and inclusion of cyberspace (Eggenschwiler & Kulesza 2020), revolutionize the communication process, and provide space for people to participate and contribute to international affairs (Adesina 2017). This disruption gave birth to the "digitalization of diplomacy" or "digital diplomacy" which explains the phenomenon of the utilization of digital tools for diplomatic and international affairs (Manor & Segev 2015). Digital diplomacy is an effective means and strategy for managing change through bottom-up, dialogical, and participatory engagement among stakeholders (Manor & Segev 2015), and it has been used by many Asian countries such as Japan, China, and South Korea to carry out cultural diplomacy strategies through preservation, conservation, and promotion of historical, artistic, and cultural heritage to build a positive national image and represent national interests (Akagawa 2015; Chan 2020).

However, the issue of "heritage diplomacy" demands bilateral or multilateral cooperation as artistic and cultural heritage, both tangible and intangible, has existed even before modern states were (Wang et al. 2020). However, the essence of digital diplomacy that should not be forgotten in this process is the existence of two-way, multi-directional, and multi-actor communication engagement which also portrays the difference between dialogical digital diplomacy from the monologic traditional diplomacy of the past (Turianskyi & Wekesa 2021).

Former Deputy Secretary-General of ASEAN Kung Phoak (2019) stated that ASEAN has a great interest in utilizing the latest digital technologies such as VR (i.e., virtual reality) and AI (i.e., artificial intelligence) as well as collaborative efforts to promote Southeast Asian arts and culture in a creative, innovative, and inclusive way possible. Through arts and culture, ASEAN seeks to promote and celebrate a culture of optimism and pluralism, as well as the values of tolerance, moderation, and mutual trust that ASEAN upholds to overcome the challenges of cultural pessimism, extremism, and intolerance (Phoak 2019).

3.3 *Museum diplomacy and regional cultural identity*

A country needs to preserve, conserve, and promote national historical, artistic, and cultural heritage at the international institutional level because it not only stores cultural and historical values but is also able to enhance and strengthen the national image for the people of that country and the world (Chan 2020). At the regional level, Southeast Asian countries need to make collaborative efforts to build a regional arts and culture ecosystem that is not only able to prosper the community through the production, consumption, and distribution of artistic and cultural works but also to spread shared noble values from local areas across Southeast Asian nations to form a regional collective identity (Farid 2020).

The development of this ecosystem can be actualized through an effort called "museum diplomacy." The museum is now seen not only as a cultural institution but also as a diplomatic actor which has great potential to empower and foster communities' aspirations, as well as bridge the local and the global through various collaborative programs such as loans, exhibitions, research, collections, people-to-people connections, traveling exhibitions, satellite museums, museum summitry, and digital initiatives (Priewe 2021). In the past, museums were only a mouthpiece for the national interest of states, but now museums are required to be more inclusive and be able to collaborate and become an inclusive international arena. The digitization of museums and online tools is helping museums around the world to be more interconnected, accessible, and efficient (Grincheva 2021). Through museum diplomacy, the identity and image of a nation can be constructed through a dialogical, bottom-up, collaborative, people-centered process from a local to a global scale (Grincheva 2019).

ASEAN as a regional organization has realized the importance of museum collaboration and cooperation to build identity and promote the shared noble values of the Southeast Asian region. The digitization of museums is therefore also an amalgamation between efforts to maintain cultural heritage while maximizing the role of museums which have traditionally been internationalized through exhibitions and research collaborations, and of course become part of cultural diplomacy efforts that cannot be underestimated, especially in increasing people-based cooperation.

4 CONCLUSION

ASEAN identity is not a new concept, but its relevance, perspective, and meaning are evolving along with the changing global strategic environment. The ASEAN leaders' agreement to adopt the same narrative in seeing their identity as a region is an effort to bring its people closer to a larger role in an ASEAN Community and, at the same time, becomes a foothold when dealing with common challenges at the global level. It is clear that the ASEAN identity is not an effort to realize singularity but rather an effort to encourage unity amidst pressing global challenges. Cultural diversity across Southeast Asian countries is a valuable asset that needs to be maintained and becomes the basis for ASEAN's efforts to forge a shared identity as a community. In this case, the existence of ACHDA is a measure made by ASEAN that strategically combines constructed values through socialization by the elite and inherited values through the existing rich cultural heritage and history of its people. The digital museum built by ASEAN through ACHDA is a combination of preserving cultural heritage while increasing the role of international museums through exhibitions and research and also using culture as a bridge to increase the role of people at the center of the formation of the ASEAN Community.

REFERENCES

Acharya, A. (2018). The Evolution and Limitations of ASEAN Identity. In *Building ASEAN Community: Political–Security and Socio-cultural Reflections.* https://www.eria.org/ASEAN_at_50_4A.2_Acharya_final.pdf

Adesina, O. S. (2017). Foreign Policy in an era of Digital Diplomacy. Http://Www.Editorialmanager.Com/Cogentsocsci, 3(1), 1297175. https://doi.org/10.1080/23311886.2017.1297175

Akagawa, N. (2015). *Heritage Conservation and Japan's Cultural Diplomacy: Heritage, National Identity and National Interest.* https://www.routledge.com/Heritage-Conservation-and-Japans-Cultural-Diplomacy-Heritage-National/Akagawa/p/book/9781138629172

ASEAN Secretariat. (2008). *The ASEAN Charter.* www.asean.org

ASEAN Secretariat. (2015). *ASEAN 2025: Forging Ahead Together.* https://www.asean.org/wp-content/uploads/2015/12/ASEAN-2025-Forging-Ahead-Together-final.pdf

ASEAN Secretariat. (2020, November 12). *The Narrative of ASEAN Identity.* https://asean.org/wp-content/uploads/2021/08/The-Narrative-of-ASEAN-Identity_Adopted-37th-ASEAN-Summit_12Nov2020.pdf

Chan, V. C. (2020). Heritage Conservation as a Tool for Cultural Diplomacy: Implications for the Sino-Japanese Relationship. In W. S.-L. and M. H. H.-H. P. Philippe (Ed.), *Heritage as Aid and Diplomacy in Asia* (pp. 167–189). ISEAS Publishing. https://doi.org/10.1355/9789814881166-008/HTML

Eggenschwiler, J., & Kulesza, J. (2020). Non-State Actors as Shapers of Customary Standards of Responsible Behaviour in Cyberspace. In *Broeders D, van den Berg B, editor, Governing Cyberspace: Behavior, Power and Diplomacy* (pp. 245–262). https://rowman.com/ISBN/9781786614940/Governing-Cyberspace-Behavior-Power-and-Diplomacy

Farid, H. (2020, May 1). Fostering ASEAN Identity Through Collaborative Efforts in Cultural Ecosystem. *The ASEAN Magazine,* 14–15. https://asean.org/wp-content/uploads/2017/09/The-ASEAN-Magazine-Issue-1-May-2020.pdf

Grincheva, N. (2019). *Glocal Diplomacy of Louvre Abu Dhabi:* Museum Diplomacy on the cross-roads of local, National and Global Ambitions. Https://Doi.Org/10.1080/09647775.2019.1683883, 35(1), 89–105. https://doi.org/10.1080/09647775.2019.1683883

Grincheva, N. (2021). *Museum Diplomacy in the Digital Age.* https://www.routledge.com/Museum-Diplomacy-in-the-Digital-Age/Grincheva/p/book/9780815369998

Mahbubani, K., & Sng, J. (2017). *The ASEAN Miracle: A Catalyst for Peace.* NUS Press.

Manor, I., & Segev, E. (2015). America's Selfie: How the US portrays itself on its social media accounts. In *Bjola, C., Holmes, M., Digital Diplomacy: Theory and Practice* (pp. 89–108). Routledge. https://doi.org/10.4324/9781315730844-7

Ministry of Foreign Affairs of The Republic of Indonesia. (2020, September 1). *Indonesian Minister Of Foreign Affairs: Narrative Of Asean Identity Crucial To Raise Awareness Toward Asean And Its Relevance*. https://kemlu.go.id/portal/en/read/1636/berita/indonesian-minister-of-foreign-affairs-narrative-of-asean-identity-crucial-to-raise-awareness-toward-asean-and-its-relevance

NTT DATA Corporation. (2020). *Creates and Makes Public a Digital Archive of the Historical Cultural Heritage of the ASEAN Region*. https://www.nttdata.com/global/en/media/press-release/2020/february/ntt-data-creates-and-makes-public-a-digital-archive-of-the-historical-cultural-heritage

Phoak, K. (2019, March). Bringing People Together Through Arts and Cultures. *ASEAN Focus*, 8–9. https://think-asia.org/handle/11540/11008

Priewe, S. (2021). *Museum Diplomacy: Parsing the Global Engagement of Museums Perspectives* (Vol. 2, p. 2021). www.figueroapress.com

Severino, R. (2008). *ASEAN*. Institute of Southeast Asian Studies. https://books.google.co.id/books?id=YC-xOF_04jQC

Turianskyi, Y., & Wekesa, B. (2021). African digital diplomacy: Emergence, evolution, and the future. *South African Journal of International Affairs*, *28*(3), 341–359. https://doi.org/10.1080/10220461.2021.1954546

Dance as the medium to develop interpersonal communication for persons with Down syndrome in a digital era

M.Y. Cobis, G.A. Mulyosantoso, K. Syahna, M. Wiguna & K.I. Septiani
Institut Komunikasi dan Bisnis LSPR, Jakarta, Indonesia

ORCID ID: 0000-0003-1074-9667

ABSTRACT: This paper is aimed to analyze dance as the medium to develop interpersonal communication for persons with Down syndrome, especially in using digital platforms or online classes. The main theory used in this research is interpersonal Communication by De Vito. The data collection used in this study was using interviews with structured questionnaires as the guide for interview flow. The questionnaire was adapted from Functional Idiographic Assessment Template-Questionnaire (FIAT-Q), and analyzed using a qualitative approach. The result has shown that dance could develop interpersonal communication skills of persons with Down syndrome.

Keywords: dance, down syndrome, interpersonal communication, performing arts

1 INTRODUCTION

In everyday life, humans cannot be separated from communication to interact with people around them. Communication is very often carried out by each individual as a process for socializing and interacting with other individuals and the surrounding environment, for instance, interpersonal communication. Interpersonal communication itself can be interpreted as the process of sending and receiving messages between two people or between a small group of people, with various effects and feedback (Sugiarto & Mutiah 2021). Through transmitting, a communication process occurs, namely the delivery of messages (both verbal and non-verbal). The process of the interpersonal communication pattern is known as one way without feedback, such as the interaction model with feedback and transactional pattern, which include the inclusion of attitudes, beliefs, self-concept, values, and communication skills.

As one of the types of communication, the non-verbal messages are manifested in a dance. Dance is a rhythmic body movement performed in a place and time to express feelings, intentions, and thoughts. Body movements and attributes used in dance are signs used to communicate. Every move in dancing has a meaning or message to be delivered. In dance, movements are made and adjusted based on a musical accompaniment.

Current digital era has enabled innovative education and digital technology-based classroom. It produces a more varied manner of performance method, and it is also impacting in more in line with the direction and goals of development according to the existing situation (Mattsson & Lundvall 2013).

Moreover, the social restriction during the COVID-19 pandemic has a significant impact on many people, such as the high level of isolation. This is due to certain health conditions and age-related risks associated with the severity of viral infection. The digital platform has played an important role to maintain communication and activities. The use of digital tools to maintain social relationships has been increasing during this pandemic (Bek *et al.*, n.d.) including for those with Down syndrome.

The particular of Down syndrome as a genetic disorder brings about when abnormal cell division results in extra genetic material from chromosome 21. Common characteristics of a person with Down syndrome are "low IQ, epicanthal folds (in which the skin of the upper eyelid forms a layer that covers the inner corner of the eye), short and broad hands, and below-average height" (Russell 2010). Many people with Down syndrome display hypotonia (low muscle tone) and are slow to develop movement skills as a result of impaired neurological development (Jobling et al. 2013). Specifically, the cerebellum of people with Down syndrome is smaller and less dense than that of their typically developing peers (Pinter et al. 2001), which results in balance, motor control, and motor learning deficits. In addition, the corpus callosum is thinner causing disrupted efficiency of bimanual movement (Ringenbach et al. 2012). These individuals may also have difficulty with goal-directed functioning as they have a smaller frontal lobe (Cebula et al. 2010) and may have hearing complications as the temporal gyrus is narrower than those with typical development (Pinter et al. 2001). Regardless of the cognitive and motor delays that accompany this chromosomal mutation, persons with Down syndrome are often able to make their own decisions, attend school/work, participate in their communities, and lead a fulfilling life of National Down Syndrome Society. For some individuals, taking part in community-based activities such as dance might be an important aspect in leading up to a fulfilling life.

Dance programs using digital platforms have shown unique advantages. Based on qualitative research, respondents showed appreciation for the new opportunities such as the variety of dance styles, the connection with others from distant locations, and the acceleration in learning of skills. This also applied to those with Down syndrome, where digital platforms provide a safe space to learn. The convenience, variety, and frequency of participation provided by digital resources resulting in the increasing desire from the respondents to continue participating in online dance (Lee et al. 2015)

Based on a study, virtual dance practice offers active learning and motivation to those with certain conditions such as Parkinson's disease. It also increases functional independence (Yavuzer et al. 2008) and daily living activities (Zhang et al. 2003). The use of dance in digital platforms can be useful for people with Down syndrome to improve their daily life activities, due to their vulnerability to decreasing daily activities.

2 LITERATURE REVIEW AND RESEARCH METHODS

2.1 *Dance*

Dance incorporates music and personal expression. Using music in a dance experience can enhance one's ability to create new memories and call upon old ones as it stimulates the medial prefrontal cortex, an area of the brain that is "involved in making prediction, paying attention, and updating the events in memory" (Alpert 2010). The combination of all of these factors makes dance a valuable activity that can be offered through community-based programs for those with and without disabilities, as well as therapeutic treatments for individuals with any combination of cognitive needs, social and physical. The Arts Council of England states that "Anyone can enjoy dancing regardless of their age or background, if they are disabled or non-disabled, whether or not they have danced before, and whatever their shape and size" (Arts council England. 2006). Many levels of development are impacted by dance, including, but not limited to, physical, social, and psychological (Duggan et al. 2009). Physical benefits include increased coordination, endurance, strength, and motor abilities such as gait speed, flexibility, and balance. Socially, dance fosters communication skills and reduces social anxiety while psychological benefits may consist of increased self-esteem, self-confidence, and quality of life (Becker & Dusing 2010). A rhythmic dance is performed to express feelings, thoughts, and messages simultaneously (Giguere 2011).

2.2 Interpersonal communication

Interpersonal communication is described as a communication between two individuals, who interact with each other, and give feedback to one another. Therefore, communication is an act of human beings born with full awareness, even human beings are actively born because there are certain objectives (Pinatih *et al.* 2018). According to Murtiningsih, Interpersonal communication is described as communication between two people where individuals physically interact with each other and provide feedback (Murtiningsih *et al.* 2019). There are elements of Interpersonal communication theory namely Openness, Empathy, Supportive, Positive Attitude, and Equality (Devito 2011).

2.3 Research methodology

Using descriptive qualitative research by analyzing events, circumstances, social phenomenon that exist in society. Documentation as evidence of events in research. Field analysis was obtained from the results of observations, interviews, and documentation. This method was phenomenology with data collection techniques in the form of interviews and observations. There were 10 participants in this study who qualified as study subjects. The criteria of the subject are persons with Down syndrome, aged 20–30 years old, accompanied by parent or family, and joined in dance class for at least 6 months and followed the online dancing class regularly. The data collection used in this study was using interviews with structured questionnaires as the guide for interview flow. The questionnaire was adapted from Functional Idiographic Assessment Template-Questionnaire (FIAT-Q) developed by Callaghan. FIAT-Q is a functional-analytic-behavioral approach for assessing interpersonal functioning for use in therapeutic modalities such as functional-analytic psychotherapy to develop interpersonal communication skills. This tool allows for standardized tracking of problems and improvements in interpersonal functioning across areas of functioning. There are five important areas/classes of behavior: Class A-E with different behavior to be observed. This study focused on Class C: disclosure of interpersonal closeness. This class observed the aspect of interpersonal functioning. Then, the questionnaire was modified to be the interview guideline to gather detailed information from participants' interpersonal communication (Callaghan 2014).

3 RESULT AND DISCUSSION

3.1 Openness

Being open here means letting others know what we are thinking and feeling. People with Down syndrome are naturally more open to their caregivers, so when learning online, people with Down syndrome are freer to express their misconceptions about the classes they are taking. Eight out of 10 respondents expressed their feelings of happiness when attending a dance class because they meet their friends and teacher. During the interview, respondents showed openness to the researcher by wanting to answer all the questions asked even though their close relatives who accompanied them during dance practices and performances. The desire to answer came spontaneously and was expressed with a happy expression. They are also happy to retell their dance class experiences to those around them. In addition, they are also willing to open a conversation with their interlocutor.

3.2 Empathy

Empathy is the ability to sense feelings from the perspective of others, to build healthy interpersonal relationships (Sugiarto & Mutiah 2021). In a dance group, each member has the ability to receive and manage different information. When asked about the response to one member who made a mistake, all respondents did not show negative expressions or responses. All respondents said that if a member made a mistake, they would be assisted to correct it and remind group members of the error. They are taught to always prioritize teamwork, to care for one another so no one falls behind, and everyone can dance happily together on stage and create great memories together.

3.3 *Supportive*

Support is important in encouraging people to communicate (Ulfah *et al.* 2018). All respondents claimed to get support from the people around them, especially support from their parents. Support from the people around them is also very influential for persons with Down syndrome because it can increase their self-confidence and opens up many opportunities. They also receive verbal support for persons with Down syndrome. In every performance they put on, the dance teacher and their companion will convey a sense of happiness and pride to them by saying constructive words such as "good job," "you are the best." The appreciation given is also expressed nonverbally by hugging and kissing an expression of pride to persons with Down syndrome. Most of the time, for the parents, they also show tears of joy and pride after many performances.

3.4 *Positive attitude*

In interpersonal communication, a positive attitude is a sense of self-respect, respect for others, and respect for the communication situation itself. All individual feelings are reflected in the words whether they are positive or negative. This feeling also affects the outcome of the conversation, just as someone is satisfied or dissatisfied with the conversation. Negative feelings such as anger, resentment, and hurt make communication difficult and ultimately turn off the communication (Ulfah *et al.* 2018). One in 11 people stated that he was annoyed because he could not follow the movement. This also fosters a sense of enthusiasm and confidence in the individual to continue to attend dance classes and tirelessly continue to try to follow the movements given by the teacher. The positive attitude that is grown in the classroom is also shown through the form of teaching which includes how to express and move the body, so that this can also have a positive impact when they appear in public, such as smile, and keep following the music that accompanies their movements.

3.5 *Equality*

Equality is a mutual understanding that both parties have an interest, and they are equally valuable and precious, and each requires the intention of equality here is the recognition or awareness (Ariyani & Hadiani 2019). Every child in the class has the same right or equal opportunity to get the best learning from their teacher. This is done so that each individual has self-awareness, in the sense of knowing the talents that exist within themselves. That way, they can accept their own strengths and weaknesses. They become confident in themselves and finally they become independent to make their choices. Finally, they can learn their own dance moves through the video given by the teacher at home.

4 CONCLUSION

The conclusion of dance as the medium of interpersonal communication with children with down syndrome apply effective communication such as, through dance. Persons with Down syndrome will have openness in communicating with others by trusting them. They have confidence in starting a conversation and develop their listening skill in receiving directions. By dance, persons with Down syndrome have self-awareness therefore they have empathy and respect for others. Supportive people who appreciate their talents surround persons with Down syndrome. Through the medium of dance, the positive attitude is shown from the way they smile and perform before audience. The dance teacher and persons build equality of communication with Down syndrome through dance. Dance using a digital platform could be the most useful tool to increase interpersonal communication for people with down syndrome.

REFERENCES

Alpert, P. T. (2010). The Health Benefits of Dance: Http://Dx.Doi.Org/10.1177/1084822310384689, *23*(2), 155–157. https://doi.org/10.1177/1084822310384689

Ariyani, E. D., & Hadiani, D. (2019). Gender Differences in Students' Interpersonal Communication. *Responsible Education, Learning and Teaching in Emerging Economies*, *1*(2), 67–74. https://doi.org/10.26710/relate.v1i2.1125

Arts council England. (2006). *Dance and Health the Benefits for People of All Ages.*

Becker, E., & Dusing, S. (2010). Participation is Possible: A Case Report of Integration into a Community Performing Arts Program. *Physiotherapy Theory and Practice*, *26*(4), 275–280. https://doi.org/10.3109/09593980903423137

Bek, J., Leventhal, D., Groves, M., Growcott, C., & Poliakoff, E. (n.d.). *Moving online: Experiences and Potential Benefits of Digital Dance for Older Adults and Individuals with Parkinson's Disease.* https://doi.org/10.31234/OSF.IO/954F2

Callaghan, G. M. (2014). The Functional Idiographic Assessment Template (FIAT) System: For Use with Interpersonally-based Interventions Including Functional Analytic Psychotherapy (FAP) and FAP-Enhanced Treatments. *The Behavior Analyst Today*, *7*(3), 357. https://doi.org/10.1037/H0100160

Cebula, K. R., Moore, D. G., & Wishart, J. G. (2010). Social Cognition in Children with Down's syndrome: Challenges to Research and Theory Building. *Journal of Intellectual Disability Research: JIDR*, *54*(2), 113–134. https://doi.org/10.1111/J.1365-2788.2009.01215.X

Devito, J. A. (2011). *Komunikasi Antar Manusia. Tangerang.* Karisma Publishing Group.

Duggan, D., Stratton-Gonzalez, S., & Gallant, C. (2009). *Dance Education for Diverse Learners: A Special Education Supplement to the Blueprint for Teaching and Learning in Dance.* New York, NY: Department of Education.

Giguere, M. (2011). Dancing Thoughts: An Examination of Children's Cognition and Creative Process in Dance. Https://Doi.Org/10.1080/14647893.2011.554975, *12*(1), 5–28. https://doi.org/10.1080/14647893.2011.554975

Jobling, A., Virji-Babul, N., & Nichols, D. (2013). Children with Down Syndrome. Http://Dx.Doi.Org/10.1080/07303084.2006.10597892, *77*(6), 34–54. https://doi.org/10.1080/07303084.2006.10597892

Lee, N. Y., Lee, D. K., & Song, H. S. (2015). Effect of Virtual Reality Dance Exercise on the Balance, Activities of Daily Living, and Depressive Disorder Status of Parkinson's Disease Patients. *Journal of Physical Therapy Science*, *27*(1), 145–147. https://doi.org/10.1589/JPTS.27.145

Mattsson, T., & Lundvall, S. (2013). The Position of Dance in Physical Education. Http://Dx.Doi.Org/10.1080/13573322.2013.837044, *20*(7), 855–871. https://doi.org/10.1080/13573322.2013.837044

Murtiningsih, M., Kristiawan, M., & Lian, B. (2019). The Correlation Between Supervision of Headmaster and Interpersonal Communication with Work Ethos of The Teacher. *European Journal of Education Studies*, *0*(0). https://doi.org/10.46827/EJES.V0I0.2398

Pinatih, I. D. S., Pratiwi, N. I., & Ekaresty, P. (2018). The Second Concert of Powers: Managing US-China Competition on the Korean Peninsula Conflict in Terms of International Communication Perspective. *International Research Journal of Management, IT and Social Sciences*, *5*(6), 17–25.

Pinter, J. D., Eliez, S., Schmitt, J. E., Capone, G. T., & Reiss, A. L. (2001). Neuroanatomy of Down's syndrome: A high-resolution MRI Study. *American Journal of Psychiatry*, *158*(10), 1659–1665. https://doi.org/10.1176/APPI.AJP.158.10.1659/ASSET/IMAGES/LARGE/J720F2.JPEG

Ringenbach, S. D. R., Mulvey, G. M., Chen, C. C., & Jung, M. L. (2012). Unimanual and Bimanual Continuous Movements Benefit From Visual Instructions in Persons With Down Syndrome. Http://Dx.Doi.Org/10.1080/00222895.2012.684909, *44*(4), 233–239. https://doi.org/10.1080/00222895.2012.684909

Russell, P. J. (2010). *iGenetics: A Molecular Approach.* 828.

Sugiarto, K. M., & Mutiah. (2021). Parents Communication to Down Syndrome in Online Learning During Pandemic. *Proceedings of the International Joint Conference on Arts and Humanities 2021 (IJCAH 2021)*, *618*, 741–744. https://doi.org/10.2991/ASSEHR.K.211223.129

Ulfah, S., Imam, U., & Padang, B. (2018). *Soft Skill of Language Teachers in the Classroom: Analysis of DeVito Humanistic Interpersonal Communication Model*. 168–173. https://doi.org/10.2991/ICLLE-18.2018.27

Yavuzer, G., Senel, A., Atay, M. B., & Stam, H. J. (2008). Playstation Eyetoy Games" Improve Upper Extremity-related Motor Functioning in Subacute Stroke: A Randomized Controlled Clinical Trial. *Eur J Phys Rehabil Med*, *44*(3), 237–244.

Zhang, L., Abreu, B. C., Seale, G. S., Masel, B., Christiansen, C. H., & Ottenbacher, K. J. (2003). A Virtual Reality Environment for Evaluation of a Daily Living Skill in Brain Injury Rehabilitation: Reliability and Validity. *Archives of Physical Medicine and Rehabilitation*, *84*(8), 1118–1124. https://doi.org/10.1016/S0003-9993(03)00203-X

The inclusive concept for designing four categories of disability education facilities

A. Akhmadi*, S.T. Virgana, A.P. Yuniati & H. Azhar
Telkom University, Bandung, Indonesia

*ORCID ID: 0000-0001-7508-1376

ABSTRACT: Based on Republic of Indonesia Law No. 20 of 2003 Chapter 5, every citizen has the right to receive an education, including physically, emotionally, mentally, intellectually, and socially disabled citizens. These individuals need comfortable and safe education facilities setting to pursue their education. This study aims to determine the space-forming factors needed in making education facilities for individuals with special needs. They are implementing a qualitative method by searching the data from scientific references, interviews, and observations on three special schools in Bandung, West Java. The results showed that every type of disability needs different interior settings, according to their specific needs.

Keywords: Interior, Inclusive Design, Disabilities

1 INTRODUCTION

Based on the Republic of Indonesia Law No. 20 of 2003 Chapter 5, every citizen has the right to receive an education, including citizens who are physically, emotionally, mentally, intellectually, and socially disabled (Indonesia 2003). These are some classifications of physical and mental disability characteristics based on the type of individual with special needs, including others:

a. Type A—Visually Impaired: According to Kurniawan (2015), Visually Impaired is classified into two categories: blind and low vision. Visually Impaired people have several cognitive characteristics, for example, relying on the sense of touch and hearing, receiving information more slowly, and having difficulty in reading and writing. In addition, having difficulties with mobility and orientation and difficulties in carrying out social behavior are other behavioral characteristics. Although physical aspects are often perfect, motoric development tends to be slower.

b. Type B—Deaf: According to Nofiaturrahmah (2018), Deaf is classified into two types: deaf and difficulty in hearing. Deafness has characteristics, such as relying on the senses of touch and sight, slow academic development, minimum vocabulary, irregular grammar, behaving like normal humans, tendency to have an aggressive gaze, walking stiffly and slightly bent, lack of balance, and irregular breathing.

c. Type C—Mental Retardation: According to Wulandari (2016), mental retardation is classified into several groups consisting of mild mental retardation, moderate mental retardation, severe mental retardation, and very severe mental retardation. Mental retardation has characteristics, such as having difficulty in learning, having short attention, having difficulty in social adaptation, lack of independence, and difficulty in interacting.

d. Type D—Quadriplegic: According to Astati (2009), it is classified into two types consisting of cerebral palsy and muscular and skeletal system disorder (physical disability). Physical disability has several characteristics, such as normal level of intelligence (except in cerebral palsy patients), often needing assistive devices for mobility (commonly wheelchairs), having a disabled body, being physically stiff and paralyzed, and having difficulty with mobility. Cerebral palsy patients have fewer cognitive abilities that affect their intelligence and sense.

According to Republic of Indonesia Law No. 20 of 2003 Chapter 5, every citizen with disabilities deserves a special education service. The education services for children with special needs among others, inclusive education services and Special Education in Special Schools (Sekolah Luar Biasa), (Munandar 2019).

According to Rahardja (2010) Special Schools (SLB) are classified into several categories: SLB A (Special School for the visually impaired), SLB B (Special School for the deaf), SLB C (Special School for mentally retarded), SLB D (Special School for Quadriplegics), SLB E (Special School for socially retarded), and SLB G (Special School for multiple disabled).

To accommodate all activities in Special Schools, referring to Indonesia Minister of National Education Law No. 33 of 2008 on Standard of Facilities and Infrastructure for Special Schools, a Special School needs to have some facilities such as classrooms, library, special learning rooms (consisting of orientation and mobility rooms, speech development rooms, sound and rhythm perception development rooms, self-development rooms, self-development and movement development rooms, personal development rooms and social, and skills room), principal's room, teacher's room, administration room, School's Health Clinic room, counseling or assessment room, toilet, and circulation room (Kemendikbud 2008).

However, Indonesia started implementing inclusive education. Inclusive education is an education system that provides opportunities for all participants with disabilities and different potentials to participate in an educational environment (Rombot 2017). This research aims to find suitable interior settings for Special Schools that started implementing an inclusive education system, which has Visually Impaired, Deaf, Mentally Retarded, and Quadriplegic students.

2 RESEARCH METHOD

In this research, qualitative methods are implemented. Creswell (2016) explains that qualitative research is a method to explore and understand individuals or groups of people that refer to social or humanitarian problems. We collected data by finding data from scientific references and doing interviews and observations. The interviews and observations were carried out in YPAC Specials School Bandung, Sumbersari Special School Bandung, and Roudhatul Jannah Special School Bandung Regency.

3 RESULT AND DISCUSSION

3.1 *Literature review*

According to Building Bulletin no. 102, Inclusive design can support and allow students with disability to live independently at school and in larger community. Inclusive school design is a design that considers all users (students, staff, and visitors) with disability without any barriers. Manley (2016) acquaint that Commission for Architecture and the Built Environment (CABE) in 2006 developed the inclusive design principle into several points: inclusive (everyone can use it safely, easily, and with dignity), responsive (takes into account what people say they need and want), flexible (different people can use the buildings and places in different ways), convenient (everyone can use it without too much effort or separation), accommodating (for all people regardless of their age, gender, mobility, ethnicity or circumstances), welcoming (with no disabling barriers that might exclude some people), realistic (offering more than one solution to help balance everyone's needs and recognizing that one solution may not work for all), and understandable (everyone knows where they are and can locate their destination).

3.2 Learning difficulties and spatial needs.

From the scientific references, interviews, and observations in three Special Schools in Bandung, we found the needs and difficulties faced by teachers and students while they are doing their education activities at school.

Table 1. Needs and difficulties faced by teachers and students at school.

Disability	Difficulties	Needs/Solutions
Visual Impaired	Having orientation and mobility difficulty while moving from one room to another	Need wayfinding, need clear circulations
	Feeling confused when the environment has too much sound and noise because it is hard to identify sound	Need a good acoustic system
	Having difficulty to determine area	Need different textures and need contrast and bright color (for low vision students)
Deaf	Too much noise causes pain in their ears while using hearing aids	Need a good acoustic system
	Seating arrangement in the classroom causes the view of the blackboard and the teacher obstructed by the students in the front, besides it is difficult to interact between students	Need seating arrangement, wide space
	Having difficulty detecting objects or people behind them, they often collide	Need a bit of reflective material, i.e., reflective material on a door or window frame
	Lack of balance	Need a rounded edge shape to prevent injury when they lost their balance
Mental Retardation	Easily to forget, so the lesson being taught must be repeated often	Need a visible learning display
	Students have a variety of focus levels, some of them are easily distracted, and some of them are too focused on their own world so they cannot follow the lessons well.	Need calming atmosphere in the classroom, so they can increase their focus
Quadriplegic	They get tantrums easily that cause self-harm, breaking things, or banging against the wall	Need soft material, i.e., wall padding, need rounded edge furniture
	Having difficulty in mobility because some of them use wheelchairs, meanwhile, sometimes the circulation is narrow	Need huge circulation
	Easily feel tired	Need resting area

3.3 Concept implementation

To find a suitable interior setting for Special Schools that have students with various special needs, we elaborate the data obtained from interviews and observations in three Special

Schools to be used as a reference in implementing the concept. The concept name is CLIFS. CLIFS stands for Calming, Legible, Intuitive, barrier-Free, and Safe.

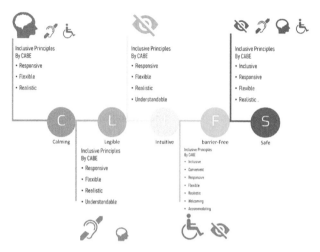

Picture 1. CLIFS concept mapping.

We know that every disability has different priority needs and different implementations of CLIFS concepts, for example, mental retardation education facilities need a more calming environment, Deaf education facilities need a more legible environment, Visual Impaired education facilities need a more intuitive environment, and Quadriplegic education facilities need more barrier-free environment. There are some implementation examples of CLIFS concepts based on some scientific references, among others:

Table 2 Implementation Examples of CLIFS Environment Concepts at education facilities.

CLIFS Concept	Disability	Implementation Examples
Calming	Mental retardation	• The seating is arranged at a distance for giving them more personal space • Using acoustic panels on the ceiling • Installing lights with in-bow position to avoid glare effect • Build a calming room for calming tantrum student • Applying soft colors for the calming atmosphere
	Deaf	• Making good acoustic by using acoustic panels on the ceiling • Using carpet on the floor to
	Quadriplegic	• Making area for short rest
Legible	Deaf	• Installing lights with in-bow position to avoid glare effect • Wider space / wider circulation for making a clear view • Letter U seating arrangement • Legible signage, i.e., bigger font on signage • Implementing a little reflective material, i.e., on a door or window frame

(*continued*)

Table 2 Continued

CLIFS Concept	Disability	Implementation Examples
	Mental retardation	• Making a display area for learning • Organizing area by color
	Visual Impaired	• Using different textures on interior elements, i.e., on the floor and on furniture surfaces • Making good acoustic by using acoustic panels on the ceiling • Implementing a guidance path on the floor • Implementing a handrail for the orientation guide • Using contrast and bright color for helping low-vision students determine the area
	Quadriplegic	• Making huge circulation and clear circulation • Minimalizing physical activity i.e. using automatic door
	Visual Impaired	• Making clear circulation
	Visual Impaired	• Applying anti-slippery material on the floor, i.e., epoxy flooring, anti-slippery ceramic tiles flooring • Applying rounded edges on furniture
	Deaf	• Applying anti-slippery material on the floor, i.e., epoxy flooring, anti-slippery ceramic tiles flooring • Applying rounded edges on furniture
	Mental retardation	• Applying soft materials, i.e., wall padding • Applying rounded edges on furniture • Applying anti-slippery material on the floor, i.e., epoxy flooring, anti-slippery ceramic tiles flooring
	Quadriplegics	• Applying rounded edges on furniture • Applying anti-slippery material on the floor, i.e., epoxy flooring, anti-slippery ceramic tiles flooring

4 CONCLUSION

From the explanation above, we conclude that every type of disability needs different interior settings, according to their specific needs. This CLIFS concept aims to specify inclusive design principles to be applied to Special Schools especially in Indonesia because the inclusive design principles that already exist are still general.

REFERENCES

Astati. (2009). *Pengantar Pendidikan Luar Biasa*. UPI.
Creswell, J. W. (2016). *Research Design: Pendekatan Metode Kualitatif, Kuantitatif dan Campuran. Edisi Keempat (Cetakan Kesatu)*. Pustaka Pelajar.

Indonesia, U.-U. (2003). *Undang-Undang Nomor 20 Tahun 2003 tentang Sistem Pendidikan Indonesia.* Sekretariat Negara.

Kemendikbud. (2008). *Peraturan Menteri Pendidikan Nomor 33 Tahun 2008 tentang Standar Sarana Dan Prasarana Untuk Sekolah Dasar Luar Biasa (SDLB), Sekolah Menengah Pertama Luar Biasa (SMPLB), dan Sekolah Menengah Atas Luar Biasa (SMALB).* Kementrian Pendidikan dan Kebudayaan Republik Indonesia.

Kurniawan, I. (2015). Implementasi Pendidikan Bagi Siswa Tunanetra Di Sekolah Dasar Inklusi. *Edukasi Islami: Jurnal Pendidikan Islam*, 16.

Manley, S. (2016). *Inclusive Design in the Built Environment Who Do We Design For?* University of the West of England.

Munandar, D. R. (2019). *Manajemen Perubahan Organisasi Sekolah Luar Biasa.* Universitas Singaperbangsa.

Nofiaturrahmah, F. (2018). Problematika Anak Tunarungu dan Cara Mengatasinya. *Quality: Journal of Empirical Research in Islamic Education*, 1–15.

Rahardja, D. (2010). Pendidikan Luar Biasa dalam Perspektif Dewasa Ini. *Jurnal Asesmen Dan Intervensi Anak Berkebutuhan Khusus*, 76–88.

Rombot, O. (2017). *Pendidikan Inklusi*. PGSD Binus. https://pgsd.binus.ac.id/2017/04/10/pendidikan-inklusi/

Wulandari, D. R. (2016). Strategi Pengembangan Perilaku Adaptif Anak Tunagrahita Melalui Model Pembelajaran Langsung. *Jurnal Pendidikan Khusus UNY*, 51–66.

Utilization of planar material for shell structure with digital tectonics approach

S.E. Indrawan
Ciputra University, Surabaya, Indonesia

ORCID ID: 0000-0001-7453-2999

ABSTRACT: In Heino Engel's book, one type of active form structure is the shell structure. The shell structure construction is composed of curve formations that have different axes. This type of structure is very flexible, but the design process is very dependent on the typing material, generally in the form of sheets or planar. On the other hand, the shell design process is quite complicated because it is very dependent on the simulation process of scale models or structural calculations. This paper explains how digital tectonics can accommodate students in their design process to a digital fabrication process and facilitates them to make simulations at accurate scale. The research method used was an experiment with object modeling process solutions using paraametric design that was simulated in design software.

Keywords: Shell structures, Material, Planar, Digital Tectonics, Digital Fabrication

1 INTRODUCTION

So far, there is still a lot of understanding that there is a domain difference between architects and engineers in civil society in designing a building where there is an understanding that the architect is in charge of designing buildings from the general concept to the details of the build (Najiyati & Muslihat 2008). In contrast, civil engineers served to take into account the stability of the structure only. Even though in existing practice the sole architect's role is in the type of multifunctional building in a social context or usually known as residential houses while the role of the sole structural engineer is on the type of building that serves the purpose of a single structure or is better known as infrastructure facilities (such as bat). However, high-rise buildings that are multifunctional and require high social consideration need an architect and engineer civil role. The more many Genre strength needs to be anticipated the more role of civil engineers is getting bigger. On the other hand, in infrastructure planning such as towers, power plants, wide-span roof structures, and wide-span bridges on a city scale a civil engineer needs a recommendation from an architect to decide the selection of materials and consider the scale of the building on the urban environment. One type of structural system that can anticipate wide spans is the shell structure (Engel 2013). The formation of the shell structure occurs due to the flow of existing forces, the formation is also the result definition of the load behavior that occurs, this system can also save building materials if it is supported by the use of local materials so it can utilize local labor and new jobs can be created.

2 LITERATURE STUDIES

2.1 *Form exploration*

Model physique for process exploration formation has already been conducted since the Renaissance era. This Model was also used by Gaudi to utilize a Suite chain that hung backward (Adriaenssens 2014), where in that era, Gaudi was able to show a simulation of the loading that occurred through simulation represented by the physical model (Huerta 2006). The results of this experiment explain the relationship between the behavior of load with the formation that occurs. Heretofore, the surface shape holds an important role in planning shell structure. Architects have used digital methods and scale models as a design and communication medium to understand the behavior of the shell structure. However, before the digital era, latex and other flexible material was also an option to study the mechanism of form-follow force in the world of structures. The use of physical models in studying structural mechanics consists of 3 main groups: drape models. The force of gravity forms a curved formation on a flexible material by pulling the material. Heinz Isler was a German architect who learned this technique and was able to apply it to the Groetzingen Naturtheater project. Meanwhile, Frei Otto understands this logic by applying it to a series of iron chains. This chain arrangement shows the concept of a mesh through a series of crossed chains so that observers can understand the logic of the construction of the shell structure. In addition to the previous experiment, Frei Otto also conducted experiments with latex and soap bubbles. Through the use of latex, observers can understand the structural logic of the tent construction. Frei Otto also used soap bubble experiments to design a train station in Stuttgart. In this method, Frei Otto succeeded in finding the concept of a minimal surface (Beesley Seebohm 2000). The soap bubble experiment also resulted in the concept of a pneumatic structure, as shown in Figure 1. The concept of a pneumatic structure is to use air as a structural element. The shell structure design concept takes the pneumatic structure principle as a reference to visualize the structural mechanics that will occur in the design (Figure 2).

 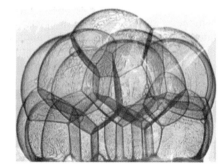

Figure 1. Eden project creation.   Figure 2. Mockup bubble soap.

The use of scale models with flexible materials is an attempt by designers to achieve the principle of form-follow-force. The architects used this concept only to recognize the mechanics of the structure and establish its basic logic. In the next stage, architects need to create conventional working drawings as media to communicate with contractors and engineers. Furthermore, digital engineering and various computer-based structural analysis methods have become popular since the 1960s.

2.2 Digital tectonics and digital fabrication

The evolution of computing technology in architectural design can link the design process more closely to the fabrication process. Based on Shelia Kennedy's explanation, two main factors influence the success of utilizing digital technology in the design process: knowledge of using software and the ability to integrate disciplines outside the field of architecture (Oxman 2006). In the digital tectonic approach, the simulation of design objects is in the form of virtual materials and natural objects. The use of digital tectonic concepts in the design process (Oxman 2012) are as a media for visualizing the use of accurate or photo-realistic materials, as a medium that can explain the deformation of material and design object, as a medium that can increase the economic value and efficiency of materials and workflow, as a medium for the exploration of more advanced formations, and as a medium that can accommodate the adaptive capabilities of advanced materials.

3 RESEARCH METHODS

This study uses a controlled or semi-structured experimental research method (Sugiyono 2015) to find the logic and basic principles of using planar materials to plan shell structures. This activity is part of the interior architecture studio of Ciputra University students. The success parameter of this activity is the ability of students to understand the basic principles and logic of shell structure design through an integrated workflow from the design process to fabrication activities with laser cutting machines. The purpose of this activity is to guide students to understand the theory and history of construction to understand the logic of the experimental process in the past and relate it to the use of computers. This process is limited to discussing the shell structure, tessellation method, dividing the structural components, production using a laser cut machine and ended with the assembly process.

Figure 3. Semi-structured experiment.

4 RESULTS AND DISCUSSION

4.1 Fabrication digital and studies case

In the evolution of architectural presentation methods, orthogonal projection is a method for identifying the formation and geometry of a design object. This orthogonal projection is a commonly used method for working drawings. This working drawing is a medium to communicate with contractors and engineers in a project. But what about complex architectural objects such as shell structures? The computer will unroll the Shell structure into smaller components with a specific arrangement in the development process at various scales. The digital fabrication approach makes this process becomes logical.

4.2 Exploration geometry

This section describes the shell structure planning process using Rhino, Grasshopper, and Lunch box software.

This paper uses a modified long-barreled dome with different heights for this type of shell structure. The designer will adjust the height of the curve according to his needs. The formation can be the basis of the design based on the desired needs. The surface has triangular-shaped fragments. The amount arrangement fragment (Sub Divide) can be set through sliders so that the designer can change it based on the design purpose. The more fragments on the surface, the smoother the surface formed. For this study, each fragment has a hole to minimize dead load and a number for guidance in assembly (Figures 4, 5). Next, each panel is parsed and sorted by number. Then during the assembly process, the guide is 3D modeling which has a serial number.

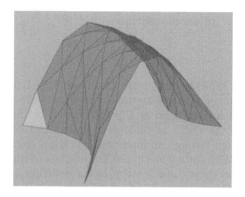
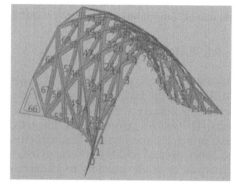

Figure 4.   Triangulation.   Figure 5.   Triangulation with serial number.

The value of each parameter can determine the surface design through the curve's shape, distance, the number of fragments, or the size of the hole in each fragment (Figure 5). The definition (Figure 7) structure of the Grasshopper software automates this series of processes (Figure 6). Finally, a laser cutting machine produces each fragment. The purpose of using a laser machine is not only to shorten the process and get good results but also to get precise results so that the process can be scaled up.

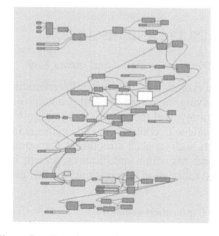

Figure 6.   Unflatten fragments.   Figure 7.   Grasshopper definition.

5 CONCLUSION

The digital tectonic and digital fabrication process is a comprehensive design process and does not just stop at the drawing process (as shown in Figures 8, 9). Through this process, students can see design practice more closely. The process of design and realization becomes real, and students can be involved in this process without involving third parties (such as contractors).

Figure 8. Realization 1.

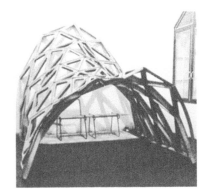

Figure 9. Realization 2.

The design process does not require a physical model with flexible modeling due to the rhino and grasshopper software definition. For future research, this process can start formation exploration, which involves simulation and even optimizing the shell structure. In future research, this process can be a prelude to formation exploration, which involves simulation of even optimization structure shell.

REFERENCES

Adriaenssens, S. (2014). Shell Structures for Architecture. In *Shell Structures for Architecture*. https://doi.org/10.4324/9781315849270

Beesley, P., & Seebohm, T. (2000). Digital Tectonic Design. *18th ECAADe Conference Proceedings*, March, 287–290.

Engel, H. (2013). *Structure System* (5th ed.). Heino Engel and Hatje Cantz Verlag.

Huerta, S. (2006). Structural design in the work of gaudi. *Architectural Science Review*, 49(4), 324–339. https://doi.org/10.3763/asre.2006.4943

Indrawan, S. E., Purwoko, G. H., & Utomo, T. N. P. (2020). The use of Minimal Surface Principles and Multiplex Joinery System for Designing Post-disaster Construction Systems. *ARTEKS: Jurnal Teknik Arsitektur*, 5(3), 347–358. https://doi.org/10.30822/arteks.v5i3.488

Iwamoto, L. (2010). Digital Fabrications:Architectural and Material Techniques. In *Architecture briefs*. http://www.papress.com/html/ book.details.page.tpl?cart=125777262860796& isbn=9781568987903

Kolarevic, B. (2004). Architecture in the Digital Age: Design and Manufacturing. In *Architecture in the Digital Age: Design and Manufacturing*. https://doi.org/10.4324/9780203634561

Najiyati, S., & Muslihat, L. (2008). Mengenal Tipe Lahan Rawa Gambut. *Seri Pengelolaan Hutan Dan Lahan Gambut*, 1–4. wetlands.or.id/PDF/Flyers/Agri05.pdf

Oxman, R. (2006). Theory and design in the first digital age. *Design Studies*, 27(3), 229–265. https://doi.org/10.1016/j.destud.2005.11.002

Oxman, R. (2012). Informed tectonics in material-based design. *Design Studies*, *33*(5), 427–455. https://doi.org/10.1016/j.destud.2012.05.005

Potter, M., & Ribando, J. (2005). Isometries, Tessellations and Escher, Oh My! *American Journal of Undergraduate Research*, *3*(4). https://doi.org/10.33697/ajur.2005.005

Sugiyono. (2015). Metode Penelitian. *Metode Penelitian*.

Woods, E. (2016). Software Architecture in a Changing World. *IEEE Software*, *33*(6), 94–97. https://doi.org/10.1109/MS.2016.149

Systematic literature review: Games for social skills development in people with Autism Spectrum Disorder (ASD)

O.D. Hutagaol, Y. Rahayu, P. Nova, M. Rafly & A. Sultani
Institut Komunikasi dan Bisnis LSPR, Jakarta, Indonesia

ABSTRACT: The use of technology, especially games in education, has recently increased, including its use for social skills development in people with autistic spectrum disorder (ASD). This study is designed to investigate existing and recommended evidence-based digital games and strategies in social skills development for people with ASD through the available published studies. This systematic literature review searched for peer-reviewed articles published in the Sage Journal, Science Direct, and Emerald Insight databases in the year 2018–2022 using specific keywords and phrases that covered games, ASD, and social skills without age restriction. The end area is composed momentarily and briefly, without the expansion of any new understandings. The consequence of the end area should answer the goals of the issue, peculiarity, and all of the work on the exploration. The outcome ought to sum up discoveries as opposed to giving information exhaustively.

Keywords: Systematic Review, Autism Spectrum Disorder, Games, Social Skills

1 INTRODUCTION

Autism spectrum disorder (ASD) is a complex neurological disorder with a wide range of symptoms The DSM-5 defines a range of diagnoses that include Autistic Disorder, Asperger's Disorder, or Asperger Syndrome (AS). (Donald w. black & E. grant. 2013) According to a global review of ASD prevalence, one out of every 160 children have ASD. As of 2018, it was estimated that the global prevalence of ASD was 7.6 people per every 100-population base (*Autism*, n.d.). People with ASD often have difficulty communicating, behaving politely, making friends, and coping with conflicts. They may also struggle to express their feelings, assertiveness, and coordinate groups. They may struggle to speak in public. The majority of studies on autism have been focused on children who are younger than 12 years. The general assumption is that when management and proper treatment address ASD identified among children in this age category, disability-related challenges are significantly reduced among the elderly. However, some of the limitations of this general assumption are that delays in the diagnosis of ASD among children have become a major global challenge, especially in emerging and developing countries. Current prevalence and incidence studies among people with ASD show that the category of adolescents is a new and emerging group that needs care and attention. In this research, although we are not limited to the ages, we are more focused on how games could assist social skills development in people with ASD recently and further research opportunities.

2 RESEARCH METHODS

This study uses a systematic literature review method with PRISMA flow as the search method. Around June and July 2022, we searched Science Direct, Sage Publishing, and

Emerald Insight databases using computerized methods. In all three, searches were limited to English-language articles published in peer-reviewed journals between the years 2018–2022 and using the following keywords: autism or ASD and games. We screened all reports, studies, and reviews that were identified by reading the titles and abstracts. In addition, we reviewed the reference list of studies that met the selection criteria to identify additional inclusion studies (Bono et al. 2016).

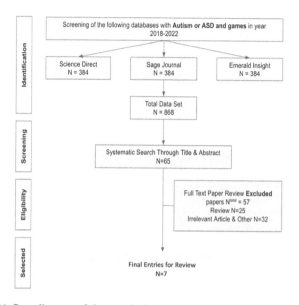

Figure 1. PRISMA flow diagram of the searched method used in this study.

3 RESULTS AND DISCUSSION

Table 1 sums up the principal qualities of the seven games that target general interactive abilities. The use of information correspondence progresses (ICTs) in treatment gives new perspectives for treating individuals with synthetic unevenness range issues (ASD) because they can be used in different ways and settings and they are charming to the patients. For getting ready on intelligent capacities, serious games are very uplifting. They can be used to plan different capacities and can lean toward correspondences in arranged settings and conditions, some of which seem to be reality. In any case, the current open serious games present a couple of cut-off points concerning the confirmation of their clinical benefits. It by and by appears to be vital to check the ampleness of ICTs and the use of serious games for remediation through extra generous assessments concerning method (immense models, control social occasions, longer treatment periods, and follow-up to assess whether changes stay stable). We in like manner need to zero in more diligently on blueprint and cultivate a specific framework for this kind of serious game to propose games that test and interface with the patients. Additionally, we really need to encourage more serious games that are acclimated to LFASD individuals and that rely upon late advances in enrolling systems that can normally answer client input. The current status of the workmanship has put a great deal of emphasis on the specific capacities associated with feeling affirmation in faces. Later, researchers should similarly progress various subsets of abilities that are critical for non-verbal social correspondence like joining thoughts or movements. Features that depict mental working in ASD should moreover be thought of. Taking into account that

Table 1. Selected research articles related to games for social skills development in people with ASD.

No	Article Details	Outcome and Findings
1.	**Title:** Online gaming, loneliness, and friendships among adolescents and adults with ASD (Sundberg 2018). **Participants:** A total of 85 people with ASD between 14 and 69 years old **Social Skills:** Making & maintaining friends **Assessment Tools:** (1) Motives for Online gaming questionnaire (2) UCLA Loneliness scale (3) Unidimensional Relationship Closeness Scale **Game Type:** Online game	**Outcome Measures:** Connection to friendships and loneliness among teenagers and adults with ASD **Findings:** People with ASD who play online games have a significant increase in the number of friends they have. Online games can provide a platform for people with ASD to make new friendships and maintain existing ones.
2.	**Title:** Dynamic difficulty adjustment technique-based mobile vocabulary learning game for children with an autism spectrum disorder (Shohieb et al. 2022) **Participants:** Seven children with ASD between 5 and 10 years old **Social Skills** Communication & Assertiveness **Assessment Tools:** (1) Pre-study & Post-Study User Interviews (2) Engagement Assessment and Usability Evaluation of Pickstar **Game Type:** Pickstar, Mobile Game with Dynamic Difficulty Adjustment (DDA)	**Outcome Measures:** A of educational and therapeutic methods for ASD and to mitigate caregivers' problems in accessing educational materials **Findings** Pickstar with DDA can be used with minimal guidance to improve the vocabulary skills of individuals with ASD & can mitigate caregiver problems in accessing good-quality, engaging, and affordable education for their children.
3.	**Title:** A pilot study on evaluating children with autism spectrum disorder using computer games. (Chen et al. 2019) **Participants:** Ninety-one children (40 ASD & 51 TD was compared) between 2 and 6 years old **Social Skills:** Communication **Assessment Tools:** Analysis of differences using t-tests **Game Type** Computer-based game	**Outcome Measures:** Children with ASD and TD children in the development fields of joint attention, ToM, visual search skills, fine motor skills, cognitive understanding, and classification **Findings:** In general, most of the children with ASD have lower development levels than the TD children with significant differences. But ASD group sometimes performed better than the TD group (in terms of raw scores) on several tasks. Computer games have great potential as support, evaluation, or intervention tools in the field of special education with appropriate support.
4.	**Title:** Does playing on a digital tablet impact the social interactions of children with autism spectrum disorder? (Paquet et al. 2022) **Participants** Four children between 3 and 5 years old and four young people between 14 and 20 years old. **Social Skills:** Social relations, communication, and restricted behaviors **Assessment Tools:** Behavioral observation with nonparametric tests and Wilcoxon test **Game Type:** Memory Card game & Snakes and Ladders Digital & Physical game.	**Outcome Measures:** Test the effect of tablets on social interactions in a multiplayer game situation involving children with ASD. **Findings:** Tablets did not improve social interaction. However, sharing a game with others seemed to contribute to the development of interactive behaviors.

(continued)

Table 1. Continued

No	Article Details	Outcome and Findings
5.	**Title:** Co-Op World: Adaptive computer game for supporting child psychotherapy (Alkalay et al. 2020) **Participants:** Four cases with different age groups of Children ranging in age from 7 to 10. **Social Skills:** Social interaction and communication **Assessment Tools:** Evaluating each case at a different type of level. **Game Type** Virtual game players, and Co-Op World	**Outcome Measures:** Evaluation of the Co-Op World game during therapy for ASD children. **Findings:** The game has great potential in therapy. Reveal high acceptance of the game by the users and demonstrate various behavioral strategies adopted by the children. By using the game, the therapists were able to hold discussions with the children about their motivations to expand their learning about their real-life social interactions.
6.	**Title:** Supporting Youth With Autism Learning Social Competence: A Comparison of Game- and Nongame-Based Activities in 3D Virtual World. (Wang & Xing 2021) **Participants:** Eleven adolescents with ASD between 10 and 13 years old **Social Skills:** Social interaction **Assessment Tools:** (1) A learning analytics approach—association rule mining, (2) Autism Diagnostic Interview-Revised, (3) Autism Diagnostic Observation Schedule **Game Type:** Game-Based Learning in 3D Collaborative Virtual World	**Outcome Measures:** Compared the social interactions of 11 adolescents with ASD in game-and non-game-based 3D collaborative learning activities in the same social competence training curriculum. **Findings:** Co-occurrence of verbal and nonverbal behaviors is much stronger in game-based learning activities. The game activities also yielded more diverse social interaction behavior patterns. the affordance of serious game-based collaborative virtual learning activities has a significant influence on learners' social interactions.
7.	**Title** MEDIUS: A Serious Game for Autistic Children Based on Decision System (Daouadji Amina & Fatima 2018) **Participants:** Ten children with ASD between 5 and 13 years old **Social Skills:** Consistency make decision & facial expression **Assessment Tools:** Picture Exchange Communication System "PECS" and Applied Behavior Analysis "ABA" **Game Type:** Medius. PECS Game-Based	**Outcome Measures:** Realize a serious game called "Medius" based on criteria of decision support to present a new mean of communication between the tutor and the autistic child in the form of a playful serious game, and in a purely educational frame. **Findings:** This work is very useful for a targeted apprenticeship and personalized to each player (autistic child) and also represents a mean of communication between the tutor and the autistic child seeing the difficulty of the exchanges of information and the communication which reign in the world of the children affected by the pervasive developmental disorders "PDD"

individuals with ASD are every now and again paid all due respects to be visual driving forces, future assessments ought to seek full advantage of the entryways dealing with the expense of visual depictions intrinsic to serious games. The general hunt system yielded a sum of 868 articles. These were every one of the articles that were caught utilizing the different watchwords and expressions listed in the procedure area.

4 CONCLUSIONS

We found a total of seven research articles related to games for social skills development in people with ASD as shown in Table 1. The end area is composed momentarily and briefly, without the expansion of any new understandings. The consequence of the end area should obviously answer the goals of the issue, peculiarity, and all of the work on the exploration. The outcome ought, to sum up (logical) discoveries as opposed to giving information exhaustively. Talk about the discoveries according to the hypothesis or the aftereffects of past examinations, so science can keep on creating. This part can likewise record the curiosity of the examination, the benefits and hindrances of the exploration, as well as proposals for additional examination. Technology, especially games, can help improve social skills in people with autism, although its significance still needs to be investigated further. The types of devices, features, interactions, and stimuli used need to be considered and adapted to the characteristics of people with ASD such as the use of dynamic difficulty adjustment and virtual characters. Technology or games in specific, can help improve social skills in people with autism although its significance needs to be investigated further. The types of devices, features, interactions, and stimuli used need to be considered and adapted to the characteristics of people with ASD such as the use of Dynamic difficulty adjustment and virtual characters. Further research related to the same research in adults and focusing on the types of devices, features, interactions, and stimuli is highly recommended.

REFERENCES

Alkalay, S., Dolev, A., Rozenshtein, C., & Sarne, D. (2020). Co-Op World: Adaptive Computer Game for Supporting Child Psychotherapy. *Computers in Human Behavior Reports, 2*, 100028. https://doi.org/10.1016/J.CHBR.2020.100028

Autism. (n.d.). Retrieved July 8, 2022, from https://www.who.int/news-room/fact-sheets/detail/autism-spectrum-disorders

Bono, V., Narzisi, A., Jouen, A. L., Tilmont, E., Hommel, S., Jamal, W., Xavier, J., Billeci, L., Maharatna, K., Wald, M., Chetouani, M., Cohen, D., Muratori, F., Bonfiglio, S., Apicella, F., Sicca, F., Pioggia, G., Cruciani, F., Paggetti, C., ... Donnelly, M. (2016). GOLIAH: A gaming Platform for Home-based Intervention in Autism – Principles and Design. *Frontiers in Psychiatry, 7*(APR), 70. https://doi.org/10.3389/FPSYT.2016.00070/BIBTEX

Chen, J., Wang, G., Zhang, K., Wang, G., & Liu, L. (2019). A Pilot Study on Evaluating Children with Autism Spectrum Disorder Using Computer Games. *Computers in Human Behavior, 90*, 204–214. https://doi.org/10.1016/J.CHB.2018.08.057

Daouadji Amina, K., & Fatima, B. (2018). MEDIUS: A Serious Game for Autistic Children Based on Decision System: Https://Doi.Org/10.1177/1046878118773891, *49*(4), 423–440. https://doi.org/10.1177/1046878118773891

donald w. black, M. d., & E. grant., J. (2013). *DSM5 Guideline*. 519. file:///C:/Users/Rose/Desktop/DSM-5(r) Guidebook – Black, Donald W., Grant, Jon E. [SRG].pdf

Paquet, A., Meilhoc, L., Mas, B., Morena, A. S., & Girard, M. (2022). Does Playing on a Digital Tablet Impact the Social Interactions of Children with Autism Spectrum Disorder? *Neuropsychiatrie de l'Enfance et de l'Adolescence*. https://doi.org/10.1016/J.NEURENF.2022.04.003

Shohieb, S. M., Doenyas, C., & Elhady, A. M. (2022). Dynamic Difficulty Adjustment Technique-based Mobile Vocabulary Learning Game for Children with Autism Spectrum Disorder. *Entertainment Computing, 42*, 100495. https://doi.org/10.1016/J.ENTCOM.2022.100495

Sundberg, M. (2018). Online Gaming, Loneliness and Friendships Among Adolescents and Adults with ASD. *Computers in Human Behavior, 79*, 105–110. https://doi.org/10.1016/J.CHB.2017.10.020

Wang, X., & Xing, W. (2021). *Supporting Youth With Autism Learning Social Competence*: A Comparison of Game-and Nongame-Based Activities in 3D Virtual World: Https://Doi.Org/10.1177/07356331211022003, *60*(1), 74–103. https://doi.org/10.1177/07356331211022003

Development of Tasikmalaya Kerancang solder design embroidery motifs for women's resort wear clothing

J. Evan, M.Y. Tanzil & Y.K.S. Tahalele
Universitas Ciputra, Surabaya, Indonesia

ABSTRACT: Indonesia is the largest archipelago nation, extending from Sabang to Merauke, with many ethnic groups, races, cultures, and traditions. A range of traditional clothing, handicrafts, accessories, and cosmetics can be used to identify ethnic groups. For example, the Sundanese ethnic in West Java province produces different handicrafts, such as designer handicrafts. However, many craftsmen have experienced a decline in sales due to the COVID-19 pandemic. This research aims to support the embroidery craftsmen in Tasikmalaya and promote the development of motifs for kerancang. It also increases public awareness of Indonesian handicrafts and fosters an appreciation for local products. The taps of resort wear are projected to be recognized and used by buyers on local and international vacations to increase their economy. This research is qualitative research with a design thinking approach. The primary data were collected directly through interviews with 6 experts and 12 extreme users, and the secondary data were obtained from literature studies of journals, books, articles, and online media on resort wear, motifs, embroidery, and designs. Kerancang soldering techniques were used to create five looks of resort wear clothes, including shirts, bottoms, and outerwear. This research reveals that traditional crafts can be developed into contemporary fashion products. The kerancang embroidery solder technique was applied to women's resort wear clothing collections as motifs and acquired validations from both the experts and the target market.

Keywords: embroidery motif, resort wear, Tasikmalaya Kerancang

1 INTRODUCTION

Indonesia is the world's biggest archipelago, with 17,504 islands stretching from Sabang to Merauke (Katadata 2021), the largest of which are Sumatra, Java, Kalimantan, Sulawesi, and Papua. This brings consequences for the diversity of culture since tribes, races, and traditions are based on ethnic distribution. One of these ethnic identities can be seen in different traditional clothes, accessories, and make-up. Furthermore, Indonesia also has its traditional clothes and handicrafts.

The Sundanese ethnicity in West Java creates a variety of handicrafts as one of Indonesia's subcultures. Each region has its special craft, including Tasikmalaya, famous as a center for producing the art of embroidery or *kerancang*, *Rajapolah* Weaving, Wooden Furniture, *Kelom Geulis*, *Tasik* Batik, and many more. The art of embroidery or *kerancang* typical of Tasikmalaya is interesting to observe (Hanjani 2015).

Tasikmalaya *kerancang* embroidery becomes the topic of this research due to interest in Indonesian handicrafts to develop motifs and foster the daily usage of kerancang by people. The goal is to care for the kerancang embroidery craftsmen in Tasikmalaya as a result of the COVID-19 outbreak. Therefore, the development of motifs on *kerancang* can help the craftsmen in Tasikmalaya. This research also aims to increase public awareness of

Indonesian handicrafts and foster appreciation for local products. For women, resort-wear clothing products can be used for vacations to showcase their crafts, and the development of motifs also allows *kerancang* to be worn as casual clothes.

2 RESEARCH METHODS

This research utilizes qualitative research with a design thinking approach. The primary data were acquired through direct and online interviews with expert panels and extreme users of the products. The secondary data were obtained from literature studies, namely journals, books, articles, and online media regarding resort wear, motifs, embroidery, and *kerancang*. Lee et al. (2020) defined design thinking as a research strategy used in the design process. The six stages of the research start with empathy, where the approach is made to the target customer to get empathy for the problem to be solved. This approach is conducted by direct observation, interviews with expert panels, and extreme users aged 25–35 who like Indonesian handicrafts and travel. The customers' desires can be determined by conducting a preliminary survey of the target market. During the empathy phase, information was gathered through interviews on the target market that enjoys resort wear with Indonesian handicrafts. The second stage is defined which reveals the market's needs and wants of contemporary Indonesian handicrafts for leisure and resort wear. The third stage ideate is the stage to generate an idea. The market desired a more modern kerancang motif from the previous stage without losing its distinctive qualities. Moreover, the market desires resort wear with added features.

Following the ideation stage, the fourth stage is the prototype. This stage is an experimental process in designing and manufacturing several products to investigate the best solution for a previously formulated problem. This stage will be accomplished by making a toile using calico and kerancang fabric, considering the silhouette, cutting, placement of the appropriate motif, and neat stitching. After the prototype is created, it will be tested, evaluated, and presented to obtain feedback on the product development at this final stage, and the test will be carried out on the target market.

2.1 *Data collection*

2.1.1 *Primary data*

Primary data were obtained directly from the informants by answering questionnaires or interviews. In addition, interviews were conducted with 6 Expert Panels and 12 Extreme Users.

3 RESULTS AND DISCUSSIONS

3.1 *Resort wear*

Resort or Cruise wear is worn when going on vacation, especially during a hot climate (Hobby 2020). As stated by Garg (2018), it has a light character, hence it is easy to carry and use. Resort wear consists of swimwear, light dresses, short and long pants, shirts, kaftans, and skirts that are suitable for leisure and travel.

3.2 *Kerancang*

Kerancang is embroidery made using a small or black sewing machine, and the motif is intertwined threads to form holes. It has two methods of making, namely using a machine or soldering. In resort wear fashion designs, *kerancang* is created manually with a sewing machine. The embroidery technique was born in 330 BC, in the Byzantine era. The decoration craftsmen made embroidery manually. As the convection industry developed, tools emerged that could be used to make embroidery decorations. Sewing machines that used feet were the first to make embroidery (Rusyaman 2013: 60). With the creativity of the Tasikmalaya people, new embroidery motifs and techniques emerged and further increased market demand. Tasikmalaya is nicknamed the City of Santri, where the people also apply embroidery with Islamic nuances on *mukena* (prayer veil), *koko* (Muslim clothing for men), robes, tunics, scarves, tablecloths, and others (Rusyaman 2013: 63).

3.3 Fashion trend Autumn/Winter 2021/2022

Figure 1. Fashion trend A/W 2021/2022. Source: WGSN, 2020.

This design used the autumn/winter 21/22 trend entitled "New Mythologies" (WGSN 2020). It is based on the idea that new meanings can be derived from what is regarded as archaic. This trend has a slow fashion theme that prioritizes handicrafts, upcycled, product quality, and natural textures. According to the WGSN, the present consumer trend cannot be avoided, and the community strongly supports local items from their different countries. The color trend in this fashion design is based on WGSN at Spinexpo Shanghai autumn/winter 21/22 with the theme "Origins." The material is based on the "New Mythologies" trend, prioritizing handicrafts that use stitches to form a symbol. Other supporting materials such as large buttons enhance the essence of minimalist clothing. Based on the conclusions drawn from the current trend of womenswear, the market prefers clothes that make them more relaxed, hence a loose or flowy silhouette is needed.

3.4 Resort wear collection design

The resort wear collection entitled "Jaringna," which means net in Sundanese, is a Resort Wear collection that uses a soldering kerancang from Tasikmalaya as a concern for Indonesian handicrafts and local craftsmen. This collection is inspired by fishermen who spread their nets in the ocean. It was designed manually using a juki machine and was combined with a feminine silhouette to show a bit of masculinity. The purpose of making day-to-night resort wear from this collection is to prevent users from using many clothes for different events. With materials that are not easily wrinkled, this collection is suitable for use as a day-to-night outfit. Additionally, it enables design craftsmen in Tasikmalaya to gain widespread recognition.

The toile and prototype made are tops from Figure 4 made of calico fabric. After several trials of making toiles, a prototype was made using the main material. Experiments were also

Figure 2. Moodboard (left), FINAL design and illustration (right). Source: Evan 2021.

conducted on fabric coloring using textile dyes to find the right baby blue color during the prototyping process.

3.5 *Product trial*

Product trials are an essential stage in carrying out research and development. First, the goal is to determine whether the product made is suitable for use (Tanzil et al 2021). The other purpose is to collect data that can be used to increase effectiveness, efficiency, and attractiveness. The prototype made is then tested on three extreme users. Data collection Instruments are fittings, materials used, comfort, and colors. Furthermore, product trials were conducted with interviews using 12 extreme users and 6 expert panels.

According to tests conducted, creating apparel is challenging due to the vast number of designs; therefore, it requires considerable time and dependable abilities. The color combination in the collection design is good, but there are suggestions to try adding combined colors evenly. The kerancang motif is consistent with the essence and philosophy, and the placement looks harmonious. However, there is input to provide some distance between the designs, hence they do not look too dense or heavy. Overall, the clothes from this collection are easy to mix and match and can be used for formal or informal events. Therefore, several design developments are carried out, including laying out patterns, color, materials, design, moodboards, and illustrations.

3.6 *Solder Kerancang techniques*

The Kerancang technique is a small cover or finishing with holes. The technique of making Kerancang is either with a sewing machine sewn manual or dikerancang directly. The

Figure 3. Final product realization. Source: Evan 2021.

process of making embroidered kerancang is starting with trace motif, small delicate motif and shearing, and start mengkerancang, then finishing with smoothed Kerancang.

4 CONCLUSION

Based on the design trials and interviews, expert panels and extreme users like modern *kerancang* motifs for resort wear with a simple and modern appearance. Accuracy and patience are needed in the manufacture and placement of *kerancang* resort wear. It should work in accordance with the patterns and directions that have been made. Furthermore, clear patterns and directions should be made for craftsmen to understand the finished product.

Resort-wear clothing is known for its comfort when worn; hence, the *kerancang* placement does not make the users uncomfortable. For example, the *kerancang* should not be placed on the neck collar and arm sleeve. The use of brown buttons with marble patterns and *kerancang* pockets gives the design a more exciting feel.

From the data described previously, the typical Tasikmalaya *kerancang* has the potential to be further developed and preserved. The resort wear in this design can also answer the needs of women to stay fashionable while on vacation, especially in the beach area, and help the economy of the *kerancang* craftsmen in Tasikmalaya. The Tasikmalaya-style resort wear clothing products with modern *kerancang* motifs are easily recognizable for their characteristics and can be used by the public, especially women.

REFERENCES

Katadata. 2021. *Berapa Jumlah Pulau di Indonesia?* 6 June 2021. https://databoks.katadata.co.id/datapublish/2018/10/16/berapa-jumlah-pulau-di-indonesia

Garg, S. 2018. *What is "Resort Wear for Women?"*. 30 October 2020. https://sandhyagarg.com/blogs/what-isresort-wear/what-is-resort-wear-for-women.

Hanjani, G. 2015. *Promosi Kerancang Bordir Khas Tasikmalaya*. 23 September 2020. http://repository.maranatha.edu/17810/

Hobby, J. 2020. *What Is Resort Wear? (with pictures)*. wiseGEEK, 30 October 2020. https://www.wisegeek.com/what-is-resort-wear.htm.

Lee, J., Ostwald, M. & Gu, N. 2020. *Design Thinking: Creativity, Collaboration, and Culture* (pp. 1–6). Switzerland: Springer Nature.

Rusyaman, M. 2013. *Pengaruh Citra Kerajinan Bordir Terhadap Motivasi Berkunjung Wisatawan Ke Sentra Kerajinan Bordir di Kota Tasikmalaya, 60–63*, 31 October 2020. http://repository.upi.edu/5145/6/S_MRL_0901512_Chapter%204.pdf.

Tanzil, M.Y., Astrid, Tjandrawibawa, P., Tahalele, Y.K., Toreh, F.R. (2021). Desain Fesyen Ready-To-Wear: Sebuah Studi Merek Fesyen Kolaborasi Akademik dan Industri, *Jurnal Idealog*, 6(2), 135–149. doi: https://doi.org/10.25124/idealog.v6i2.4281

Utilization of waste fabric for ready-to-wear deluxe unisex clothing design with fabric manipulation technique on Kamisado brand

C.A.P. Kurniadi, D.M.W. Githapradana & Soelistyowati
Universitas Ciputra, Surabaya, Indonesia

ABSTRACT: In Indonesia, national waste reaches 170 thousand tons per day, and about 5% includes cloth waste. The rapid growth of fast fashion will also increase continuous production. In this age of globalization, clothing continues to evolve to the point where it is used not only to cover the body but also as a lifestyle or a status symbol. Dresses continue to evolve in terms of design, materials, construction patterns, and the concept of gender. Kisfaludy stated that fashion is more developed in ready-to-wear clothes. As a result, clothing with new revolutionary designs is a style that easily attracts attention because feelings and thoughts are open to new and exciting ideas. This challenge can be addressed by upcycling, combining fabric manipulation skills with sewing techniques, and arranging combinations of these materials in various ways. Therefore, ready-to-wear deluxe clothing can be developed using tweed, suede, and corduroy materials. The Kamisado brand is expected to provide clothing alternatives for unisex fashion consumers while also being environmentally friendly. This qualitative study is based on observations, interviews with experts and extreme users, as well as data from literature, journals, and publications. The upcycling design concept has the potential to reduce environmental waste while also becoming an invention with a beneficial impact on fashion enthusiasts.

Keywords: fabric manipulation techniques, fabric waste, ready-to-wear deluxe, unisex

1 INTRODUCTION

The gender-neutral design concept of unisex clothes has widespread application. Jeans and t-shirts were popularized in the 50s by Hollywood cinema, and since then, clothing has undergone a revolution, with women and men traditionally wearing skirts and pants. The popularity of boyfriend jeans in the women's fashion category is a result of the lifestyles of many who wear their partner's pants as fashion accessories.

Beauty in contemporary art is made to appear old-fashioned by modernism. In this fast-fashion era, the times focus more on ready-to-wear clothes (Kisfaludy 2008:59). Fast fashion is a term used by the fashion industry to describe a collection of affordable clothing that follows trends from high-end or designer brands produced in a quick time. Due to the rapid turnover of fashion products to meet consumers' demands, many issues appear regarding environmental pollution. In Indonesia, national waste reaches 170 thousand tons/day, 25% including fabric, glass, and metal waste. Textile scraps, also known as patchwork, significantly contribute to environmental pollution in the fashion sector. This will negatively impact the environment because it is difficult to reunite with nature. According to this background, the Kamisado brand, which manufactures unisex clothing by upcycling fabric waste based on a sustainable concept, aims to create public awareness of environmental problems and appreciate works of art of high quality to avoid being trapped in the era of fast fashion. The problem is how does the Kamisado brand recycle fabric waste to design ready-to-wear upscale unisex garments using fabric manipulation techniques? This research purpose is to design ready-to-wear deluxe unisex clothes under the Kamisado brand using recycled fabric waste and fabric manipulation techniques.

1.1 Literature study

a. Fabric waste

The fashion industry is an example of the largest waste contributor, such as fabric/textile waste. According to statistical data, this type of waste is most commonly found in landfills. Reprocessing textile waste into functional items is one method for addressing this issue. In addition, unutilized waste should be converted into more valuable textile waste craft products such as bags, blankets, and displays. Textile waste is also one of the focuses in the creative industry sector. According to the 2016 Indonesia Economic Forum website, the use of recycled materials, systems for more ethical production processes, and increasing environmental awareness through design are some of the innovations needed for eco-sustainable fashion for the growing industry.

b. Ready to wear deluxe clothing

Ready-to-wear deluxe is clothing that uses better quality materials and embellishments in the manufacturing process and requires good skills (Bel et al. 2008). It is still wearable and has more extra details or models than ready-to-wear clothes but does not reach haute couture clothing. Ready-to-wear deluxe clothes are also often called demi-couture.

c. Unisex

Unisex clothing is designed to suit both sexes and for styles where men and women look and dress in the same fashion (Callan 1998). It was popular from 1960 to 1970 because of the androgyny fashion style (Park 2015). However, unisex fashion was limited to men and women who borrowed from each other's closets or dressed the same way. This implies the unisex approach has become the standard for designing and manufacturing clothing for men and women.

d. Fabric manipulation technique

Fabric manipulation is a method that tries to make the fabric more three-dimensional (Przybylek 2019), incorporating sculpture and embellishing techniques to create a distinctive texture and appearance. It transforms clothing into a work of art with a high level of craftsmanship. There are three basic ways to produce fabric manipulation: fabric texturing, stitching, and bling.

2 RESEARCH METHODS

The data collection method that will be used is a qualitative study. It is shown to understand better an object being studied, which aims to elaborate a concept on the problem to be faced and apply a passion related to grounded theory (Mulyadi 2013). Data were collected through observation, literature study, and interviews. This research also use design thinking methods:

1. Empathize: The observations focus on the problem of cloth waste in Indonesia regarding the processing of materials of high value to meet the target market through quirky and unique fashion styles.
2. Define: In this stage, market observations are carried out to obtain information related to their needs. For example, the market requires "additional" clothing to complement their outstanding appearance.

3. Ideate: It focuses on designing deluxe ready-to-wear clothing using fabric waste materials processed with fabric manipulation techniques as the focused detail. This clothing uses design principles and elements to produce products worth being sold.
4. Prototype: In this process, a sample of fabric waste processing ideas will be made and tested on lecturers and the target market.
5. Test: The results of the prototype that has been tested are useful for getting feedback, which will later be retested to become a solution design.

3 DISCUSSION AND DESIGN SOLUTION

The title of the Autumn–Winter 2021/2022 collection is Rasavada. This design aims to design ready-to-wear deluxe unisex clothing by utilizing fabric waste processed through fabric manipulation techniques to reduce the problem of national waste and the negative impact of fast fashion. The target market of this brand is men and women with a unique and quirky style, as well as an age range of 20–30 years with middle to upper social status with the Psychological Experiences group. The material used is fabric waste from the production of Indonesian designers, namely Yogie Pratama, which is supported by complementary materials such as suede, tweed, and corduroy. "Rasasvada" is a collection from the Kamisado brand with a theme inspired by basic shapes such as rectangles and circles, which are then combined into new shapes that give a unique impression on cutting and creative fabrics.

The following are some cloth wastes obtained from Yogie Pratama. It was in the form of pieces of cloth that are no longer intact, and their parts have been reduced to form patterns. The characters and types of fabrics also vary from color to texture. Not all waste can be directly used to make fabric manipulation. For example, tulle waste that is very light and thin needs to be balanced in weight, and when combined with other fabrics, it is not easily torn. A layer of sugar cloth and plaid motif was given with suede ribbon to make the fabric even more attractive and not plain. Fabric manipulation is made with a patchwork technique, and the shape is made abstract and irregular. Furthermore, the good parts are stitched neatly inside, and the remaining seam is not more than 1 cm for maximum results. The fabric should be sewn slowly because the position may change due to the condition of the Lsicn fabric.

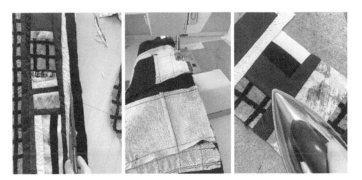

Figure 1. Fabric manipulation process.

According to experts, the concept is intriguing since it has an immediate value, particularly regarding the fabric waste issue, which can be applied to production and save material costs. Fabric waste processing with fabric manipulation technology is deemed appropriate, and it can later become a focal point in the collection and the brand's qualities. The men's M-XL and women's large sizes of unisex clothing sizing can be worn by both genders.

In addition, the grouping of fabric types needs to be considered for maintenance problems. Before purchasing, it is important to consider whether certain materials can be cared for at a fabric care facility. For the target market, it is believed that the market opportunity in Indonesia is relatively high. Since most artworks are hand-crafted, they are limited and more expensive than the average market pass. External factors influence products that sell better, as seen during fashion weeks, where unique and quirky designs will increase their sales. Demographics are also a determining factor, for example, South Jakarta is more daring in dressing than other areas of the city.

Extreme users are starting to like the fashion world with the quirky-edgy style used to express their character. This distinctive style is also used daily when traveling, but during events or photoshoots, they will wear clothing that is even more outlandish. The model is the most crucial consideration when purchasing clothing, and the brand is of no concern. To reduce the prevalence of fast fashion, they carefully consider the durability of each item purchased, and the most frequently bought items are outer and vest.

The Figure below is the final design from the Rasasvada collection, with developments in design, color, and fabric manipulation. These are the results of a pre-final project in which five designs were made, two of which were maintained, and three were developed. There is also a color development for shirts and turtlenecks (designs 3 and 5), which were previously terracotta and now maroon. Therefore, the collection consists of four tops (sleeveless blazer, shirt, long sleeve, turtleneck), three outer (oversized long outer, oversized outer, crop jacket), three vests (full vest, half vest, pocket vest), two trousers (skinny fit chino pants, asymmetrical pants), and three short pants (tweed short pants, fabric manipulation short pants, corduroy short pants).

Figure 2. Final design on male illustration.

4 CONCLUSION

The results showed that the national waste reaches 170 thousand tons/day in Indonesia, 5% of which includes cloth waste. Although the fast fashion industry also contributes to waste, patchwork fabric scraps account for most textile waste. This will harm the ecology when not handled appropriately, as it is hard to reunite with nature.

The fabric manipulation technique in processing waste is patchwork because it can maximize its usage. It can also increase the product's value and high craftsmanship value. The design of this final project is based on interviews and product trials with experts and extreme users. Based on interviews and initial trials, the ideas and products to be produced received positive feedback. However, several factors such as pattern construction and bright color need to be considered to maximize the product. The test results carried out development starting from design, material fabric manipulation, color placement, and construction patterns during production. As a result, the fabric manipulation is more engaging, and the portion is significant. The lime green color is also only on small elements such as ropes to make them pop out. In addition, the manufacturing pattern is improved, resulting in a more defined clothing form.

REFERENCES

Bel, L., Daerah, P., & Maluku, P. (2008). *1 Bab I.* 1–102.

Bestari, Astrid. 2019, Juni 3. *Arti Sustainable Fashion yang Perlu Anda Ketahui Sekarang.* Accessed from https://www.harpersbazaar.co.id/articles/read/6/2019/6987/Arti-Sustainable-Fashion-yang-Perlu-Anda-Ketahui-Sekarang

Callan, G. O. 1998. *The Thames and Hudson Dictionary of Fashion and Fashion Designers.* London: Thames and Hudson Ltd.

Davis, Fred. *Fashion, Culture, and Identity.* Chicago and London: The University of Chicago Press, 1992.

Firdhaussi. 2018, Oktober 10. *Memahami Fast Fashion dan Sustainable Fashion.* Accessed from https://medium.com/@setali/memahami-fast-fashion-dan-sustainable-fashion-c467de1d5a2a

Kisfaludy M. (2008), *Fashion and Innovation, Acta Polytechnica Hungarica*, ISSN 1785-8860, 06/2008, Vol. 5, No. 3, p. 59–64

Kurniawan, A., & Ashfahani, S. (2018). Jurnal Ilmiah Ilmu Komunikasi. *Jurnal Ilmiah Ilmu Komunikasi*, 130. doi: http://dx.doi.org/10.38041/jikom1.v10i03.37

Mulyadi, M. (2013). Penelitian Kuantitatif Dan Kualitatif Serta Pemikiran Dasar Menggabungkannya. *Jurnal Studi Komunikasi Dan Media*, 15(1), 128.

Nirmala, Andini, T dan Aditya Pratama, 2003, *Kamus Pintar Bahasa Indonesia.* Surabaya, Prima Media.

Park, J. 2015. *Unisex Clothing* (online). Accessed from https://fashionhistory.lovetoknow.com/clothing-types-styles/unisex-clothing

Przybylek, S. (2019). *Fabric Manipulation Techniques*, Accessed from https://study.com/academy/lesson/fabricmanipulation-techniques.html

Pujiriyanto, 2005, *Desain Grafis Komputer.* Yogyakarta, Andi.

Sari, Endah. 2019, Oktober 31. *Sejarah Kain Perca dan Perkembangannya.* Accessed from https://5news.co.id/artikel/2019/10/31/sejarah-kain-perca-dan-perkembangannya/

Sarkar, P. (2017, June 14). *What is a Specification Sheet in Fashion?* Retrieved from Online Clothing Study: https://www.onlineclothingstudy.com/2017/06/what-is-specification-sheet-in-fashion.html

Steele, Valerie. *Fifty Years of Fashion: New Look to Now. New Haven, Conn.*: Yale University Press, 1997.

Susilo, R. (2012). Pemanfaatan Limbah Kain Perca Untuk Pembuatan Furnitur. *Jurnal Tingkat Sarjana Senirupa Dan Desain*, (No. 1), 1.

Strategy for revitalizing city parks in the city of Bandung, West Java, Indonesia. Case study: City Hall Park

I. Sudarisman
Telkom University, Bandung, Indonesia

M. Mustafa & M. Hafizal
Universiti Sains Malaysia, Penang, Malaysia

ABSTRACT: City Hall Park is the oldest park in the city of Bandung. This park was built since the Dutch East Indies era in 1885. This park was originally built as a place of recreation for European businessmen, a place to rest, hold music concerts and get closer to nature. The function of the park continues to grow and has become a place for gathering, exercising, relaxing, playing, recreation, interacting with nature and holding various events. In 2013, the City Hall Park underwent a revitalization process in order to facilitate the development of this function. Based on the results of a survey in 2018, park users, namely city residents, think that revitalization has had a positive impact. This study aims to identify the implementation in park design of the important factors for the success of a city park by the government in revitalizing the City Hall Park so that it can be used as a reference in the process of new design or revitalization of city parks in the future. The research was conducted with a qualitative descriptive method through data collection methods in the form of documentation, observation and mapping. The results showed that the success of revitalization was due to the fulfillment of important factors in the success of a park, namely accessibility, fun and attractiveness for users, as well as comfort and safety.

Keywords: Public Open Spaces, City Parks, Park Users and Activities

1 INTRODUCTION

The city of Bandung is the largest city in the province of West Java. The city of Bandung has around 1700 hectares of green open space, including city parks. "Town of Parks" or Tuinstad was the development concept applied by the Dutch East Indies Government in Bandung during the colonial period (Husodo *et al.* 2014).

City parks as a form of urban green open space (RTHKP) as stated in the Minister of Home Affairs Regulation no. 1 of 2007, has a function as a public open space in an urban area that is used as a place for people to meet, gather and interact, and also the identity of a city (Helmi 2018). Green open space has space elements in the form of shady trees, shrubs and neatly arranged ornamental plants, park benches, paths, ponds, fountains, and children's playgrounds (Dewiyanti 2011).

City Hall Park is the oldest park in the city of Bandung. Initially, this park was built in 1885 to commemorate the Assistant Resident of Priangan, Pieter Sijthoff, who is considered to have contributed greatly to the development of the city of Bandung so it was named Pieter Sijhoff Park (Astri *et al.* 2013). According to Kunto (1984), the existence of Pieter Sijhoff Park is used as a place to unwind. Due to its proximity to Kweekschool, this park is often used by students during their lunch breaks. The Dutch army drum band players often practice at Pieter Sijhoff Park. For pre-anger planters, Pieter Sijhoff Park is also used as a place to listen to music orchestras while eating (Budiman 2015).

City Park with an area of 14,700 m² is a landmark of the city of Bandung (Helmi 2018). Bandung City Hall Park is located in the downtown area of Bandung, which is bordered by Jalan Merdeka on the east side, Jalan Watukencana on the west side, and Jalan Aceh on the north. This park is integrated with the Merdeka Building area which is an office building where the Bandung city government is located. City Hall Park is close to several shopping centers and other commercial facilities (shopping and culinary) which are always crowded with Bandung residents both on weekdays and holidays.

City Hall Park is crowded with residents of the city of Bandung, especially on holidays. City residents can do various activities such as gathering, sports, playing, and relaxing in the park. City Hall Park is often used to hold various events both by the city government and by other parties, such as bazaars, competitions, art performances, and so on. This park is also used by the city government as an office parking area and to hold ceremonial activities.

Initially, Bandung City Park had several areas, namely the office area (Gedung Merdeka), parking area, motor vehicle circulation, pedestrian circulation, greenery, white rhino pond, gazebo, and plaza. This park area is limited by a guardrail that separates it from the pedestrian circulation area outside the park. The main entrance access is on Jalan Merdeka and Wastukencara. In addition, there is also one entry from the office area (Gedung Merdeka) which is located on Jalan Aceh.

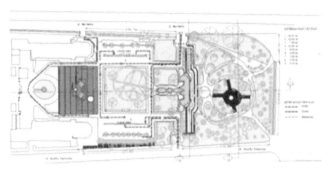

Figure 1. Map of Bandung City Hall Park before revitalization. Source: Astri Z, Wiguna M. H, Rismunanda D, Brahmana R. B. (2013).

Initiated by the Mayor of Bandung, Ridwan Kamil, the City Hall Park underwent a revitalization that began in 2013 until 2017 (Helmi 2018). The purpose of the revitalization is to make the City Hall Park more attractive and to accommodate the needs of the citizens of Bandung for outdoor activities such as gathering, playing, relaxing, exercising, and nature tours.

Based on the results of research conducted in 2018, the responses of most of the City Hall Park visitors to the results of the revitalization of the park are as follows: (Helmi 2018)

1. The park is easily accessible either on foot or by motor vehicle
2. The existence of the City Hall Park is considered to have a good impact
3. City Hall Park can be stress-healing for them
4. City Hall Park is comfortable for them

Based on the data above, park users believe that the park revitalization efforts carried out by the city government are successful and are able to facilitate their needs. These needs include the need for a place for stress-healing through gathering activities, exercising, relaxing, recreation, and playing.

According to Aulia Hariz (2013), several criteria can be used to measure the success rate of a city park as a public open space. These criteria include accessibility and fun; it can attract users, safety and comfort, and can bind the community (Helmi 2018). Project for

Public Spaces (2000) states that access is important in assessing the success of public open spaces, namely the ease of public open spaces to be visited and easy to see. Carr (1992) states that public open space must be democratic, which means that the space must be accessible to all groups, both physically and visually (Helmi 2018). Carr (1992) states that public space must be responsive, which can answer the needs of users in this case active and passive activities that can provide pleasure. Project for Public Spaces (2000) makes activities and uses the criteria for successful public open spaces. The existence of activities makes the space can provide pleasure and can attract users (Helmi 2018). Miller (2009) considers that public spaces must be safe. The safe condition of the space makes the space able to be visited and function properly. A safe space is characterized by easy visibility into and out of the garden, easy access, and openness. Meanwhile, the Project for Public Spaces (2000) states that a comfortable space is characterized by clean and safe space conditions (Helmi 2018).

This research was conducted to examine the strategy of the Bandung city government in revitalizing the park so that it can be felt to be successful and have a positive impact on park users, namely residents of the city of Bandung. The results of the study can be used as a reference in the process of revitalizing city parks in the future for the city government and city planners.

2 RESEARCH METHODS

The research method used is the descriptive qualitative method. The data collection method is carried out by three events, namely documentation, observation, and mapping.

(a) Documentations are carried out on the park area and facilities, the condition of the park, and the activities that occur in the park by using photos. Documentations were mainly made on important factors in park design, namely accessibility, fun, and can attract users' interest, safety, and comfort which were implemented in the design of the City Hall Park.
(b) Observations were made on the park area and facilities, the condition of the park, as well as the activities that occur in the park. Observations were mainly made on important factors in park design, namely accessibility and fun; it can attract users' interest, safety, and comfort which were implemented in the design of the City Hall Park.
(c) Mappings area is carried out on park areas and facilities, as well as activities that occur in the park. Mappings were mainly made on important factors in park design, namely accessibility and fun and can attract users' interest, safety, and comfort which were implemented in the design of the City Hall Park.

The data obtained is then analyzed descriptively and then conclusions are drawn as research results.

3 RESULT AND DISCUSSION

In this chapter, the discussion is divided based on the factors that are important to be implemented into park design based on the theory described in the previous chapter, namely accessibility, fun, and can attract users' interest, safety, and comfort.

- *Accessibility*

Before the revitalization process, City Hall Park only had three entrances. In addition, access to Jalan Merdeka and Wastukencana is guarded by security forces. However, after the revitalization process, the City Hall Park has access to more parks and most of them are not guarded by security forces so that they are more accessible to the citizens of Bandung. One of the strategies taken by the city government to make access to the park easier is to remove most of the fences that separate the park area from the pedestrian circulation path on Jalan Merdeka, Jalan Wastukencana, and Jalan Aceh. In addition, efforts are being made to unite the pedestrian circulation path with the pedestrian circulation path in the park.

The loss of most of the guardrails around the garden makes the inside of the garden easier to see from outside the garden. Meanwhile, users who are in the park find it easier to see the conditions outside the park. Park users make easy access visually both inside and outside the park as part of relaxing and recreational activities, namely in the form of sitting and relaxing activities while observing the surrounding conditions.

Figure 2. The park fences on Jalan Merdeka and Jalan Aceh are removed, and the pedestrian circulation path outside the park is integrated with the circulation path inside the park. Source: (TripAdvisor 2022).

- *Fun and can attract users*

Before the revitalization process, City Hall Park had motorized vehicle parking facilities, motorized vehicle circulation lanes, pedestrian circulation paths, greenery, plazas, white rhino ponds, and gazebos. These facilities are used by park users for sports, gathering, playing, and relaxing activities. After the revitalization process, City Hall Park has more facilities that park users can use to accommodate their activities. These facilities include flower gardens, labyrinth gardens, children's play areas, waterfalls, water play areas (pools and rivers with shallow depths so that children can play in water), more sitting facilities, and circulation paths. Spacious and attractively designed pedestrian area, additional motorized vehicle parking area, additional plaza area, bicycle parking area, mural wall, historical park, fountain, public art which is used as a photo spot, mini zoo, waterfront, bandros which is used for walking around, as well as an attractive park sign system (see Figures 3 and 4).

Figure 3. Flower garden and maze garden facilities.

Figure 4. Facilities for children's playgrounds and children's play pools.

These new facilities can accommodate the various needs of park users and can be an attraction for tourists, both local and from outside the city of Bandung.

- *Safety and comfort*

Park security is still related to accessibility both physically and visually as previously discussed. Physical and visual ease of access is obtained by removing the fence that limits the garden area and connecting the pedestrian circulation path from inside to outside the park. Ease of visual access at night is facilitated by an adequate amount of artificial lighting.

Meanwhile, comfort related to the cleanliness of the park is achieved by providing garbage disposal facilities scattered in various areas of the park. These waste disposal facilities are distinguished by the type of waste (organic and non-organic) using color markers (see Figure 5).

Figure 5. Garden lighting at night and garbage disposal facilities are differentiated by the type of waste using different colors.

4 CONCLUSION

The results of the revitalization of the Bandung City Hall Park are considered successful and have a positive impact on the citizens of Bandung because they have fulfilled the accessibility, comfort and security factors, as well as an attractive design and the availability of various facilities to accommodate the activity needs of the city residents.

Accessibility, comfort, and safety factors are achieved in park design by eliminating the park area barrier and making the area outside the park as if it were part of the park. These factors are also achieved in park design by connecting the pedestrian circulation path inside the park with the pedestrian circulation path outside the park. The safety factor is also achieved in park design by providing adequate lighting in the garden so as to facilitate visual access. Meanwhile, the comfort factor is also achieved in park design by providing a garbage disposal that is distinguished by the type of waste so that the park remains clean and well-maintained.

The fun and interesting factor is achieved in park design through an attractive physical garden design (not monotonous but more organic) and the provision of various facilities to accommodate the activities of city residents. The facilities provided are not only intended for adults, but also for children. Facilities for gathering, exercising, playing, relaxing, and enjoying the atmosphere are provided in various forms such as plazas, pedestrian circulation paths, children's playgrounds, ponds and rivers, seating and gazebos, as well as greening, flower gardens, sign systems, public art, memorial statues, fountains, and waterfalls.

REFERENCES

Astri Z., Wiguna M. H., Rismunanda D., Brahmin R. B. 2013. Study of Open Space Patterns in Bandung City Hall Park Area. *Reka Karsa Journal* 2(1): 1–12.

Budiman H.G. 2015. City Park Developments in Bandung in Dutch East Indies Era (1918–1942). *Patanjala* 7 (2): 185–200.

Carr S., Francis M., Rivlin L. G., Stone A. M. 1992. *Public Space*. New York: Cambridge University Press.

Dewiyanti D. 2011. Green Open Spaces of Bandung City, A Preliminary Review of City Parks on the Concept of a Child Friendly City. *Unikom Scientific Magazine Journal* 7(1): 13–26.

Hariz, A. (2013). Evaluation of the Success of Environmental Parks in Dense Housing as Public Open Spaces, Case Study: Environmental Parks in Galur Village, Central Jakarta. *Journal of Urban and Regional Planning* 24(2): 109–124.

Helmi S. 2018. *Community Perception About Bandung City Hall Park*. Bandung: Biomanagement, Bandung Institute of Technology.

Husodo S., Irawan B., Wulandari I., Dasanova W. M. 2014. *Trees in Bandung City Park*. Bandung: Bandung City Environmental Management Agency.

Miller, L. B. 2009. *Parks, Plants, and People Beautifying the Urban Landscape*. New York: W.w. Norton and Company.

Project for Public Spaces. 2000. *How to Turn a Place Around: A Handbook for Creating Successful Public Spaces*. New York: Project for Public Spaces.

Reflection of sustainability concept on the use of resources in the 19th-century plantation

R. Wulandari
Telkom University, Indonesia
The University of Western Australia, Australia

L. Nuralia
Research Center for Prehistoric and Historical Archaeology, BRIN

ABSTRACT: The World Green Building Council established sustainable architecture guidance, which is Net Zero Healthy, consisting of six assessment categories, including material resources and cycle. Several research papers studied the relationship between culture and sustainability with respect to the environment. This research aimed to show the architecture–environment ecological relationship in cinchona plantation architecture in the context of sustainability. The research was conducted qualitatively through fieldwork on two plantations in Pangalengan and Cilengkrang, data digging, interviews, and theoretical review. It was found that the cinchona trunk is being used as a building material as structural elements, infill walls, window frames, roof structures, entrance stairs, and fences. It is concluded that initial buildings in cinchona plantations are efficient in material extraction, zero energy mobilization, and nontoxic and purely natural materials. Plantation architecture as part of Indonesian architecture has the potential to be researched from various perspectives, including sustainable or green architecture.

Keywords: sustainable architecture, plantation architecture, building materials

1 INTRODUCTION

The concept of sustainability has been introduced for more than two decades with the peak in the year 2000 (Donovan 2020). A critical text on sustainable architecture/building was done by Sara Cook and Bryn Golton in 1994 (Donovan 2020). Donovan (2020) also mentioned differences in understanding the definition of sustainable architecture, that there is not one definition used largely, but rather multiple understandings based on context, which are all valid. This also applies to Indonesia where the notion of sustainability is dynamic and diverse. It is all because different regions have different needs and different resources, thus different criteria of sustainable architecture may apply.

The World Green Building Council established guidance for sustainable/green building/architecture criteria. Every national green building association describes and develops a criterion that is suitable for the condition and needs of each nation. In Indonesia, Green Building Council Indonesia sets up Net Zero Healthy criteria (reduction of carbon emission) for building certification. This certification helps to give an understanding and a standard for sustainable architecture based on standardized criteria that consist of six assessment categories. One of the categories is material resources and cycle. However, the issue of material resources and cycle has not been an intensive issue, especially in Indonesia where the use of fabricated or industrialized materials is in major use. On the other hand, studies on material sustainability in Indonesian green architecture mostly took examples from vernacular architectures and not yet looking at other typologies.

An observation of architecture in the cinchona plantations in Pangalengan and Cilengkrang exposed an example of an architecture–environment ecological relationship in terms of green building materiality. This research aimed to show a strong architecture–environment ecological relationship in cinchona plantation architecture, in the context of sustainability on building material use, extraction, material impact, and afterlife. As most criteria of green building or green architecture are fitted to the urban context where the material is almost 100% fabricated, and other examples are taking on vernacular architecture, it will be a good reference to see an example from another context, in this case from an agro-industrial (rural) context, which is the plantation.

2 SUSTAINABILITY AND ARCHITECTURE

2.1 *Sustainable architecture and materials*

One of the key goals that humanity has made as the ultimate model for all their operations is to create sustainable and eco-friendly architecture. As a result, the primary objective of the current architecture of our time is thoughtfully working toward a greener architecture (Mohammadjavad *et al.* 2014). According to Burcu (2015), "green architecture" is defined as "an understanding of environment-friendly building under all classes and contains some universal consent." Nonsynthetic, nontoxic materials? Locally sourced stone and wood ethically harvested forests. The utilization of minimally processed, abundant, or renewable resources, as well as those that, when recycled or salvaged, create healthy living conditions and maintain indoor air quality, are key components of natural building methods that promote sustainability (Ragheb *et al.* 2016). The last has also been a concern of Indonesian designers in designing buildings or interior design.

Shafii (2006) mentioned that the biggest factor influencing climate change is concrete and steel. More to that, construction, operation, and demolition of buildings share 40% of the energy used, the same amount of greenhouse emissions, which makes the construction and building sector a target to reduce carbon emissions (Shafii 2006).

The School of Architecture and Urban Planning of the University of Michigan recapped concepts of sustainable architecture in 1998 (Gil-Piqueras & Rodríguez-Navarro 2021). The three concepts concern environmental consideration within the principles. On the other hand, there is cultural and ecological context to be considered in a built environment to achieve sustainability in architecture (Fabbri & Tronchin 2004).

Baweja (2021), while discussing the need to see architectural sustainability from an environmental history perspective to see the relationship between sustainability, architecture and urbanism, proposed a survey on architectural building materials. In Baweja's opinion, the city and architecture are ecosystems. To establish an ecological relationship between architecture and the environment, the information needed on the method of building is material extraction, the environmental impact caused by the extraction who extracted them, their rate of consumption, and their afterlife (Baweja 2014).

2.2 *Sustainable architecture and vernacular/traditional architecture*

Several research studies discuss the relationship between culture/tradition and the concept of sustainability. In between the studies, researchers studied the elements of vernacular (or traditional) architecture under sustainable architecture principles. While sustainability parameters are different in every country, there are similarities in vernacular architecture with respect to the environment, use of nearby materials, traditional labor and construction techniques, and culture in dealing with climatic problems (Al Tawayha *et al.* 2019; Asadpour

2020; Gil-Piqueras & Rodríguez-Navarro 2021; Lahji & Lakawa 2017; Malhotra 2020; Nasir & Arif Kamal 2021; Riany et al. 2014). Respect for local resources, technology, and traditional knowledge of a place is implied by the cultural context (Gülmez & Uraz 2007).

The recycling of energy by passive or renewable energy is implied by an ecological environment. Additionally, it needs to be in sync with regional economies and have evidence of biological diversity. Vernacular architecture is frequently used to demonstrate how to build with inherent sustainable features such as energy, materials, and local resources. It is widely accepted that an ecological building should use local materials whenever possible. The simplicity, minimalism, operativeness, honesty, coherence, and durability of these buildings throughout generations, as well as their overwhelming genius loci, were made possible by the vernacular way of thinking about architecture (Alves 2017). In a variety of fields, local knowledge systems have been found to support sustainability. Additionally, useful for ecosystem restoration, local knowledge frequently contains components for adaptive management (Pandey 2002).

3 METHODOLOGY

The research was conducted in a qualitative manner through observation and theoretical review. During field surveys of Priangan plantations, field surveys were conducted in two plantations in Pangalengan and Cilengkrang. Priangan plantation was well known for its coffee, tea, and cinchona commodities. The two Pangalengan and Cilengkrang plantations were known for their commodity and history. The architecture of the first construction was listed, the typology of the architecture was examined, and its construction method and structural system were observed. During the field survey, visual documentation and documentation through redrawing and measuring were also conducted. Data on the type of buildings, function, style, building position, and material utilization were observed and documented. Concerning sustainability, the focus was on the use of resources, while energy consumption was not yet measured. Interviews were also conducted with people in the area and plantations' staff to verify what was seen on site. Literature research was conducted to understand whether the issue has been written previously. Historical data digging was also conducted to review the primary situation of the area and its buildings in their early state. The data were analyzed using a checklist of building materials on the type of materials used on each building element, extraction method, environmental impact caused by the extraction, who extracted them, and their afterlife.

4 RESULT AND DISCUSSION

The plantation in Pangalengan (Kertamanah) is a multi-commodity plantation. It produced tea, coffee, and cinchona in the past. At present, the commodity production focuses on tea and coffee, while cinchona production has stopped. The plantation in Cilengkrang (Bukit Unggul) has a single commodity, which is cinchona. The second plantation also manages a plantation nursery that grows cinchona plant seeds.

In the surveyed plantation site, there exist employee houses, offices, and factories in various forms—modern and traditional styles. According to an interviewer, it was said that these buildings have been existing since 1800, but no one has an exact estimation of their construction. It was found that in these two plantations that grow cinchona, the cinchona trunk is being used as a building material. The cinchona trunk is used as a structural element and infill wall. The elements using cinchona trunks are columns, walls, window frames, roof structures, entrance stairs, and fences. The use of the cinchona trunk as a column was found in two styles. The first style is a form of a circular column where the main structure is placed on the structural point and then covered with wood panels. The model is reflected in

Figures 1.a and 1.h. Another form is shown in Figure 1.e where the cinchona trunk is placed as a pole in a structural point uncovered. All columns are only finished with paint. However, there is no sign of corrosion from the cinchona trunks used as columns.

Figure 1. Images on the use of cinchona trunk in plantation architecture, Source: Authors, 2019.

Another use of cinchona trunk was found in walls. The cinchona trunks were cut in half and arranged horizontally to make a wall. The cut trunks are put after a stone wall that is usually placed on the lower part of the wall. To keep the log attached to one another to form a wall, a cinchona trunk is also used as a wall frame (Figure 1.b, 1.e, 1.f, 1.g, 1.h). The use of cinchona trunks for wall material was found not only in housing but also in office buildings (Figure 1.f). As a structural material, smaller cinchona trunks are being used as part of the roof structure as joists or rafters (Figure 1.d). Lastly, cinchona branches are being used as fences and posts for footsteps structures at entrances of housing buildings. The use of cinchona trunk and branches as building materials was only found on plantations that grow cinchona. In this use, the material also does not show a sign of decay.

A data-digging activity was conducted to find data concerning the plantation's initial building situation. The focus of data digging was on the search of building photographs, site information, maps, and images that may be related to the research issues. However, the data digging did not succeed in finding image collection of cinchona plantation buildings in this area in the period of the 19th to 20th century. Very few images found came from other plantations and they did not inform any use of cinchona as a building material.

One of the criteria for green or sustainable architecture is the resources used; in this case, building materials and their characteristics. Sustainable architecture views the way material was extracted, and how it is reused. Besides material extraction, material finishes are also an issue in sustainable architecture. The material should be nontoxic, easily harvested, and have a short life-cycle. The use of cinchona trunks and branches in the case of cinchona plantation architecture reflected this issue. Cinchona trunks and branches used in buildings were extracted from cinchona trees that had been harvested in the past. The 'leftovers' of cinchona trees were not disposed of but were used as building materials. The use of cinchona as a building material has shown no sign of decay so far. It implies that cinchona trunks have good quality and strength as a wood substitute. If the interview was valid, that these buildings have existed since the 1800s, then the quality of cinchona as a building material is very high. However, the statement needs further study because the data digging did not result in any evidence of these housing in the 19th century. Roughly, it can be said that initial buildings in cinchona plantations have almost a net zero carbon footprint. It is because the extraction of material was done at the same time as crop harvesting, zero energy mobilization of material to the building site, the material is nontoxic and purely natural, and zero waste of material because the material itself is leftover. In this case, the green concept of "reduce" and "reuse" is well applied. The cinchona plantation architecture reduces the waste from harvesting and reuses it as a building material. However, this does not apply in other plantations without cinchona plants.

On energy consumption issues, buildings in cinchona plantations have a characteristic that similes vernacular architecture and Indo-European architecture. The form and building orientation mimic traditional architecture, while the use of large windows copies Indo-European architecture. It gives the architecture the ability to control microclimate and use of daylight. The interior temperature is cool during hot days and large windows help penetrate the interior with sufficient daylight that the building does not need extra lighting unless during bad weather. A structure measurement is needed further to calculate and confirm these issues.

In accordance with previous studies, this research can be considered a form of architectural survey that reviews an architectural example from the sustainable perspective (Baweja 2014). It discovered the potential of nonvernacular architecture potentials, in this case, the plantation architecture as a model of sustainable architecture that can be further measured through performance-based paradigms through empirical measures such as energy consumption, greenhouse gas emissions, resource management, life cycle assessment, indoor air quality control, and waste management (Baweja 2014). It indirectly reviews the plantation architecture as an example of anthropogenic transformation of the environment (Baweja 2014).

5 CONCLUSION

Plantation architecture as part of Indonesian architecture has the potential to be researched from various perspectives. From the perspective of sustainable or green architecture, plantation buildings reflect values of sustainable/green architecture in general, especially on the use of material. The way cinchona trunks and branches, which are used as building materials for various functions and purposes, have shown the creativity and technical capability of the people designing and constructing the building within the context. Cinchona trunks and branches are uncommon materials used in building construction, yet with an aesthetic value. Hence, they are well used and well applied in cinchona plantation architecture. This reflection is a source of knowledge on site management, design, and sustainability. Further research on plantation architecture is needed to value plantation architecture from a sustainable perspective. It also opens an opportunity to research cinchona as a building material and a substitute for wood. However, this study did not review the plantation architecture as Indo-European architecture because of the various style it possesses (vernacular, Indo-European, in between, or both). In fact, an anthropo-architecture study of plantation architecture may add important findings to improve the Indo-European architecture theory.

REFERENCES

Al Tawayha, F., Braganca, L., & Mateus, R. (2019). Contribution of the Vernacular Architecture to the Sustainability: A Comparative Study Between the Contemporary Areas and the Old Quarter of a Mediterranean City. *Sustainability (Switzerland)*, *11*(3). https://doi.org/10.3390/su11030896

Alves, S. (2017). The Sustainable Heritage of Vernacular Architecture: the Historic Center of Oporto. *Procedia Environmental Sciences*, *38*, 187–195. https://doi.org/10.1016/j.proenv.2017.03.105

Asadpour, A. (2020). Studies Defining the Concepts. *National Academic Journal of Architecture*, *7*, 241–255.

Baweja, V. (2014). Sustainability and the Architectural History Survey. *Enquiry The ARCC Journal for Architectural Research*, *11*(1), 12. https://doi.org/10.17831/enq:arcc.v11i1.207

Donovan, E. (2020). Explaining Sustainable Architecture. *IOP Conference Series: Earth and Environmental Science*, *588*(3). https://doi.org/10.1088/1755-1315/588/3/032086

Fabbri, K., & Tronchin, L. (2004). Level of Comfort in a Traditional Architectural Typology: Study the Change to the Level of Comfort to Know a Gap-energy-need in a Sustainable Perspective. *Plea2004 – The 21th Conference on Passive and Low Energy Architecture, September*, 19–22.

Gil-Piqueras, T., & Rodríguez-Navarro, P. (2021). Tradition and Sustainability in Vernacular Architecture of Southeast Morocco. *Sustainability (Switzerland)*, *13*(2), 1–18. https://doi.org/10.3390/su13020684

Gülmez, N. Ü., & Uraz, T. U. (2007). Vernacular Urban Fabric as a Source of Inspiration for Contemporary Sustainable Urban Environments: Mardin and the case of "Mungan House." *ENHR International Conference*.

Lahji, K., & Lakawa, A. R. (2017). Sustainability Concept: Cultural Based Method in Building Gurusina Sao in Flores-Nusa Tenggara Timur. *International Journal on Livable Space*, *2*(2), 97–108. https://doi.org/10.25105/livas.v2i2.4698

Malhotra, B. (2020). Living Footprints Sustainability and traditional wisdom. *International Journal of Scientific & Engineering Research Volume*, *11*(10), 5–13.

Nasir, O., & Arif Kamal, M. (2021). Vernacular Architecture as a Design Paradigm for Sustainability and Identity: The Case of Ladakh, India. *American Journal of Civil Engineering and Architecture*, *9*(6), 219–231. https://doi.org/10.12691/ajcea-9-6-2

Pandey, D. N. (2002). *Indigenous Sustainability Science*.

Ragheb, A., El-shimy, H., & Ragheb, G. (2016). Green Architecture: A Concept of Sustainability. *Procedia – Social and Behavioral Sciences*, *216*(October 2015), 778–787. https://doi.org/10.1016/j.sbspro.2015.12.075

Saha, K., Sobhan, R., Nahyan, M., & Mazumder, S. A. (2021). Vernacular Architecture as Cultural Heritage: An Interpretation of Urban Vernacular 'Bangla Baton' Houses of Sylhet City, Bangladesh. *Journal of Settlements and Spatial Planning*, *12*(1), 35–49. https://doi.org/10.24193/JSSP.2021.1.04

Samalavičius, A., & Traškinaitė, D. (2021). Traditional Vernacular Buildings, Architectural Heritage and Sustainability. *Journal of Architectural Design and Urbanism*, *3*(2), 49–58. https://doi.org/10.14710/jadu.v3i2.9814

Shafii, F. (2006). *Achieving Sustainable Construction In The. September 2002*, 5–6.

Digital skills in education: Perspective from teaching capabilities in technology

R.R. Wulan*, D.A.W. Sintowoko & I. Resmadi
Telkom University, Bandung, Indonesia

Y. Siswantini
Binus University, Bandung, Indonesia

ORCID ID: 0000-0001-6827-0144

ABSTRACT: During the pandemic, the new culture of teaching has changed massively. Teachers' digital skills are one of the essential skills in the era of digital learning. Unfortunately, after two years of the learning process was implemented the digital skills, and the competence of teachers in Indonesia is in a moderate, meaning that skills in using and utilizing digital media for learning are still mediocre. This condition will certainly have an impact on the quality of education received by students. This article aims to provide an overview of this phenomenon. Literature review in the past decade will be analyzed systematically.

Keywords: digital, teaching, technology

1 INTRODUCTION

Digital literacy is one of the essential skills in today's 21st century. Through digital literacy, teachers and students will have the ability to utilize technology, information and communication, social skills, learning and attitude skills, critical thinking, and creativity in the context of developing digital competencies (Rahmawati et al. 2021). In recent years, especially during the Covid-19 pandemic, it is undeniable that digital transformation has become an unavoidable necessity for the online learning process between teachers and students. In addition, Rahmawati et al. (2021) showed the correlation between the ability of teachers to create learning content related to improving students' capabilities in digital literacy. Teachers' contribution to making course content can improve students' digital literacy skills (Rahmawati et al. 2021). Thus, there is a mutual relationship between the creation of course content and the digital literacy skills of students. Teachers' ability to process various social media platforms is also eminent in developing students' digital literacy skills, such as building-integrated classrooms with social media platforms such as Facebook and WhatsApp. Digital literacy is closely related to the ability of teachers to develop learning technology capabilities (Lathipatud Durriyah & Zuhdi 2018).

The challenges faced in digital literacy in Indonesia are related to culture, habits, technological capabilities, and digital discrepancies. Data from the Indonesian Teachers Association (Ikatan Guru Indonesia/IGI) shows that around 60% of teachers in schools have poor skills in using information technology during the teaching process (Indahri 2020). In addition to competency issues, other problems also arise between public schools and private schools, schools in urban areas and schools in rural areas, or schools located in border areas.

---

*Corresponding Author: rorowoelan@telkomuniversity.ac.id

According to the 2012 PISA report, several factors created discrepancies in digital literacy skills like the problem of socio-economic background in schools, advantaged and disadvantaged students, immigrant and non-immigrant students, or between those attending rural and urban schools (Student & Ii 2013).

2 RESEARCH METHODS

The present study examines 21 papers in total published in the last two decades. These papers were organized according to the publication year and were investigated based on several different attributes, namely the digital literacy index in Indonesia and online learning practice in technology. All selected papers were based on the year of publication as mentioned previously.

3 RESULT AND DISCUSSION

The concept of digital literacy is identically considered to media literacy, some of them refer to digital competency due to its scope of improvement in digital skill including access-capability, information, and communication (Suwarto *et al.* 2022). Digital literacy has become more than just educational institution regulation. It has become an example of two different skills, critical thinking, and ability of digital skills (Diskominfo 2018). Martinez strongly mentioned that digital literacy refers to various basic abilities or skills such as ICT skills, information literacy, and technological literacy (Martínez-Bravo *et al.* 2020). While others mentioned curriculum of digital literacy centered on personal reflection, combining both cognitive and emotional learning, especially in digital tools, and making innovative process designs (Hobbs & Coiro 2018). Although digital literacy is traditionally highlighted in the user's skill, however, the term "digital" is a core competence where the teacher achieved digital competencies for lifelong learning as well. Even though the concept of digital literacy commonly focuses on digital competency regarding software and hardware in 2000, results from the present studies might be diverse due to 21st-century competencies. Therefore, some methods have been developed to explore issues about more independent concepts related to the ability of users. Though digital literacy has expanded given the possibility of users creating and distributing media content products on their own. Teachers added their technological knowledge in order to understand their digital skills for enhancing subject learning outcomes qualities (e.g., Harris *et al.* (2009); Falloon, (2020). However, different countries might differ in digital competency. Research on the demands of digital literacy has been criticized for their lack of digital knowledge and competence among teachers and students in the classrooms. The two reasons are as follows. First, teachers have the requisite expertise to engage in digital to address student enthusiasm (e.g., Jenson, (Hébert *et al.* 2021). Teachers have different approaches to teaching students through more challenging media as in their age. Thus, the condition of teachers in digital competence might describe digital literacy in Indonesia as well.

3.1 *Digital literacy index*

Digital Literacy Index is at the medium level at 3.49 score in Indonesia. It is measured with four pillars: digital skill, digital ethics, digital safety, and digital culture (Kominfo 2021). Studies show that demographic indicators are varied such as gender, education, and area of living. Those living in the urban area have a digital literacy index above the national average. From the 34 provinces in Indonesia, Yogyakarta has the highest digital literacy index at a 3.71 score on a scale of 1–5 in 2021. Meanwhile, North Maluku is the lowest index score at 3.18. Indonesians traditionally receive information from social media where *WhatsApp* and

Facebook are dominant. In Java, the digital literacy of teachers must posit a lifelong learning process. Teachers learned from some media institution programs to create content media. However, there is a potential gap in learning expectations between teachers and students because of Covid-19 where online learning was applied in various ways.

3.2 *Online learning practice in technology*

Online learning is a capability to support the teaching process using web media and communication technology to encourage individual or group communication processes. Online learning encourages collaborative learning, activities, and assignments related to access and interaction (Garrison 2009). In simple terms, online learning practices are learning methods that actively utilize information technology in the teaching process such as online learning, e-learning, web-based learning, distance learning, and cyberlearning, and the use of other formats such as self-instructional packaged, web-based subjects, two-way audio, or video conferencing (Zalon 2000). According to Korkmaz (2020), the definition of online learning is also closely related to digital technology used in the context of web-based learning, blended learning, e-learning, learning management system (LMS), computer-assisted instruction (CAI), massive open online courses (MOOCs), and virtual learning environments (VLE) Korkmaz (2020). However, research conducted by Sit *et al.* (2004) shows that the use of technology alone will not be sufficient in the online learning process, but also needs to be focused on aspects of teaching and learning. In fact, the biggest challenge in online learning should be emphasized in aspects of human interaction to facilitate a peer support environment, create academic dialogue, and build socialization. During the Covid-19 pandemic and online learning, problems are usually related to internet connection problems, lack of teacher and student interaction, lack of skills in learning assessments, lack of understanding related to learning evaluation, lack of feedback from students, lack of teacher understanding of students' interest in learning materials, lack of motivation to learn, and behavior of school admins to teachers regarding online learning obligations. Referring to some of the definitions and contexts above, online learning is not entirely about proficiency in aspects of digital technology, but it is also very important to see social patterns between teachers and students in dealing with online learning in Indonesia as well.

4 CONCLUSION

The conclusion section is written briefly and concisely, without the addition of any new interpretations. The result of the conclusion section must clearly answer the objectives of the problem, the phenomenon, and all of the work on the research. The result should summarize (scientific) findings rather than provide data in great detail. Discuss the findings in relation to theory or previous research results so that science can continue to develop. This section can also write down the novelty of the research, the advantages, and disadvantages of the research, as well as recommendations for further research.

REFERENCES

Diskominfo. (2018, May 31). *Pengertian Literasi Media*. (Diskominfo) Retrieved January 8, 2022, from https://diskominfo.badungkab.go.id/artikel/17916-pengertian-literasi-media

Falloon. (2020). From Digital Literacy to Digital Competence: The Teacher Digital Competency (TDC) framework. *Educational Technology Research and Development, 68*(5), 2449–2472.

Garrison. (2009). Implications of Online Learning for the Conceptual Development and Practice of Distance Education. *Journal Of Distance Education Revue De L'éducation À Distance 2009, Vol. 23*(No. 2), 93–104.

Harris, Mishra, & Koehler. (2009). Teachers' Technological Pedagogical Content Knowledge and Learning Activity Types: Curriculum-based Technology Integration Reframed. *Journal of Research on Technology in Education*, *41*(4), 393–416.

Hébert, Jenson, & Terzopoulos. (2021). Access to Technology is the Major Challenge": Teacher Perspectives on Barriers to DGBL in K-12 Classrooms. *E-Learning and Digital Media*, *18*(3), 307–324.

Hobbs & Coiro. (2018). Design Features of a Professional Development Program in Digital Literacy. *Journal of Adolescent and Adult Literacy*, *62*(4), 401–409. DOI:10.1002/jaal.907.

https://berkas.dpr.go.id/puslit/files/info_singkat/Info Singkat-XII-12-II-P3DI-Juni-2020-201.pdf

Indahri, Y. (2020). Permasalahan Pembelajaran Jarak Jauh di Era Pandemi. *Info Singkat: Kajian Singkat Terhadap Isu Aktual Dan Strategis*, *12*(2), 13–18.

Kominfo. (2021). *Status Literasi Digital di Indonesia 2021*. Jakarta: Katadata Insight Center.

Korkmaz & Toraman. (2020). Are we Ready for the Post-COVID-19 Educational Practice? An Investigation into what Educators Think as to Online Learning. *International Journal of Technology in Education and Science (IJTES)*, *4*(4), 293–309.

Lathipatud Durriyah, T., & Zuhdi, M. (2018). Digital Literacy With EFL Student Teachers: Exploring Indonesian Student Teachers' Initial Perception About Integrating Digital Technologies Into a Teaching Unit. *International Journal of Education and Literacy Studies*, *6*(3), 53. https://doi.org/10.7575/aiac.ijels.v.6n.3p.53

Martínez-Bravo, Sádaba-Chalezquer & Serrano-Puche. (2020). Fifty Years of Digital Literacy Studies: A Meta-research for Interdisciplinary and Conceptual Convergence. *Profesional de la Información*, *29*(4).

Rahmawati, A. Z., Haryanto, Z., & Sulaeman, N. F. (2021). Digital Literacy of Indonesian Prospective Physics Teacher: Challenges Beyond the Pandemic. *Journal of Physics: Conference Series*, *2104*(1), 012004. https://doi.org/10.1088/1742-6596/2104/1/012004

Student, G. E., & Ii, S. V. (2013). *PISA 2012 Results: Excellence Through Equity: Vol. II*. https://doi.org/10.1787/9789264201132-en

Sit. (2004). Experiences of Online Learning: Student's Perspective. *The Hong Kong Polytechnic University, School of Nursing, Hong Kong*.

Suwarto, Setiawan, & Machmiyah. (2022). Developing Digital Literacy Practices in Yogyakarta Elementary Schools. *Electronic Journal of E-learning*, *20*(2), pp101–111.

Zalon. (2000). A Prime-time Primer for Distance Education. *Nurse Educator*, *25*(1), 28–33.

Author index

Abdalloh, A.T.Z. 87
Abdulhadi, R.H.W. 42, 135, 290
Abdullah, S. 144
Adi, A.E. 107
Adino, T.F.L. 226
Aditia, P. 170
Aditya, D. 107
Aditya, D.K. 242, 349
Adji, R.B. 91
Adler, J. 96
Agung, L. 113
Ahada, K.S. 354
Akhmadi 321
Akhmadi, A. 393
Alam, S. 1
Alfarizy, M.A. 343
Alissa, A.R. 149
Amansyah, R. 59
Amelia, K.P. 186, 302
Andhyka Satria Putra, M. 311
Andrianto 164
Anis, S. 107
Anwar, A.A. 327
Anwar, H. 102, 290
Apsari, D. 70
Ar-Rasyid, M.D.D. 237
Assegaf, A.H. 381
Atamtajan, A.S.M. 164
Aulia, G. 381
Aulia, R. 59
Auliarahman, A.M. 237
Azhar, H. 307, 321, 393
Azhari, I. 118

Balqis, F. 149
Barlian, Y.A. 257

Bashori, M. 321
Bastari, R.P. 159, 196
Basten, G.P.S. 19
Bismo, A. 59
Buana, M. 354
Budi, A.P. 332
Budiman, A. 1
Budiono, I.Z. 290

Cardiah, T. 7
Chalik, C. 164
Christian 338
Ciptandi, F. 14, 24, 311
Cobis, M.Y. 387

Desintha, S. 87
Dijiwa, D.G. 343
Dolah, J.B. 76
Dwija Putra, I.D.A. 349

Evan, J. 410

Fadilla, A.N 170
Farida, A. 247
Fauzia, I.F. 231
Fawwaz, R. 247
Febriani, R. 19, 316
Firmansyah, R. 7, 343

Githapradana, D.M.W. 415
Gumilar, G. 208
Gunawan, A.N.S. 343

Hadiansyah 360
Hafizal, M. 420
Hambali, R. 102
Hanom, I. 237
Hapsoro, N.A. 302

Harditya, A. 332
Haristianti, V. 42, 273, 296
Haswati, S.M.B. 354
Hendiawan, T. 220
Hidayat, D. 1
Hidayat, S. 87
Humaira, H. 307
Hutagaol, O.D. 405

Ilhamsyah 91, 263
Indrawan, S.E. 399

Jawak, J.B.W. 30

Koesoemadinata, M.I.P. 81, 118, 154
Kurniadi, C.A.P. 415
Kurniawan, F.V. 338
Kurniawan, M.R. 220
Kurniawan, Y.W. 381
Kusumowidagdo, A. 365

Laksitarini, N. 7, 252, 268
Larasaty, P.N. 370
Lase, L.C. 252, 268
Lemona, M. 370
Lionardi, A. 242
Longi, D.Y.P. 140

Mahendra Nur 360
Machfiroh, R. 192
Maharani, P. 70
Mahdiyah, F. 192
Mawuntu, C. 381
Mayor, E. 180
Mentari, E. 263
Miraj, I.M. 123, 129
Mulya Raja, T. 102

Mulyosantoso, G.A. 387
Murdowo, D. 42, 296
Mustafa, M. 279, 420
Mustikawan, A. 48
Muttaqien, T.Z. 135

Nafi'ah, U. 154
Narwasti, I.P. 149
Naufalina, F.E. 48
Nova, P. 405
Nugaraha, N.D. 107
Nuralia, L. 426
Nurbani, S. 257
Nurkamilah, A.M. 231

Pangaribuan, N.P. 370
Pangestu, D.P. 54
Pinasti, A.D. 202
Prahara, G.A. 96
Prameswari, D.D. 186
Pratiwi, D.S. 81
Purnomo, A.D. 252, 268
Puruhito, E.A. 376
Putra, G.M. 14
Putra, I.D.A.D. 144

Rachmawanti, R. 285
Rafly, M. 405
Rahadiyanti, M. 365
Rahayu, Y. 405

Raja, T.M. 135, 186, 273
Ramadhan, M.S. 175
Ramli, R.R.F. 354
Rangga Lawe, I.G.A. 242
Resmadi, I. 159, 196, 432
Riyanto, B. 381
Rosandini, M. 140, 214
Rusdy, S.D.S. 24

Sabrani, Y. 186
Sagala, A.J.C. 376
Salayanti, S. 107
Salsabila Az-zahra, S. 7
Santosa, I. 180
Sari, S.A. 118
Sarihati, T. 343
Septiani, D.M. 65
Septiani, K.I. 387
Simanjuntak, A.H. 376
Simanjuntak, M.B. 370
Sintowoko, D.A.W. 307, 321, 432
Siswanto, R.A. 76, 159
Soelistyowati 415
Soewardikoen, D.W. 54, 91, 202
Subagio, I. 247
Sudarisman, I. 231, 420
Sultani, A. 405
Sunarto 1

Suskarwati, S.U. 376
Swasty, W. 48, 279
Syafikarani, A. 149
Syahid, K.M. 37
Syahna, K. 387

Tahalele, Y.K.S. 410
Tanzil, M.Y. 410
Tifany Fahira, C. 214

Utami, D.D.A. 231

Viniani, P. 321
Virgana, S.T. 393

Wardaya, M. 226
Widiandari, A. 175
Widowati, D. 370
Wiguna, M. 387
Wirasari, I. 30, 37, 65, 107, 263
Wismoyo, E.A. 273
Wiwaha, L.R. 113
Wulan, R.R. 432
Wulandari, R. 426

Yuniati, A.P. 393
Yuningsih, C.R. 123, 285

Zen, A.P. 123, 129